T0198432

Introduction to Particle Technology

Introduction to Particle Technology

THIRD EDITION

Martin Rhodes
Monash University, Australia

Jonathan Seville
University of Birmingham, UK

Registered Offices
John Wiley &a Sons, Inc., 111 River Street, Hoboken, NJ 07030, USA
John Wiley & Sons Ltd, The Atrium, Southern Gate, Chichester, West Sussex, PO19 8SQ, UK

For details of our global editorial offices, customer services, and more information about Wiley products visit us at www.wiley.com.

Wiley also publishes its books in a variety of electronic formats and by print-on-demand. Some content that appears in standard print versions of this book may not be available in other formats.

Library of Congress Cataloging-in-Publication Data

Names: Rhodes, M. J. (Martin J.), author. | Seville, J. P. K. (Jonathan P. K.), author.
Title: Introduction to particle technology / Martin Rhodes, Monash University, Australia, Jonathan Seville, University of Birmingham, UK.
Description: Third edition. | Hoboken, NJ, USA : John Wiley & Sons, Inc., 2024. | Includes index.
Identifiers: LCCN 2024005304 (print) | LCCN 2024005305 (ebook) | ISBN 9781119931102 (paperback) | ISBN 9781119931157 (adobe pdf) | ISBN 9781119931164 (epub)
Subjects: LCSH: Particles.
Classification: LCC TP156.P3 R48 2024 (print) | LCC TP156.P3 (ebook) | DDC 620/.43–dc23/eng/20240304
LC record available at https://lccn.loc.gov/2024005304
LC ebook record available at https://lccn.loc.gov/2024005305

Cover Design: Wiley
Cover Image: © Eugene Mymrin/Getty Images

Set in 10/12pt Palatino by Straive, Pondicherry, India
SKY10074391_050424

Contents

About the Authors

Martin Rhodes has a bachelor's degree in chemical engineering and a PhD in particle technology from Bradford University in the UK, industrial experience in chemical and combustion engineering and many years' experience as an academic at Bradford and Monash Universities. He has research interests in various aspects of gas fluidization and particle technology, areas in which he has many refereed publications in journals and international conference proceedings. Martin was co-founder of the Australasian Particle Technology Society. He is Emeritus Professor and former Head of the Department of Chemical and Biological Engineering at Monash University, Australia.

Jonathan Seville is Professor of Formulation Engineering in the School of Chemical Engineering at the University of Birmingham, UK, and a former Dean of Engineering and Physical Sciences at the University of Surrey. He was the editor-in-chief of the Elsevier journal *Powder Technology* for 20 years and is the author of several books and over 200 journal publications on particle technology. He has championed research into product-based particle technology and co-founded the Birmingham Positron Imaging Centre, which carries out radiation-based measurements on particulate systems. He is a Fellow of the Royal Academy of Engineering and was President of the Institution of Chemical Engineers (IChemE) in 2016–2017.

Kit Windows-Yule gained a PhD in Nuclear Physics from the University of Birmingham, UK, before working in the University of Twente's Thermal and Fluid Engineering department and the Friedrich-Alexander-Universität Erlangen-Nürnberg's Engineering of Advanced Materials Excellence Cluster, ultimately returning to Birmingham to take up a faculty position in the School of Chemical Engineering. He is a two-time Royal Academy of Engineering Industrial Fellow and a Turing Fellow and has applied his knowledge of particle technology and numerical modelling to diverse industrial problems spanning the aerospace, agriculture, chemical, defence, fast-moving consumer goods, food, green energy, and pharmaceutical sectors.

Preface to the Third Edition

I was delighted when Jonathan Seville accepted my invitation to join me as co-author of the 3rd edition of Introduction to Particle Technology. This permitted us to broaden the scope of the text introducing a chapter on engineering of particles as products together with supporting chapters on aerosols and colloids and the mechanical properties of particles. Aware that discrete element modelling of particulate systems is increasingly important, we invited Kit Windows-Yule to contribute a chapter on the subject. Although pitched at an introductory level, the coverage is thorough and describes the state of the art as it stands today. We intend that this chapter should be a reference for people starting out in discrete element modelling and for practising engineers and scientists wishing to understand something that may at first seem inaccessible to many. Other chapters have been refined and updated. We have aimed to maintain the popular style of the first and second editions, with updated worked examples, test yourself questions and exercises for each chapter.

Martin Rhodes
Cairns, October 2023

Preface to the Second Edition

It is 10 years since the publication of the first edition of Introduction to Particle Technology. During that time many colleagues from around the world have provided me with comments for improving the text. I have taken these comments into consideration in preparing the second edition. In addition, I have broadened the coverage of particle technology topics – in this endeavour, I am grateful to my co-authors Jennifer Sinclair Curtis and George Franks, who have enabled the inclusion of chapters on Slurry Transport and Colloids and Fine Particles, and Karen Hapgood, who permitted an improved chapter on size enlargement and granulation. I have also included a chapter on the Health Effects of Fine Powders – covering both beneficial and harmful effects. I am also indebted to colleagues Peter Wypych, Lyn Bates, Derek Geldart, Peter Arnold, John Sanderson and Seng Lim for contributing case studies for Chapter 16.

Martin Rhodes
Balnarring, December 2007

Preface to the First Edition

Particle Technology

Particle technology is a term used to refer to the science and technology related to the handling and processing of particles and powders. Particle technology is also often described as powder technology, particle science and powder science. Powders and particles are commonly referred to as bulk solids, particulate solids and granular solids. Today particle technology includes the study of liquid drops, emulsions and bubbles as well as solid particles. In this book only solid particles are covered and the terms particles, powder and particulate solids will be used interchangeably.

The discipline of particle technology now includes topics as diverse as the formation of aerosols and the design of bucket elevators, crystallization and pneumatics transport, slurry filtration and silo design. A knowledge of particle technology may be used in the oil industry to design the catalytic cracking reactor which produces gasoline from oil or it may be used in forensic science to link the accused with the scene of crime. Ignorance of particle technology may result in lost production, poor product quality, risk to health, dust explosion or storage silo collapse.

Objective

The objective of this textbook is to introduce the subject of particle technology to students studying degree courses in disciplines requiring knowledge of the processing and handling of particles and powders. Although the primary target readership is amongst students of chemical engineering, the material included should form the basis of courses on particle technology for students studying other disciplines including mechanical engineering, civil engineering, applied chemistry, pharmaceutics, metallurgy and minerals engineering.

A number of key topics in particle technology are studied giving the fundamental science involved and linking this, wherever possible, to industrial practice. The coverage of each topic is intended to be exemplary rather than exhaustive. This is not intended to be a text on unit operations in powder technology for chemical engineers. Readers wishing to know more about the industrial practice and equipment for handling and processing are referred to the various handbooks of powder technology which are available.

The topics included have been selected to give coverage of broad areas within easy particle technology: characterization (size analysis), processing (fluidized beds granulation), particle formation (granulation, size reduction), fluid-particle separation (filtration, settling, gas cyclones), safety (dust explosions), transport (pneumatic transport and standpipes). The health hazards of fine particles or dusts are not covered. This is not to suggest in any way that this topic is less important than others. It is omitted because of a lack of space and because the health hazards associated with dusts are dealt with competently in the many texts on Industrial or Occupational Hygiene which are now available. Students need to be aware however, that even chemically inert dusts or 'nuisance dust' can be a major health hazard. Particularly where products contain a significant proportion of particles under 10 μm and where there is a possibility of the material becoming airborne during handling and processing. The engineering approach to the health hazard of fine powders should be strategic wherever possible, aiming to reduce dustiness by agglomeration, to design equipment for containment of material and to minimize exposure of workers.

The topics included demonstrate how the behaviour of powders is often quite different from the behaviour of liquids and gases. Behaviour of particulate solids may be surprising and often counter-intuitive when intuition is based on our experience with fluids. The following are examples of this kind of behaviour:

When a steel ball is placed at the bottom of a container of sand and the container is vibrated in a vertical plane, the steel ball will rise to the surface.

A steel ball resting on the surface of a bed of sand will sink swiftly if air is passed upward through the sand causing it to become fluidized.

Stirring a mixture of two free-flowing powders of different sizes may result in segregation rather than improved mixture quality.

Engineers and scientists are used to dealing with liquids and gases whose properties can be readily measured, tabulated and even calculated. The boiling point of pure benzene at one atmosphere pressure can be safely relied upon to remain at 80.1 °C. The viscosity of water at 20 °C can be confidently predicted to be 0.001 Pa s. The thermal conductivity of copper at 100 °C is 377 W / m K. With particulate solids, the picture is quite different. The flow properties of sodium bicarbonate powder, for example, depend not only on the particle size distribution, the particle shape and surface properties, but also on the humidity of atmosphere and the state of the compaction of the powder. These variables are not easy to characterize and so their influence on the flow properties is difficult to predict with any confidence.

In the case of particulate solids it is almost always necessary to rely on performing appropriate measurements on the actual powder in question rather than relying on tabulated data. The measurements made are generally measurements of bulk properties, such as shear stress, bulk density, rather than measurements of fundamental properties such as particle size, shape and density. Although this is the present situation, in the not too distant future, we will be able to rely on sophisticated computer models for simulation of particulate systems. Mathematical modelling of particulate solids behaviour is a rapidly developing area of research around the world, and with increased computing power and better visualization software, we will soon be able to link fundamental particle properties directly to bulk powder behaviour. It will even be possible to predict, from first principles, the influence of the presence of gases and liquids within the powder or to incorporate chemical reaction.

Particle technology is a fertile area for research. Many phenomena are still unexplained and design procedures rely heavily on past experience rather than on fundamental understanding. This situation presents exciting challenges to researchers from a wide range of scientific and engineering disciplines around the world. Many research groups have websites which are interesting and informative at levels ranging from primary schools to serious researchers. Students are encouraged to visit these sites to find out more about particle technology.

Martin Rhodes
Mount Eliza, May 1998

Acknowledgements

The authors are indebted to a large number of people who have helped them in various ways to prepare this edition, by advising, reading, commenting and in some cases supplying material which we have included. In particular, we would like to thank the following:

Professor Karen Hapgood, now of Swinburne University, Melbourne, Australia, who prepared the chapter on Size Enlargement for the 2nd edition, upon which the 3rd edition chapter on this subject is based.

Professor George V. Franks of the University of Melbourne, Australia, who prepared the chapter on Colloids and Fine Particles for the 2nd edition, upon which the 3rd edition chapter on Colloids and Aerosols is based.

Professors Mike Adams, Bettina Wolf and Zhibing Zhang and Dr. Andy Ingram at the University of Birmingham, UK.

Professor Peter Knight, formerly of Unilever Research.

Professor Chuan-yu Wu at the University of Surrey, UK.

Professor Mojtaba Ghadiri at the University of Leeds, UK.

Dr. Ted Knowlton of Particulate Solids Research Inc., USA.

Professor Mark Jones at Newcastle University, Australia.

Dominik Werner, Andrei Leonard Niçusan, Ben Jenkins, PhD candidates at the University of Birmingham, UK.

Allen Forbes of the Collaborative Teaching Laboratory at the University of Birmingham, UK, for help with microphotography.

Professor Hsiu-Po Kuo of National Taiwan University.

Granutools for permission to use Figure 2.2.

GEA Group for permission to use Figure 12.20.

Winkworth Machinery for permission to use Figure 10.7.

Hosokawa Micron for permission to use Figure 10.8.

Freeman Technology for permission to use Figure 9.8c.

About the Website

This book is accompanied by the website:

www.wiley.com/go/rhodes/particle3e

A website with laboratory demonstrations in particle technology and interactive code examples, designed to accompany this text, is available. This easily navigated resource incorporates many video clips of particle and powder phenomena with accompanying explanatory text. The videos bring to life many of the phenomena that we have tried to describe here in words and diagrams. For example, you will see: fluidized beds (bubbling, non-bubbling, spouted) in action; core flow and mass flow in hoppers, size segregation during pouring, vibration and rolling; agglomeration of fine powders, a coal dust explosion; a cyclone separator in action; dilute and dense phase pneumatic conveying. The interactive online notebooks will provide the reader with a deeper understanding of the techniques used to model particulate and multiphase systems. As a whole, the website will aid the reader in understanding particle technology and is recommended as a useful adjunct to this text.

Introduction

Particulate materials, powders or bulk solids are used widely in all areas of the process industries, for example in the food processing, pharmaceutical, biotechnology, energy, chemical, mineral processing, metallurgical, detergent, paint, plastics and cosmetics industries. These industries involve many different types of professional scientists and engineers, such as chemical engineers, chemists, biologists, physicists, pharmacists, mineral engineers, food technologists, metallurgists, material scientists/engineers, environmental scientists/engineers, mechanical engineers, combustion engineers and civil engineers. Numerous surveys have suggested the importance of particle-based products to the world's economy and new products and applications are emerging every day.

Some examples of the processing steps involving particles and powder include particle formation processes (such as crystallization, precipitation, granulation, spray drying, tabletting, extrusion and grinding), transportation processes (such as pneumatic and hydraulic transport, mechanical conveying and screw feeding) and mixing, drying and coating processes. In addition, processes involving particulates require reliable storage facilities and give rise to health and safety issues, which must be satisfactorily handled. Design and operation of these many processes across this wide range of industries require a knowledge of the behaviour of powders and particles. This behaviour is often counterintuitive, when intuition is based on our knowledge of liquids and gases. For example, actions such as stirring, shaking or vibrating, which would result in mixing of two liquids, are more likely to produce size segregation in a mixture of free-flowing powders of different sizes. A storage hopper holding 500 tonnes of powder may not deliver even 1 kg when the outlet valve is opened unless the hopper has been correctly designed. When a steel ball is placed at the bottom of a container of sand and the container is vibrated in the vertical plane, the steel ball will rise to the surface. This steel ball will then sink swiftly to the bottom again if air is passed upwards through the sand causing it to be fluidized.

Engineers and scientists are used to dealing with gases and liquids, whose properties can be readily measured, tabulated or even calculated. The boiling point of pure benzene at atmospheric pressure can be safely assumed to remain at 80.1 °C. The thermal conductivity of copper can always be relied upon to be 377 W/mK at 100 °C. The viscosity of water at 20 °C can be confidently expected to be 0.001 Pa s. With particulate solids, however, the situation is quite different. The flow properties of sodium bicarbonate powder, for example, depend not only on the particle size distribution, but also on particle shape and surface properties, the humidity of the surrounding atmosphere and the state of compaction of the powder. These variables are not easy to characterize and so their influence on the flow properties of the powder is difficult to predict or control with any confidence. Intriguingly, powders appear to have some of the behavioural characteristics of all three phases: solids, liquids and gases. For example, like gases, powders can be compressed;

like liquids, they can be made to flow; and like solids, they can withstand some deformation.

The importance of knowledge of the science of particulate materials (often called particle or powder technology) to the process industries cannot be overemphasized. Very often, difficulties in the handling or processing of powders are ignored or overlooked at the design stage, with the result that powder-related problems are the cause of an inordinate number of production stoppages. However, it has been demonstrated that the application of even a basic understanding of the ways in which powders behave can minimize these processing problems, resulting in less downtime, improvements in quality control and reduced environmental emissions.

This text is intended as an introduction to particle technology. The topics included have been selected to give coverage of the broad areas of particle technology: characterization (size analysis, surface area), processing (granulation, fluidization), particle formation (granulation, crystallisation, tableting, size reduction), storage and transport (hopper design, pneumatic conveying, standpipes), separation (filtration, settling, cyclones), safety (fire and explosion hazards, health hazards), engineering the properties of particulate systems to achieve desired product performance, discrete element modelling. For each of the topics studied, the fundamental science involved is introduced and this is linked, where possible, to industrial practice. In each chapter there are worked examples and exercises to enable the reader to practice the relevant calculations and a 'Test Yourself' section, intended to highlight the main concepts covered.

Martin Rhodes
Jonathan Seville

1

Particle Analysis

Particle size and size distribution are fundamentally important in determining how a powder will behave in its bulk form. Measuring the particle size distribution, describing it in graphical and mathematical form and comparing it with other distributions are therefore important tasks for the particle technologist and are introduced in this chapter. Particle "size" can be described unambiguously by a single number only for a distribution of monosized spheres, but real particles are neither spherical nor monosized. We therefore need to understand how to describe particle shape and what effects this has on measurement and calculation. In many industrial applications it is common to represent an entire distribution by some sort of averaged single number. Calculation methods and choices for this are shown. Finally, we also introduce methods for measurement of the surface area of particle distributions.

1.1 PARTICLE SIZE

Size is the most fundamental of particle properties. We will see throughout this book that the size of a particle affects all of its properties. For example, larger particles usually flow freely whereas smaller particles do not. Larger particles dissolve slowly and smaller ones more quickly, resulting in different pharmaceutical effectiveness. Light is scattered strongly from small particles but much less so from larger ones, resulting in different atmospheric effects and a different appearance of painted surfaces, for example.

The objects we describe as particles can cover a wide range of sizes, as shown in Figure 1.1, from large molecules (of order $0.01\,\mu m$ or $10^{-8}\,m$) to bricks (of order $10\,cm$ or $10^{-1}\,m$). Particles almost always come in large numbers (there are hundreds of millions of salt particles on your dinner table right now!) and as a *distribution* of sizes. Mono-sized distributions, containing only one particle size or a very narrow distribution of sizes, are very rare (pollen is an example in nature). Distributions can be wide, often very wide. For example, your container of salt will include not only salt crystals of around $0.3\,mm$ or $300\,\mu m$ but also broken salt dust particles down to $1\,\mu m$ in size. In many processes

Introduction to Particle Technology, Third Edition. Martin Rhodes and Jonathan Seville.
© 2024 John Wiley & Sons Ltd. Published 2024 by John Wiley & Sons Ltd.
Website: www.wiley.com/go/rhodes/particle3e

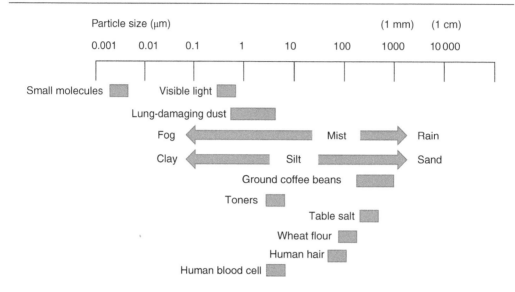

Figure 1.1   Ranges of particle size

and products the entire distribution is important; those very fine particles can have a big effect on how the whole distribution behaves.

What do we mean by *particle size*? This might seem like a simple question but in general it is not. If the particle is a sphere, the obvious answer is that its size is the same as its diameter. What if the particle is a cube? (Crystalline particles usually have an angular shape and crystals of common salt – sodium chloride – are roughly cubic.) In this case, it might seem logical to choose the side length of the cube to represent its size but as shown in Figure 1.2, there are other choices and the maximum dimension is actually $\sqrt{3}$ times the side length a.

Another way of looking at the problem of selection of a representative or equivalent diameter is to calculate the size of a sphere which has the same property as the non-spherical particle we are interested in. Two widely used possibilities are:

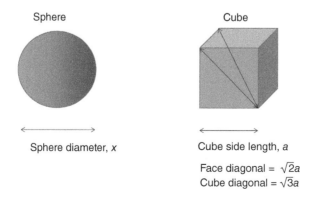

Figure 1.2   Sizes of spheres and cubes

the diameter of a sphere of equivalent projected area A

$$x_A = \sqrt{\frac{4A}{\pi}} \tag{1.1}$$

or the diameter of a sphere of equivalent volume V

$$x_V = \sqrt[3]{\frac{6V}{\pi}} \tag{1.2}$$

For the cube considered earlier, Figure 1.3 shows how the projected area may take the value a^2 or $\sqrt{2}\,a^2$ or $\sqrt{3}\,a^2$ according to the orientation of the particle, or, of course, values in between these for other orientations. The equivalent projected area diameter will then depend upon the orientation and in these three cases will take values of $a\sqrt{4/\pi}$, $2a\sqrt{1.414/\pi}$ and $2a\sqrt{1.732/\pi}$.

Note that the equivalent volume diameter x_V, does not depend on the orientation.

The example of a cube is relatively easy to deal with. Real particles seldom have a regular shape but approximation to a spheroid, plate or rod may sometimes be useful.

For irregular particles, the projected image can be used to obtain a number of types of *shape factors*, as illustrated in Figure 1.4.

Roundness S_R, is a measure of how closely the particle outline resembles a circle:

$$S_R = \frac{4\pi A}{P^2} \tag{1.3}$$

where P is the perimeter length and A is the area. Roundness is defined such that the value for a circle is 1. The roundness of other particles is then less than 1.

The Feret diameter is obtained by taking the distance between two parallel lines on either side of the particle perimeter, as shown. By varying the angle φ it is possible to find the minimum and maximum Feret diameters; a shape factor can be simply obtained from the ratio between them.

Many other measures of shape can be defined.

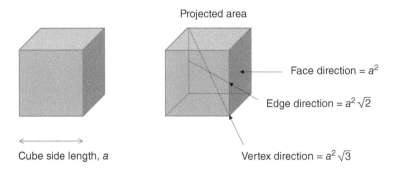

Projected area

Face direction = a^2

Edge direction = $a^2\sqrt{2}$

Cube side length, a

Vertex direction = $a^2\sqrt{3}$

Figure 1.3 Equivalent areas for a cube

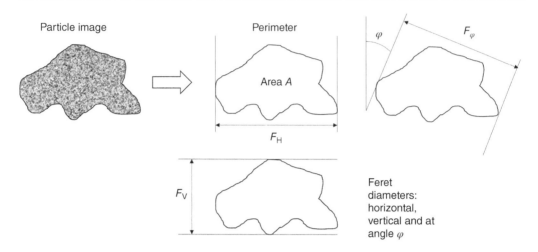

Figure 1.4   Shape and the Feret diameter

1.2   DESCRIPTION OF POPULATIONS OF PARTICLES

A population of particles is described by a particle size distribution. Particle size distributions may be expressed as frequency distributions or cumulative distributions. These are illustrated in Figure 1.5. The two are related mathematically in that the cumulative distribution is the integral of the frequency distribution; i.e. if the cumulative distribution is denoted as F, then the frequency distribution is dF/dx. For simplicity, dF/dx is often written as $f(x)$. The distributions can be by number, surface, mass or volume (where particle density does not vary with size, the mass distribution is the same as the volume distribution). Incorporating this information into the notation, $f_N(x)$ is the frequency distribution by number, $f_S(x)$ is the frequency distribution by surface, F_S is the cumulative distribution by surface and F_M is the cumulative distribution by mass. In reality, for many particles these distributions are smooth continuous curves. However, size measurement methods usually divide the size spectrum into size ranges or classes and the size distribution becomes a histogram.

For a given population of particles, the distributions by mass, number and surface can differ dramatically, as can be seen in Figure 1.6.

1.3   CONVERSION BETWEEN DISTRIBUTIONS

Many modern size analysis instruments measure particles individually and therefore produce a number distribution, which is rarely the one which is of most practical use. These instruments include software to convert the measured distribution into more practical distributions by mass, surface, etc.

Relating the size distributions by number $f_N(x)$, and by surface $f_S(x)$ for a population of particles having the same geometric shape but different size:

$$\text{Fraction of particles in the size range } x \text{ to } x + dx = f_N(x)dx$$

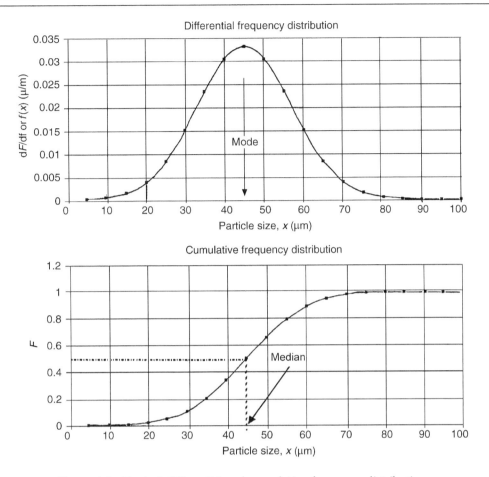

Figure 1.5 Typical differential and cumulative frequency distributions

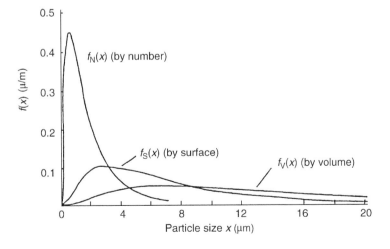

Figure 1.6 Comparison between distributions

Fraction of the total surface of particles in the size range x to $x + dx = f_S(x)dx$

If N is the total number of particles in the population, the number of particles in the size range x to $x + dx = Nf_N(x)dx$ and the surface area of these particles $= (x^2\alpha_S)Nf_N(x)dx$, where α_S is the factor relating the linear dimension of the particle to its surface area.

Therefore, the fraction of the total surface area contributed by these particles $[f_S(x)dx]$ is:

$$\frac{(x^2\alpha_S)Nf_N(x)dx}{S}$$

where S is the total surface area of the population of particles.

For a given population, the total number of particles, N, and the total surface area, S are constant. Also, assuming particle shape is independent of size, i.e. all particles have the same shape, α_S is constant, and so

$$f_S(x) \propto x^2 f_N(x) \quad \text{or} \quad f_S(x) = k_S x^2 f_N(x) \tag{1.4}$$

where

$$k_S = \frac{\alpha_S N}{S}$$

Similarly, for the distribution by volume

$$f_V(x) = k_V x^3 f_N(x) \tag{1.5}$$

where

$$k_V = \frac{\alpha_V N}{V}$$

where V is the total volume of the population of particles and α_V is the factor relating the linear dimension of the particle to its volume.

And for the distribution by mass

$$f_m(x) = k_m x^3 f_N(x) \tag{1.6}$$

where

$$k_m = \frac{\alpha_V \rho_p N}{V}$$

assuming particle density ρ_p is independent of size, i.e. all the particles have the same density.

The constants k_S, k_V and k_m may be found by using the fact that

$$\int_0^\infty f(x)dx = 1 \tag{1.7}$$

Thus, when we convert between distributions it is necessary to make assumptions about the constancy of shape and density with size. If these assumptions are not valid, the conversions are likely to be in error. For example, this approach would not be valid in the case where the population consists of whole spheres and broken pieces of spheres, because shape will then vary with size. Also, calculation errors are introduced into the conversions. For example, imagine that we used an electron microscope to produce a number distribution of size with a measurement error of ±2%. Converting the number distribution to a mass distribution we triple the error involved (i.e. the error becomes ±6%). For these reasons, conversions between distributions are to be avoided wherever possible. This can be done by choosing the measurement method which gives the required distribution directly.

1.4 DESCRIBING THE POPULATION BY A SINGLE NUMBER

In most practical applications, we require to describe the particle size of a population of particles (millions of them) by a single number. There are many options available: the mode, the median, and several different means including arithmetic, geometric, quadratic, harmonic, etc. Whichever expression of central tendency of the particle size of the population we use needs to reflect the property or properties of the population of importance to us. We are, in fact, modelling the real population with an artificial population of mono-sized particles. This section deals with calculation of the different expressions of central tendency and selection of the appropriate expression for a particular application.

The *mode* is the most frequently occurring size in the sample (Figure 1.5). We note, however, that for the same sample, different modes would be obtained for distributions by number, surface and volume. The mode has no practical significance as a measure of central tendency and so is rarely used in practice.

The *median* is easily read from the cumulative distribution as the 50% size (Figure 1.5): the size which splits the distribution into two equal parts. In a mass distribution, for example, half of the particles by mass are smaller than the median size. Since the median is easily determined, it is often used. However, it has no special significance as a measure of central tendency of particle size.

Various *means* can be defined for a given size distribution. However, they can all be described by:

$$g(\bar{x}) = \frac{\int_0^1 g(x)\mathrm{d}F}{\int_0^1 \mathrm{d}F} \quad \text{but} \int_0^1 \mathrm{d}F = 1 \quad \text{and so} \quad g(\bar{x}) = \int_0^1 g(x)\mathrm{d}F \tag{1.8}$$

where $\bar{x}$ is the mean and g is the weighting function, which is different for each mean definition. Examples are given in Table 1.1.

Equation (1.8) tells us that the mean is the area between the curve and the $F(x)$ axis in a plot of $F(x)$ versus the weighting function $g(x)$ (Figure 1.7).

Each mean can be shown to retain two properties of the original population of particles. For example, the arithmetic mean of the surface distribution retains the surface and volume of the original population. This is demonstrated in Worked Example 1.3. This mean

Table 1.1   Definitions of means

$g(x)$	Mean and notation
x	Arithmetic mean $\bar{x}_a$
x^2	Quadratic mean $\bar{x}_q$
x^3	Cubic mean $\bar{x}_c$
$\log x$	Geometric mean $\bar{x}_g$
$1/x$	Harmonic mean $\bar{x}_h$

is commonly referred to as the *surface-volume mean* or the *Sauter mean*. The arithmetic mean of the number distribution $\bar{x}_{aN}$ retains the number and length of the original population and is known as the number-length mean $\bar{x}_{NL}$:

$$\text{Number-length mean } \bar{x}_{NL} = \bar{x}_{aN} = \frac{\int_0^1 x \, dF_N}{\int_0^1 dF_N} \tag{1.9}$$

As another example, the quadratic mean of the number distribution $\bar{x}_{qN}$ conserves the number and surface of the original population and is known as the number-surface mean $\bar{x}_{NS}$:

$$\text{Number-surface mean } \bar{x}_{NS}^2 = \bar{x}_{qN}^2 = \frac{\int_0^1 x^2 \, dF_N}{\int_0^1 dF_N} \tag{1.10}$$

A comparison of the values of the different means and the mode and median for a given particle size distribution is given in Figure 1.8. This figure highlights two points: (a) that

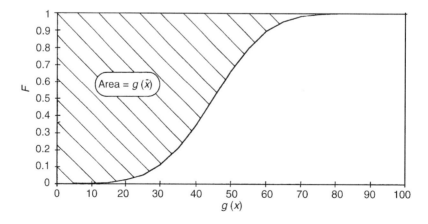

Figure 1.7   Plot of cumulative frequency against weighting function $g(x)$. Shaded area is $g(\bar{x}) = \int_0^1 g(x) dF$

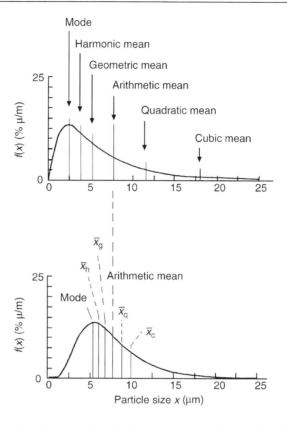

Figure 1.8   A comparison between measures of central tendency

the values of the different expressions of central tendency can vary significantly for the same distribution and (b) that two quite different distributions could have the same arithmetic mean or median, etc. If we select the wrong one for our design correlation or quality control, we may be in serious error.

So how do we decide which mean particle size is the most appropriate one for a given application? Worked Examples 1.3 and 1.4 indicate how this is done.

1.5   EQUIVALENCE OF MEANS

Means of different distributions can be equivalent. For example, as shown below, the arithmetic mean of a surface distribution is equivalent (numerically equal) to the harmonic mean of a volume (or mass) distribution:

$$\text{Arithmetic mean of a surface distribution } \overline{x}_{aS} = \frac{\int_0^1 x \, dF_S}{\int_0^1 dF_S} \tag{1.11}$$

The harmonic mean $\bar{x}_{hV}$ of a volume distribution is defined as:

$$\frac{1}{\bar{x}_{hV}} = \frac{\int_0^1 \left(\frac{1}{x}\right) dF_V}{\int_0^1 dF_V} \tag{1.12}$$

From Equations (1.4) and (1.5), the relationship between surface and volume distributions is:

$$dF_v = x dF_s \frac{k_v}{k_s} \tag{1.13}$$

hence

$$\frac{1}{\bar{x}_{hV}} = \frac{\int_0^1 \left(\frac{1}{x}\right) x \frac{k_v}{k_s} dF_s}{\int_0^1 x \frac{k_v}{k_s} dF_s} = \frac{\int_0^1 dF_s}{\int_0^1 x \, dF_s} \tag{1.14}$$

(assuming k_s and k_v do not vary with size) and so

$$\bar{x}_{hV} = \frac{\int_0^1 x \, dF_s}{\int_0^1 dF_s}$$

which, by inspection, can be seen to be equivalent to the arithmetic mean of the surface distribution $\bar{x}_{aS}$ [Equation (1.11)].

Recalling that $dF_s = x^2 k_s dF_N$, we see from Equation (1.11) that

$$\bar{x}_{aS} = \frac{\int_0^1 x^3 \, dF_N}{\int_0^1 x^2 \, dF_N}$$

which is the surface-volume mean $\bar{x}_{SV}$ [see Worked Example 1.3].

Summarizing, the surface-volume mean may be calculated as the arithmetic mean of the surface distribution or the harmonic mean of the volume distribution. The practical significance of the equivalence of means is that it permits useful means to be calculated easily from a single size analysis.

1.6 COMMON METHODS OF DISPLAYING SIZE DISTRIBUTIONS

1.6.1 Normal Distribution

In this distribution, shown in Figure 1.9, particle sizes with equal differences from the arithmetic mean occur with equal frequency. Mode, median and arithmetic mean coincide. The distribution can be expressed mathematically by:

$$\frac{\mathrm{d}F}{\mathrm{d}x} = \frac{1}{\sigma\sqrt{2\pi}}\exp\left[-\frac{(x-\bar{x})^2}{2\sigma^2}\right]$$

(1.15)

where σ is the standard deviation.

A normal distribution gives a straight line when plotted on a probability axis.

The normal distribution is commonly used to represent natural variations in, for example, human height, but is not often useful in representing particle sizes, because (a) particle size distributions tend to be skewed and (b) they often cover many orders of magnitude.

1.6.2 Log-normal Distribution

This distribution is more commonly used for particle populations. An example is shown in Figure 1.10. If plotted as $\mathrm{d}F/\mathrm{d}(\log x)$ versus x, rather than $\mathrm{d}F/\mathrm{d}x$ versus x, an arithmetic-normal distribution in log x results (Figure 1.11). The mathematical expression describing this distribution is:

$$\frac{\mathrm{d}F}{\mathrm{d}z} = \frac{1}{\sigma_z\sqrt{2\pi}}\exp\left[-\frac{(z-\bar{z})^2}{2\sigma_z^2}\right]$$

(1.16)

where $z = \log x$, $\bar{z}$ is the arithmetic mean of log x and σ_z is the standard deviation of log x.

A log-normal distribution gives a straight line when plotted on a log-probability axis.

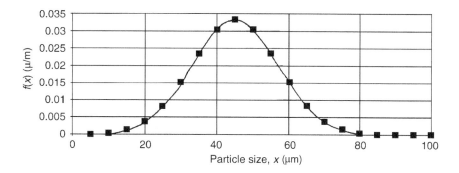

Figure 1.9 Normal distribution with an arithmetic mean of 45 and standard deviation of 12

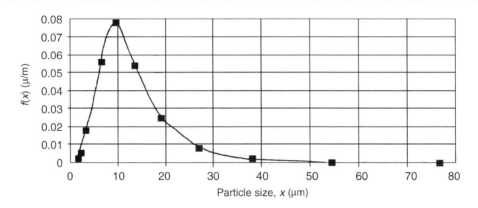

Figure 1.10   Log-normal distribution plotted on linear coordinates

1.7   *METHODS OF PARTICLE SIZE MEASUREMENT*

There are three general points to be considered when choosing a method for size analysis:

Direct observation of the particles through a microscope is always a good place to start, in order to get a good idea of what kind of particle is being considered.

Different methods and different instruments can give different numerical results for the particle size of the same sample. This is because they will usually measure different types of equivalent diameters. The more extreme the shape of the particles, the more the difference.

It is best to use the method of size measurement which gives directly the particle size which is relevant to the situation or process of interest.

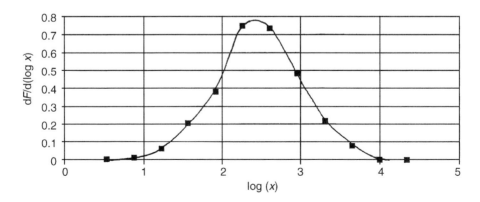

Figure 1.11   Log-normal distribution plotted on logarithmic coordinates

1.7.1 Image Analysis

It should be apparent from the earlier parts of this chapter that there is no substitute for viewing the particles of interest directly, whatever method is then chosen for measuring their size distribution. How else will it be possible to answer questions such as the following:

Are the particles of a single material or a mixture of materials?

Are the particles all of similar shape?

Does the distribution contain broken pieces or fine particles resulting from prior processing?

Are the particles actually agglomerates of smaller particles?

Viewing will certainly require some visual enlargement, either using a traditional optical microscope (for particles down to about 5 µm in size) or an electron microscope. Electron microscopy is often preferred, even for larger particles, because it does not suffer from the restricted depth of focus of optical microscopy.

The resulting digital image can then be used to obtain a distribution by image analysis, for which there are many commercial and open-source software codes. These are usually very versatile and capable of determining most if not all of the measures of size distribution introduced above, including measures of particle shape. Standard outputs include Feret lengths, perimeter, projected area, aspect ratio, roundness, sphericity and size distributions based on different size definitions. However, two issues need to be considered:

These are inherently 2D measurements of projections of particle images, so any of the 3D size distributions derived from them will require assumptions or approximations to be made about particle shape.

For the same reason, the presentation of particles to the camera or viewer needs to be considered, as in Section 1.1. Is the orientation of particles random or biased in some way, perhaps to present their maximum cross section? Is the lighting of the image biasing the measurement?

The images used for size distribution measurement may be static, in microscopy, for example, or dynamic, in which a high-speed camera is used to capture digital images of the particles in motion (for example in free fall or in a stream of air or liquid). From these images (each one a projection of a particle in a certain orientation), specialized image analysis software determines the distribution of size and shape characteristics of the particles. Commercial instruments for dynamic image analysis are available for laboratory use or they can be integrated directly into the process to give online measurements for quality control.

1.7.2 Light Scattering

As noted earlier, the way in which particles scatter light depends on their size, and this has given rise to a variety of measuring instruments, including (a) single particle counters, (b) extinction meters and (c) diffraction sizers (Figure 1.12).

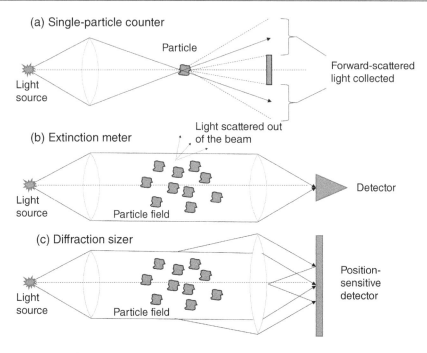

Figure 1.12   Types of light-scattering instruments. (a) Single particle counter, (b) extinction meter and (c) diffraction sizer

The single-particle counter, as its name implies, counts single particles passing through the focus of a light beam, usually suspended in a gas. The forward-scattered light is collected and converted into an electrical pulse from which the information about the particle size can be derived and a size distribution by number can be obtained over time. It is important that only one particle is in the measuring volume at a time, so these instruments tend to be used in low concentration applications, to ensure air cleanliness in clean rooms for example.

Both the extinction meter and the diffraction sizer rely on passing a parallel beam of light through a suspension of many particles simultaneously. The extinction meter measures the amount of light passing through the suspension while the diffraction instrument measures the light which is scattered. The extinction meter is designed for higher concentration measurements in applications such as effluent monitoring and is able to give limited information about size but responds rapidly to changes in concentration. The diffraction sizer is one of the most commonly used instruments for measuring particle size distribution and is considered in more detail below.

Diffraction sizing instruments convert the forward-scattered light intensity from the particle suspension into a size distribution. The suspension can be in either a liquid or a gas and is illuminated by a parallel beam of laser light. As noted earlier, the angular distribution of scattered light is very sensitive to the particle size, with smaller particles scattering light more intensely and over a wider angle. The scattered light is collected with a position-sensitive detector, usually a set of concentric rings, and its distribution can be converted into a size distribution approximating to the distribution of equivalent projected area diameter. The range of operation is wide: from approximately 1 to 2000 μm. However, over part of the size range, the inversion calculation which is used to obtain the size distribution is sensitive to the refractive index of the suspension medium (which is usually known) and the particles (which is usually not known) so some experimentation is required in order to establish the best operating conditions. This measurement method is quick and easy to perform and so has become very widely used in laboratories. Diffraction sizers can also be used *in situ* on processes to give real-time size information which is useful in process control.

1.7.3   Dynamic Light Scattering

Dynamic light scattering (DLS) is a commonly used method for measuring the size of sub-micron particles. It makes use of the fact that such particles, when dispersed in a liquid, are in constant *Brownian motion*, due to their collisions with solvent molecules, as described in Section 5.2. The rate of diffusion of particles by Brownian motion is dependent on particle size – large particles diffuse more slowly than small particles. In the DLS technique, laser light is shone into a suspension of particles in a liquid and the resulting fluctuations in scattered light analysed by autocorrelation between time-shifted traces in order to estimate the diffusion distance and hence to give a mean diameter, weighted by intensity of the light. This is not a direct measurement of particle size but responds to the hydrodynamic diameter. The method finds applications in formulation development in a wide range of industries. It has the disadvantages that it cannot be used for dry measurement and sample preparation requires considerable care to avoid contamination by dust, for example.

1.7.4   Scanning Mobility Particle Sizer

This device operates on the principle that the mobility of a particle in an electrical field depends on its size.

Electrical mobility is defined as the ratio of drift velocity V_d to the strength of the electrical field E:

$$\text{Mob} = \frac{V_d}{E}$$

The drift velocity is a terminal velocity reached by the particle moving in air under the influence of the electrical field. It is therefore the velocity at which the electrical and drag forces balance (assuming gravity plays no part). For the small particles in question, Stokes' law gives us the drag force (refer to Chapter 3).

A device called a differential mobility analyser (DMA) selects particles with a narrow range of electrical mobility (i.e. drift velocity in the applied field). These are particles of a narrow range of size, a size which is sometimes referred to as the *electrical mobility diameter*, a derived diameter specific to this method. Particles of the selected mobility are passed to a condensation particle counter which determines the number concentration of particles of that mobility and size.

The applied electrical field in the DMA is varied incrementally over time, allowing the mobility distribution and hence particle size distribution of the sample to be measured.

Differential mobility techniques can measure the size of particles in the submicron range, down to a few nanometres and are used extensively in the study of aerosols.

1.7.5 Other Methods

This section briefly covers several other measurement methods which readers may encounter, most of which have been largely superseded by the modern methods described above.

Sieving: Dry sieving using woven wire mesh sieves is a simple, cheap method of size analysis suitable for particle sizes greater than 45 μm. Particles are introduced to the top of a stack of sieves arranged in decreasing size. After a period of shaking, the mass on each sieve is weighed. Sieving therefore gives a mass distribution directly and a size known as the sieve diameter. Since the length of the particle does not hinder its passage through the sieve apertures (unless the particle is extremely elongated), the sieve diameter is dependent on the maximum width and maximum thickness of the particle. The most common sieve sets are available such that the ratio of adjacent sieve sizes is the fourth root of two (e.g. 45, 53, 63, 75, 90, and 107 μm). If standard procedures are followed and care is taken, sieving gives reliable and reproducible size analysis.

Sedimentation: In this method, the rate of sedimentation of a sample of particles in a liquid is followed. The suspension is dilute and so the particles are assumed to fall at their single particle terminal velocity in the liquid (usually water). Stokes' law is assumed to apply ($Re_p < 0.3$) and so the method using water is suitable only for particles typically less than 50 μm in diameter. The rate of sedimentation of the particles is followed by plotting the suspension density at a certain vertical position against time. The suspension density is directly related to the cumulative undersize and the time is related to the particle diameter via the terminal velocity.

Permeametry: This is a method of size analysis based on fluid flow through a packed bed (see Chapter 6), making use of the Carman–Kozeny equation [Equation (6.9)] for laminar flow through a randomly packed bed of uniformly sized spheres. In this method, the pressure gradient across a packed bed of known void fraction is measured as a function of flow rate. The diameter we calculate from the Carman–Kozeny equation is the surface-volume diameter $\bar{x}_{SV}$. This method has the advantage that it gives surface–volume mean results directly, which may be useful for processes involving particle–fluid interaction.

Electrozone sensing (sometimes known as the Coulter counter, after its inventor): Particles are held in suspension in a dilute electrolyte which is drawn through a tiny orifice with a voltage applied across it. As particles flow through the orifice a voltage pulse is recorded. The amplitude of the pulse can be related to the volume of the particle passing the orifice. Thus, by electronically counting and classifying the pulses according to amplitude this technique can give a number distribution of the equivalent volume sphere diameter. For process industry use, this method has been largely superseded by light-scattering methods.

1.8   SURFACE AREA MEASUREMENT

1.8.1   Adsorption Isotherms

Adsorption isotherms are graphical representations of the adsorption of gas molecules onto a material surface as the gas pressure is increased. The shape of the isotherm is determined by the nature of the surface (e.g. smooth, porous, or microporous), the interactions between the gas molecules and the surface and the interaction between the gas molecules.

The common adsorption isotherms, known as Type I, II, III, IV and V, are shown in Figure 1.13.

Type I is also known as the Langmuir isotherm and indicates that a monolayer of molecules has formed on the surface and further increase in pressure causes no increase in adsorption.

Type II describes the adsorption processes for both non-porous and macroporous solids. The similarity to the Type I isotherm at low pressures suggests monolayer formation. Increase in pressure results in multilayer adsorption.

Similar considerations explain the forms on Types III, IV and V isotherms.

1.8.2   The BET Technique

The specific surface area of powders and other materials is commonly measured by the Brunauer, Emmett and Teller (BET) technique, which is based on the BET theory of physical adsorption of gas molecules onto the surface of the material of the particles. The BET

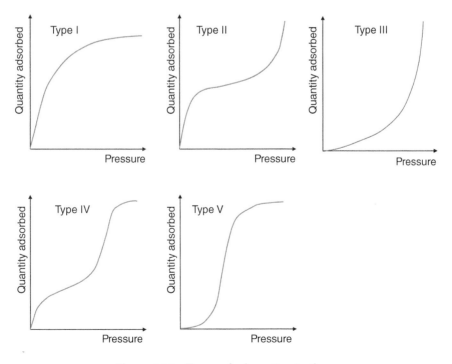

Figure 1.13   Types of adsorption isotherm

theory is based on several assumptions including physical adsorption only (i.e. no chemical adsorption) and multilayer adsorption of molecules on the material surface. The technique models the adsorption isotherm with the equation:

$$\frac{n_{\text{ads}}}{n_{\text{m}}} = \frac{CP/P_0}{(1 - P/P_0)(1 + (C-1)P/P_0)} \tag{1.17}$$

where n_{ads} is the adsorbed quantity of vapour, n_{m} is the quantity of vapour adsorbed if a monolayer were formed (theoretical only), C is a constant which describes the relative strength of the interaction between the material surface and the first layer of molecules compared with subsequent layers, P is the applied pressure and P_0 is the saturated vapour pressure at the applied temperature

The BET measurement of specific surface area involves measurement of n_{ads} for a series of values of P/P_0. The values of C and n_{m} are not known *a priori*. For ease of analysis, Equation (1.17) is rearranged to give

$$\frac{P/P_0}{n_{\text{ads}}(1 - P/P_0)} = \frac{1}{n_{\text{m}}C} + \frac{(C-1)}{n_{\text{m}}C}P/P_0 \tag{1.18}$$

Based on Equation (1.18), values for $\dfrac{P/P_0}{n_{\text{ads}}(1 - P/P_0)}$ are plotted against P/P_0 in the region where the BET model is valid (typically in the range of P/P_0 from 0.05 to 0.30); the BET plot gives a straight line of slope $\dfrac{C-1}{n_{\text{m}}C}$ and intercept $\dfrac{1}{n_{\text{m}}C}$. From the slope and intercept in that region, n_{m} and C can be calculated. From n_{m}, knowing the adsorption cross section of the adsorbed vapour molecules, we can determine the surface area per unit mass of material.

Figure 1.14 shows the BET model prediction as a function of the value of C. One can see that the BET theory might model Type III isotherms (at low values of C), Type II

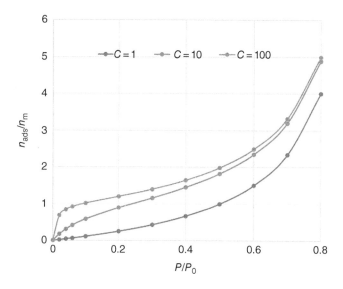

Figure 1.14   Isotherms predicted by the BET model as a function of the value of C [axis labels as for Equation (1.17)]

isotherms (mid-range values of C) and Type IV over some of its range, but does not come close to modelling the other types of isotherms. So, although this method is used for most materials, it is most suitable for materials with a Type II or Type IV isotherm. Results for materials with other types of isotherms should be treated with caution. It should also be borne in mind that any linear portion is small and that the analysis usually relies on discarding a considerable amount of data.

BET most commonly uses nitrogen gas at $-196°C$ (boiling point of liquid nitrogen at atmospheric pressure), but other gases may be used; for example, argon at $-186°C$, krypton at $-196°C$, and carbon dioxide (CO_2) at $0°C$ or at $25°C$.

Sample preparation is important in order to produce accurate and repeatable results. By applying elevated temperature with vacuum or inert gas flow in a process called degassing, impurities are removed from the surface of the sample.

1.9 SAMPLING

In practice, the size distributions of many tonnes of powder are often assumed from an analysis performed on just a few grams or milligrams of sample. The importance of that sample being representative of the bulk powder cannot be overstated. However, as pointed out in Chapter 10 on mixing and segregation, most powder handling and processing operations (pouring, belt conveying, handling in bags or drums, motion of the sample bottle, etc.) cause particles to segregate according to size and to a lesser extent density and shape. This natural tendency towards segregation means that extreme care must be taken in sampling.

There are two golden rules of sampling:

(1) The powder should be in motion when sampled.

(2) The whole of the moving stream should be taken for many short time increments.

Since the eventual sample size used in the analysis may be very small, it is often necessary to split the original sample in order to achieve the desired amount for analysis. These sampling rules must be applied at every step of sampling and sample splitting.

Splitting of samples without segregation can be achieved by flow through a splitter box (a series of knives mounted in a frame) or by use of a spinning *riffler*, in which the flow is deposited into a succession of containers rotating on a turntable.

Detailed description of the many devices and techniques used for sampling in different process situations and sample dividing are outside the scope of this chapter. However, Allen (1990) gives an excellent account, to which the reader is referred.

1.10 WORKED EXAMPLES

WORKED EXAMPLE 1.1

Calculate the equivalent volume sphere diameter x_v and the surface-volume equivalent sphere diameter x_{sv} of a cuboid particle of side length 1, 2, and 4 mm.

Solution

The volume of cuboid = $1 \times 2 \times 4 = 8 \, \text{mm}^3$.

The surface area of the particle = $(1 \times 2) + (1 \times 2) + (1 + 2 + 1 + 2) \times 4 = 28 \, \text{mm}^2$.

The volume of sphere of diameter x_v is $\pi x_v^3 / 6$.

Hence, diameter of a sphere having a volume of $8 \, \text{mm}^3$, $x_v = 2.481 \, \text{mm}$.

The *equivalent volume sphere diameter* x_v of the cuboid particle is therefore $x_v = 2.481 \, \text{mm}$.

The surface–volume ratio of the cuboid particle $= \dfrac{28}{8} = 3.5 \, \text{mm}^2/\text{mm}^3$.

The surface–volume ratio for a sphere of diameter x_{sv} is therefore $6/x_{sv}$.

Hence, the diameter of a sphere having the same surface–volume ratio as the particle = $6/3.5 = 1.714 \, \text{mm}$.

The *surface-volume equivalent sphere diameter* of the cuboid, $x_{sv} = 1.714 \, \text{mm}$.

WORKED EXAMPLE 1.2

Convert the surface distribution described by the following equation to a cumulative volume distribution:

$$F_S = (x/45)^2 \quad \text{for} \quad x \leq 45 \, \mu\text{m}$$
$$F_S = 1 \quad \text{for} \quad x > 45 \, \mu\text{m}$$

Solution

From Equations (1.4) and (1.5),

$$f_v(x) = \frac{k_v}{k_s} x f_s(x)$$

Integrating between sizes 0 and x:

$$F_v(x) = \int_0^x \left(\frac{k_v}{k_s}\right) x f_s(x) \, \mathrm{d}x$$

Noting that $f_s(x) = \mathrm{d}F_s/\mathrm{d}x$, we see that

$$f_s(x) = \frac{\mathrm{d}}{\mathrm{d}x} \left(\frac{x}{45}\right)^2 = \frac{2x}{(45)^2}$$

and our integral becomes

$$F_v(x) = \int_0^x \left(\frac{k_v}{k_s}\right) \frac{2x^2}{(45)^2} \, dx$$

Assuming that k_v and k_s are independent of size,

$$F_v(x) = \left(\frac{k_v}{k_s}\right)\int_0^x \frac{2x^2}{(45)^2} \, dx$$

$$= \frac{2}{3}\left[\frac{x^3}{(45)^2}\right]\frac{k_v}{k_s}$$

k_v/k_s may be found by noting that $F_v(45) = 1$; hence

$$\frac{90}{3}\frac{k_v}{k_s} = 1 \quad \text{and so} \quad \frac{k_v}{k_s} = 0.0333$$

Thus, the formula for the volume distribution is

$$F_v = 1.096 \times 10^{-5}x^3 \quad \text{for} \quad x \leq 45 \, \mu m$$
$$F_v = 1 \quad \text{for} \quad x > 45 \, \mu m$$

WORKED EXAMPLE 1.3

What mean particle size do we use in calculating the pressure gradient for flow of a fluid through a packed bed of particles using the Carman–Kozeny equation (see Chapter 6)?

Solution

The Carman–Kozeny equation for laminar flow through a randomly packed bed of particles is:

$$\frac{(-\Delta p)}{L} = K\frac{(1-\varepsilon)^2}{\varepsilon^3}S_v^2\mu U$$

where S_v is the specific surface area of the bed of particles (particle surface area per unit particle volume) and the other terms are defined in Chapter 6. If we assume that the bed void fraction is independent of particle size, then to write the equation in terms of a mean particle size, we must express the specific surface, S_v, in terms of that mean. The particle size we use must give the same value of S_v as the original population or particles. Thus, the mean diameter $\bar{x}$ must conserve the surface and volume of the population; that is, the mean must enable us to calculate the total volume from the total surface of the particles. This mean is the surface-volume mean $\bar{x}_{sv}$

$$\bar{x}_{sv} \times (\text{total surface}) \times \frac{\alpha_v}{\alpha_s} = (\text{total volume})\left(\text{e.g. for spheres,} \frac{\alpha_v}{\alpha_s} = \frac{1}{6}\right)$$

$$\text{and therefore } \bar{x}_{sv}\int_0^\infty f_s(x)dx \cdot \frac{k_v}{k_s} = \int_0^\infty f_v(x)dx$$

$$\text{Total volume of particles } V = \int_0^\infty x^3 \alpha_V N f_N(x) dx$$

$$\text{Total surface area of particles } S = \int_0^\infty x^2 \alpha_S N f_N(x) dx$$

$$\text{Hence, } \overline{x}_{sv} = \frac{\alpha_S}{\alpha_v} \frac{\int_0^\infty x^3 \alpha_V N f_N(x) dx}{\int_0^\infty x^2 \alpha_S N f_N(x) dx}$$

Then, since α_V, α_S and N are independent of size x,

$$\overline{x}_{sv} = \frac{\int_0^\infty x^3 f_N(x) dx}{\int_0^\infty x^2 f_N(x) dx} = \frac{\int_0^1 x^3 \, dF_N}{\int_0^1 x^2 \, dF_N}$$

This is the definition of the mean which conserves surface and volume, known as the surface-volume mean $\overline{x}_{SV}$.

So

$$\overline{x}_{SV} = \frac{\int_0^1 x^3 \, dF_N}{\int_0^1 x^2 \, dF_N}$$

The correct mean particle diameter is therefore the surface-volume mean as defined above. (We saw in Section 1.5 that this may be calculated as the arithmetic mean of the surface distribution $\overline{x}_{aS}$, or the harmonic mean of the volume distribution.) Then in the Carman–Kozeny equation we make the following substitution for S_v:

$$S_v = \frac{1}{\overline{x}_{SV}} \frac{k_s}{k_v}$$

e.g. for spheres, $S_v = 6/\overline{x}_{SV}$.

WORKED EXAMPLE 1.4

A gravity settling device processing a feed with size distribution $F(x)$ and operates with a grade efficiency $G(x)$. Its total efficiency is defined as:

$$E_T = \int_0^1 G(x) dF_M$$

How is the mean particle size to be determined?

Solution

Assuming plug flow (see Chapter 3), $G(x) = U_T A/Q$ where A is the settling area, Q is the volume flow rate of suspension and U_T is the single particle terminal velocity for particle size x, given by (in the Stokes region):

$$U_T = \frac{x^2 \left(\rho_p - \rho_f \right) g}{18 \mu} \quad \text{(Chapter 2)}$$

hence

$$E_T = \frac{A g \left(\rho_p - \rho_f \right)}{18 \mu Q} \int_0^1 x^2 \, dF_M$$

where $\int_0^1 x^2 \, dF_M$ is seen to be the definition of the quadratic mean of the distribution by mass $\bar{x}_{qM}$ (see Table 1.1).

This approach may be used to determine the correct mean to use in many applications.

WORKED EXAMPLE 1.5

Analysis of a cracking catalyst sample using an electrozone sensing device gives the following cumulative volume distribution:

Channel	1	2	3	4	5	6	7	8
Lower size of range (µm)	3.17	4.0	5.04	6.35	8.0	10.08	12.7	16
Upper size of range (µm)	4.0	5.04	6.35	8.0	10.08	12.7	16	20.16
% Volume cumulative	0	0.5	1.0	1.6	2.6	3.8	5.7	8.7

Channel	9	10	11	12	13	14	15	16
Lower size of range (µm)	20.16	25.4	32	40.32	50.8	64	80.63	101.59
Upper size of range (µm)	25.4	32	40.32	50.8	64	80.63	101.59	128
% Volume cumulative	14.3	22.2	33.8	51.3	72.0	90.9	99.3	100

(a) Plot the cumulative volume distribution versus size and determine the median size.

(b) Determine the surface distribution, giving assumptions. Compare with the volume distribution.

(c) Determine the harmonic mean diameter of the volume distribution.

(d) Determine the arithmetic mean diameter of the surface distribution.

Solution

(a) The cumulative undersize distribution is determined from the volume differential distribution data supplied and is shown numerically in column 5 of Table 1.W5.1 and graphically
in Figure 1.W5.1. By inspection, we see that the median size is 50 μm (b), i.e. 50% by volume
of the particles is less than 50 μm.

Table 1.W5.1   Size distribution data associated with Worked Example 1.5

1 Channel number	2 Lower size of range (μm)	3 Upper size of range (μm)	4 Cumulative percent undersize	5 F_v	6 $1/x$	7 Cumulative area under F_v versus $1/x$	8 F_s	9 Cumulative area under F_S versus x
1	3.17	4.00	0	0	0.2500	0.0000	0.0000	0.0000
2	4.00	5.04	0.5	0.005	0.1984	0.0011	0.0403	0.1823
3	5.04	6.35	1	0.01	0.1575	0.0020	0.0723	0.3646
4	6.35	8.00	1.6	0.016	0.1250	0.0029	0.1028	0.5834
5	8.00	10.08	2.6	0.026	0.0992	0.0040	0.1432	0.9480
6	10.08	12.70	3.8	0.038	0.0787	0.0050	0.1816	1.3855
7	12.70	16.00	5.7	0.057	0.0625	0.0064	0.2299	2.0782
8	16.00	20.16	8.7	0.087	0.0496	0.0081	0.2904	3.1720
9	20.16	25.40	14.3	0.143	0.0394	0.0106	0.3800	5.2138
10	25.40	32.00	22.2	0.222	0.0313	0.0134	0.4804	8.0942
11	32.00	40.32	33.8	0.338	0.0248	0.0166	0.5973	12.3236
12	40.32	50.80	51.3	0.513	0.0197	0.0205	0.7374	18.7041
13	50.80	64.00	72	0.72	0.0156	0.0242	0.8689	26.2514
14	64.00	80.63	90.9	0.909	0.0124	0.0268	0.9642	33.1424
15	80.63	101.59	99.3	0.993	0.0098	0.0277	0.9978	36.2051
16	101.59	128.00	100	1	0.0078	0.0278	1.0000	36.4603

Note: Based on arithmetic means of size ranges.

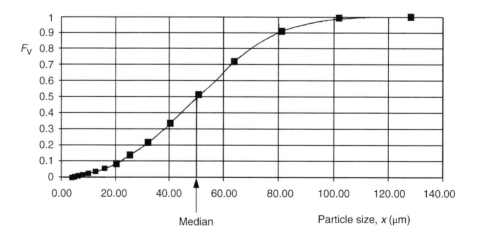

Figure 1.W5.1   Cumulative volume distribution

(b) The surface distribution is related to the volume distribution by the expression:

$$f_s(x) = \frac{f_v(x)}{x} \times \frac{k_s}{k_v} \quad \text{[from Equations (1.4) and (1.5)]}$$

Recalling that $f(x) = dF/dx$ and integrating between 0 and x:

$$\frac{k_s}{k_v} \int_0^x \frac{1}{x} \frac{dF_v}{dx} \, dx = \int_0^x \frac{dF_s}{dx} \, dx$$

or

$$\frac{k_s}{k_v} \int_0^x \frac{1}{x} \, dF_v = \int_0^x dF_s = F_s(x)$$

(assuming particle shape is invariant with size so that k_s/k_v is constant).

So, the surface distribution can be found from the area under a plot of $1/x$ versus F_v multiplied by the factor k_s/k_v (which is found by noting that $\int_{x=0}^{x=\infty} dF_s = 1$).

Column 7 of Table 1.W5.1 shows the area under $1/x$ versus F_v. The factor k_s/k_v is therefore equal to 0.0278. Dividing the values of column 7 by 0.0278 gives the surface distribution F_S shown in column 8. The surface distribution is shown graphically in Figure 1.W5.2. The shape of the surface distribution is quite different from that of the volume distribution; the smaller particles make up a high proportion of the total surface. The median of the surface distribution is around 35 µm, i.e. particles under 35 µm contribute 50% of the total surface area.

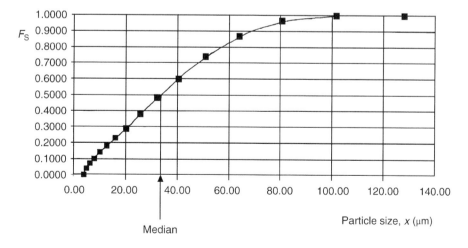

Figure 1.W5.2 Cumulative surface distribution

(c) The harmonic mean of the volume distribution is given by:

$$\frac{1}{\bar{x}_{hV}} = \int_0^1 \left(\frac{1}{x}\right) dF_v$$

This can be calculated graphically from a plot of F_v versus $1/x$ or numerically from the tabulated data in column 7 of Table 1.W5.1. Hence,

$$\frac{1}{\bar{x}_{hV}} = \int_0^1 \left(\frac{1}{x}\right) dF_v = 0.0278$$

and so, $\bar{x}_{hV} = 36\,\mu m$.

We recall that the harmonic mean of the volume distribution is equivalent to the surface-volume mean of the population.

(d) The arithmetic mean of the surface distribution is given by:

$$\bar{x}_{aS} = \int_0^1 x dF_s$$

This may be calculated graphically from our plot of F_s versus x (Figure 1.W5.2) or numerically using the data in Table 1.W5.1. This area calculation as shown in column 9 of the table shows the cumulative area under a plot of F_s versus x and so the last figure in this column is equivalent to the above integral.

Thus:

$$\bar{x}_{as} = 36.4\,\mu m$$

We may recall that the arithmetic mean of the surface distribution is also equivalent to the surface-volume mean of the population. This value compares well with the value obtained in (c) above.

WORKED EXAMPLE 1.6

Consider a cuboid particle $5.00 \times 3.00 \times 1.00$ mm. Calculate for this particle the following diameters:

(a) the volume diameter (the diameter of a sphere having the same volume as the particle);

(b) the surface diameter (the diameter of a sphere having the same surface area as the particle);

(c) the surface-volume diameter (the diameter of a sphere having the same external surface–volume ratio as the particle);

(d) the sieve diameter (the width of the minimum aperture through which the particle will pass);

(e) the projected area diameters (the diameter of a circle having the same area as the projected area of the particle resting in a stable position).

Solution

(a) Volume of the particle = $5 \times 3 \times 1 = 15\,mm^3$

Volume of a sphere = $\dfrac{\pi x_v^3}{6}$

Thus, volume diameter $x_v = \sqrt[3]{\dfrac{15 \times 6}{\pi}} = 3.06\,mm$

(b) Surface area of the particle = $2 \times (5 \times 3) + 2 \times (1 \times 3) + 2 \times (1 \times 5) = 46\,mm^2$

Surface area of sphere = πx_s^2

Therefore, surface diameter $x_s = \sqrt{\dfrac{46}{\pi}} = 3.83\,mm$

(c) Ratio of surface–volume of the particle = $46/15 = 3.0667$.

For a sphere, surface–volume ratio = $\dfrac{6}{x_{sv}}$

Therefore, $x_{sv} = \dfrac{6}{3.0667} = 1.96\,mm$

(d) The smallest square aperture through which this particle will pass is 3 mm. Hence, the sieve diameter $x_p = 3\,mm$

(e) This particle has three projected areas in stable positions:

Area 1 = $3\,mm^2$; area 2 = $5\,mm^2$; area 3 = $15\,mm^2$

Area of circle = $\dfrac{\pi x^2}{4}$

hence, projected area diameters:

Projected area diameter 1 = 1.95 mm;
Projected area diameter 2 = 2.52 mm;
Projected area diameter 3 = 4.37 mm;

WORKED EXAMPLE 1.7

BET adsorption data in the approximately linear range for Krypton on the surface of a powder sample is presented below:

P/P_0	n_{ads} (mmol/g)
0.063	0.0014
0.12	0.00173
0.202	0.00202

Given that the cross-sectional area of an adsorbed Krypton atom is 0.21 nm^2, determine the BET-specific surface area of the powder sample (Avogadro's number = 6.02×10^{23} mol^{-1}).

Solution

BET theory assumes that the linear range is described by Equation (1.18)

$$\frac{P/P_0}{n_{ads}(1-P/P_0)} = \frac{1}{n_m C} + \frac{(C-1)}{n_m C} P/P_0 \qquad (1.18)$$

Calculating $(P/P_0)/(n_{ads}(1-P/P_0))$:

P/P_0	n_{ads} (mmol/g)	$(P/P_0)/(n_{ads}(1-P/P_0))$ (g/mmol)
0.063	0.00140	48.03
0.120	0.00173	78.81
0.202	0.00202	125.31

From this, the slope of the straight-line region $\dfrac{C-1}{n_m C}$ = 556.8 g/mmol

Intercept on the P/P_0 axis, $\dfrac{1}{n_m C}$ = 12.60 g/mmol

From the values for slope and intercept we can solve for C and n_m

Hence, C = 7016 mmol/g powder and n_m = 0.00180 mmol/g powder

So, for every gram of powder, 0.00180 mmol of Krypton has been adsorbed in forming a monolayer.

Hence, the number of Krypton adsorbed per gram of powder:

$$= 0.00180 \times 6.02 \times 10^{23} \text{ molecules per gram} = 1.08 \times 10^{18} \text{ molecules per gram}$$

Since each adsorbed Krypton atom occupies an area of 0.21 nm^2, the specific surface area of the powder sample is:

$$\text{Specific surface area} = \left(1.08 \times 10^{18}\right) \times \left(0.21 \times 10^{-18}\right) = 0.227 \text{ m}^2/\text{g}$$

TEST YOURSELF

1.1 Define the following equivalent sphere diameters: *equivalent volume diameter, equivalent surface diameter, and equivalent surface-volume diameter*. Determine the values of each one for a cuboid of dimensions 2 mm × 3 mm × 6 mm.

1.2 List three types of distribution that might be used in expressing the range of particle sizes contained in a given sample.

1.3 If we measure a number distribution and wish to convert it to a surface distribution, what assumptions must be made?

1.4 Write down the mathematical expression defining (a) the quadratic mean and (b) the harmonic mean.

1.5 For a given particle size distribution, the mode, the arithmetic mean, the harmonic mean and the quadratic mean all have quite different numerical values. How do we decide which mean is appropriate for describing the powder's behaviour in a given process?

1.6 What are the golden rules of sampling?

1.7 When using the DLS method for the determination of particle size distribution, what assumptions are made?

1.8 Image analysis is used to determine the size of a distribution of cubic particles. What equivalent volume diameter does it determine if the particles are presented in their most stable orientation and how does this differ from the real equivalent volume diameter?

EXERCISES

1.1 For a regular cuboid particle of dimensions $1.00 \times 2.00 \times 6.00$ mm, calculate the following diameters:
(a) the equivalent volume sphere diameter;

(b) the equivalent surface sphere diameter;

(c) the surface–volume diameter (the diameter of a sphere having the same external surface–volume ratio as the particle);

(d) the sieve diameter (the width of the minimum aperture through which the particle will pass);

(e) the projected area diameters (the diameter of a circle having the same area as the projected area of the particle resting in a stable position).

[Answer: (a) 2.84 mm; (b) 3.57 mm; (c) 1.80 mm; (d) 2.00 mm; (e) 2.76, 1.60 and 3.91 mm.]

1.2 Repeat Exercise 1.1 for a regular cylinder of diameter 0.100 mm and length 1.00 mm.

[Answer: (a) 0.247 mm; (b) 0.324 mm; (c) 0.142 mm; (d) 0.10 mm; (e) 0.10 mm (unlikely to be stable in this position) and 0.357 mm.]

1.3 Repeat Exercise 1.1 for a disc-shaped particle of diameter 2.00 mm and length 0.500 mm.

[Answer: (a) 1.44 mm; (b) 1.73 mm; (c) 1.00 mm; (d) 2.00 mm; (e) 2.00 and 1.13 mm (unlikely to be stable in this position).]

1.4 Estimate the (a) arithmetic mean, (b) quadratic mean, (c) cubic mean, (d) geometric mean and
(e) harmonic mean of the following distribution.

Size	2	2.8	4	5.6	8	11.2	16	22.4	32	44.8	64	89.6
Cumulative % undersize	0.1	0.5	2.7	9.6	23	47.9	73.8	89.8	97.1	99.2	99.8	100

[Answer: (a) 13.6; (b) 16.1; (c) 19.3; (d) 11.5; (e) 9.8.]

1.5 The following volume distribution was derived from a sieve analysis:

Size (µm)	37–45	45–53	53–63	63–75	75–90	90–106	106–126	126–150	150–180	180–212
Volume % in range	0.4	3.1	11	21.8	27.3	22	10.1	3.9	0.4	0

(a) Estimate the arithmetic mean of the volume distribution.

From the volume distribution derive the number distribution and the surface distribution,
giving assumptions made. Estimate:

(b) the mode of the surface distribution;

(c) the harmonic mean of the surface distribution.

Show that the arithmetic mean of the surface distribution conserves the surface–volume ratio
of the population of particles.

[Answer: (a) 86 µm; (b) 70 µm; (c) 76 µm.]

1.6 BET adsorption data in the approximately linear range for Krypton on the surface of a powder
sample is presented below:

P/P_0	n_{ads} (mmol/g)
0.05	0.02240
0.11	0.02758
0.195	0.03232

Given that the cross-sectional molecular area of an adsorbed Krypton molecule is $0.21 \, \text{nm}^2$,
determine the BET-specific surface area of the powder sample (Avogadro's number =
$6.02 \times 10^{23} \, \text{mol}^{-1}$).

[Answer: $3.4 \, \text{m}^2/\text{g}$.]

1.7 The data below is extracted from the results of a BET experiment for nitrogen adsorption on the surface of a powder sample:

P/P_0	n_{ads} (STP cm^3/g)
0.1004	0.2806
0.2000	0.3424
0.2994	0.3934

Given that the cross-sectional molecular area of an adsorbed nitrogen molecule is 0.162 nm^2, determine the BET-specific surface area of the powder sample (Avogadro's number = 6.02 × 10^{23} mol^{-1}).

[Answer: 1.24 m^2/g.]

2
Mechanical Properties of Particles

This chapter deals with the mechanical properties of particles and the ways in which they affect the behaviour of particles in processes.

Consider a single particle surrounded by other particles in movement (Figure 2.1). What forces act upon it and what is its response?

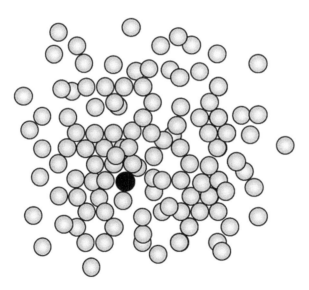

Figure 2.1   A single particle interacting with others

The particles are surrounded by a continuous fluid – a gas or a liquid. The drag and other forces which act as the particles move relative to the surrounding fluid are considered in detail in Chapter 3. It is sufficient here to note that the influence of the drag forces is greater as particles become smaller in size.

Introduction to Particle Technology, Third Edition. Martin Rhodes and Jonathan Seville.
© 2024 John Wiley & Sons Ltd. Published 2024 by John Wiley & Sons Ltd.
Website: www.wiley.com/go/rhodes/particle3e

It is obvious that particles will interact with each other through particle–particle contacts or collisions. What kind of contacts these are will depend on the nature of the *flow*. In dense, low-velocity flows in a storage bin or silo (see Chapter 9), for example, particles are in more or less continuous contact with each other and may be subjected to considerable stress. On the other hand, in rapid aerated flows in fluidization (see Chapter 6), for example, collisions may occur at high relative velocities but particles are under smaller compressive stresses. Sometimes both types of flow occur in the same piece of equipment, such as the rotating drum shown in Figure 2.2, for example.

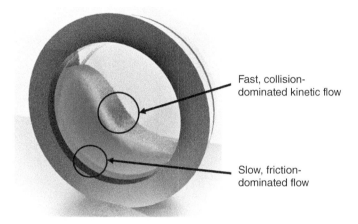

Fast, collision-
dominated kinetic flow

Slow, friction-
dominated flow

Figure 2.2 Both slow, friction-dominated flow and rapid, inertia-dominated flow occur in a rotating drum if the speed is sufficiently high. Image reused with permission from Herald et al. 2022.

Collisions result in dissipation and exchange of energy between particles and the resulting change in the energy and direction of motion, multiplied over all the particles in the system, results in the bulk motion, which may range from slow friction-dominated flow to rapid inertia-dominated *kinetic* flow. Collisions can also produce permanent changes in the particles themselves: damage due to major fractures or localized abrasion (see Chapter 11) or, if the particles are sufficiently energy-absorbing ('sticky'), aggregation or agglomeration into larger entities can occur (see Chapter 12).

This chapter considers the physics of particle–particle contact, while Chapter 3 considers the physics of particle–fluid interaction. Together these form the underpinning science behind computational methods for simulating the behaviour of particulate systems, which are presented in Chapter 4.

2.1   INTRODUCTION TO MATERIAL PROPERTIES

Particles can be made up of an almost infinite variety of materials, from very hard, brittle substances such as quartz, a major constituent of what is often called 'sand', to the easily deformable polymers known as 'plastics'. Not surprisingly, these behave in very different ways when their particles come into contact.

When a force is applied to a material, how does it behave? The simplest form of material behaviour is *elasticity*, which is easiest to visualize and to measure in a simple tensile test, as illustrated in Figure 2.3.

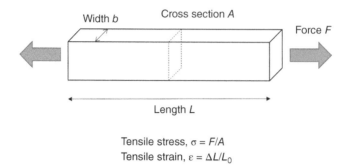

Tensile stress, $\sigma = F/A$
Tensile strain, $\varepsilon = \Delta L/L_0$

Figure 2.3   A simple tensile test

The bar in Figure 2.3 is subjected to a tensile force F. The force per unit cross-sectional area, or the *tensile stress* $\sigma = F/A$. In response the length of the bar increases by an amount ΔL, so that the *tensile strain* ε[1], is equal to $\Delta L/L_0$, where L_0 is the original length.

Figure 2.4(a) shows perfect linear elastic behaviour: the bar lengthens in proportion to the stress applied[2]. The gradient of this line is E, the Young's[3] modulus, which is a material property characterizing elasticity and of great importance in any consideration of particle contact. The behaviour shown in Figure 2.4(a) is *reversible*; that is, the elongation goes back to zero along the same line as the load is reduced and the energy stored in the stretched solid is recovered again. This is an important feature of perfectly elastic behaviour and is approximately the case for an elastic rubber band, for example.

Values of the Young's modulus vary widely. Examples for some common materials are given in Table 2.1.

Most solid materials show some initial elastic deformation in a tensile test, as in Figure 2.4(a), but it is obvious that extension cannot continue indefinitely. As the load is increased, many materials show an *elastic limit* at a *yield stress*, Y, beyond which they yield and become *plastic*, as shown in Figure 2.4(b). In effect, above the yield point they *flow*, showing deformation which is not reversible. In this case, the energy put into

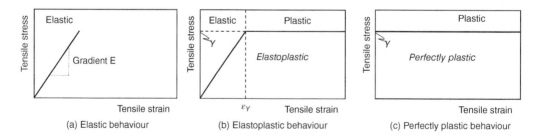

Figure 2.4   Stress–strain curves for (a) elastic, (b) elastoplastic and (c) perfectly elastic behaviour

---

[1] Note that ε is commonly used for strain as well as void fraction. Be careful not to confuse the two!
[2] Linear-elastic behaviour is sometimes described as Hookean, after Robert Hooke (1635–1703), English scientist and architect, who first documented it.
[3] After Thomas Young (1773–1829), English scientist and polymath.

Table 2.1 Indicative values of material properties for some common solids

Material	Density (kg/m³)	Young's modulus (GPa)	Poisson's ratio (–)
Quartz	2650	75–100	0.17
Glass	2400–3100	70	0.22
Steel	7600–7900	200	0.3
MMC[a]	500–1600	0.5–8.5	0.02–0.29
Polyethylene (HDPE)	930–970	1.0–1.4	0.45

[a] MMC = microcrystalline cellulose, a common pharmaceutical excipient; values represent the range from 66% internal porosity to almost fully dense.

extending the bar is not all recovered as the stress is reduced to zero. Materials that behave in this way are termed *elastoplastic*.

Figure 2.4(c) shows the extreme case in which yield occurs as soon as the load is imposed. This can apply to very soft materials such as some suspensions of particles in a fluid, which are considered in Chapter 5.

Stretching of a bar of material, as shown in Figure 2.5, has another effect, which is that as the length increases, the width b, will generally decrease. For incompressible materials the total volume is conserved. This effect is quantified by the use of Poisson's ratio[4] ν, as follows:

The tensile strain $\varepsilon_{11} = \Delta L/L_0$ where the indices 1,1 indicate the direction in which the stress is applied and the direction in which the strain is measured.

The lateral strain $\varepsilon_{22} = \varepsilon_{33} = -\Delta b/b_0$ where b_0 is the initial lateral width.

Poisson's ratio $= -\varepsilon_{22}/\varepsilon_{11}$

Values of Poisson's ratio of some common materials are given in Table 2.1. The maximum value of Poisson's ratio, for an incompressible material, is 0.5, as shown in Worked

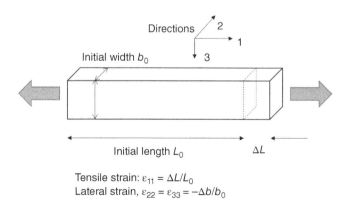

Directions   2
                1
Initial width b_0 3

Initial length L_0 ΔL

Tensile strain: $\varepsilon_{11} = \Delta L/L_0$
Lateral strain, $\varepsilon_{22} = \varepsilon_{33} = -\Delta b/b_0$

Figure 2.5 Poisson's ratio for a bar under tensile stress

[4] After Simeon Denis Poisson (1781–1840), French mathematician and physicist.

Example 2.1. For compressible materials ν is less than 0.5, and in some extreme cases, such as that of cellular materials like cork, ν can be negative.

In all of this section, the geometry considered is in tension. In general, all of the same concepts apply in compression, so that the Young's modulus, the yield stress and the Poisson's ratio are the same in tension and in compression.

2.2   PARTICLE–PARTICLE CONTACT FOR ELASTIC MATERIALS

Clearly, the geometry of particles in contact is very different from that of the tensile test considered in Section 2.1, but all the material properties so far introduced are also applicable and useful in the case of particle–particle contact and particle–wall contact. That is, a material parameter is a constant irrespective of how it is measured.

Consider two equally sized spherical elastic particles in contact under an applied normal load W, as shown in Figure 2.6. They will deform elastically to form a circular contact spot of radius c. This section concerns the prediction of the behaviour of the contact – particularly the extent of the deformation and the stress–strain relationship – so that we can use this in simulations such as those introduced in Chapter 4.

The most widely used analysis of this problem is due to Hertz[5], who made the following assumptions:

- Linear elastic behaviour of both spheres.

- Frictionless contact (the effects of friction are considered in Section 2.4).

- Small displacements, such that the two particles remain largely spherical.

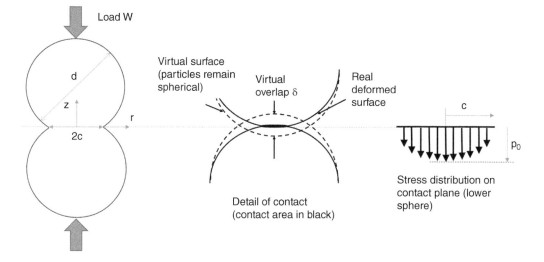

Figure 2.6   Elastic spheres in contact (exaggerated); normal stress distribution at the contact

---

[5] Heinrich Hertz (1857–1894), German physicist, also known for his work on electromagnetics.

Hertz assumed a stress distribution in the contact area of the form:

$$p(r) = p_0 \left(1 - \frac{r^2}{c^2} \right)^{1/2} \tag{2.1}$$

where p_0 is the maximum pressure on the centre line of the contact and r is the radial distance from that line. It is beyond the scope of this book to show the full derivation, but the important results arising from the Hertz analysis are as follows, for equal spheres of diameter x:

$$c^3 = \frac{3}{8} \frac{(1 - \nu^2)}{E} Wx \tag{2.2}$$

giving the dependence of contact area A, on load W:

$$A = \pi c^2 \propto W^{2/3} \tag{2.3}$$

The deformation δ, is then,

$$\delta^3 = 9 \left(\frac{1 - \nu^2}{E} \right)^2 \frac{W^2}{x} \tag{2.4}$$

giving the dependence of deformation on load:

$$\delta \propto W^{2/3} \tag{2.5}$$

This is an important result which shows how contact between linear-elastic solids can produce a nonlinear response because of the loading geometry, in this case sphere–sphere contact.

In the same way as for a bar under tension, there must be a limit to the elastic region for sphere–sphere contact and this occurs when one of the spheres yields. The shear stress distribution beneath the contact area is shown in Figure 2.7 for the similar case of a

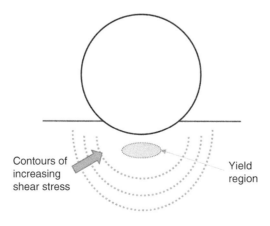

Figure 2.7   Yielding under contact (sphere–plane geometry)

sphere–plane contact. The maximum stress occurs slightly below the surface and the volume of material in the initial yield zone grows as the load is further increased. As for the case of a bar under tension, if the yield stress is exceeded then the energy of collision is not all recovered.

2.3   CONTACT IN THE PRESENCE OF SURFACE FORCES

2.3.1   Cohesion and Adhesion

Particles are sometimes described as *cohesive* (they show attraction to each other and readily form clumps or agglomerates) or *adhesive* (they show attraction to other materials such as the walls of containers). Figure 2.8 shows how cohesion and adhesion are defined. *Cohesion* is defined as the energy per unit area of interface required to separate two parts of the same substance (to infinity, in theory, although in practice they do not have to be separated by very much because of the relatively short range of the forces involved; see Chapter 5).

Adhesion is similarly defined as the energy per unit area of interface required to separate two different substances (to infinity). In each case, separation requires one interface to be destroyed while two new interfaces are created. The change in surface energy in each case is:

$$E_C/A = \gamma_{11} - 2\gamma_{13} \tag{2.6}$$

where E_C/A is the energy of cohesion and γ_{ab} is the interfacial energy associated with an interface between materials a and b, and

$$E_A/A = \gamma_{12} - \gamma_{13} - \gamma_{23} \tag{2.7}$$

where E_A/A is the energy of adhesion.

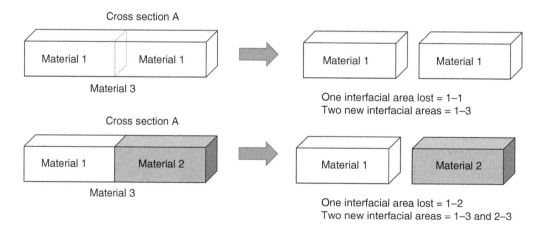

Figure 2.8   Cohesion and adhesion (note that in each case the bar is surrounded by material 3)

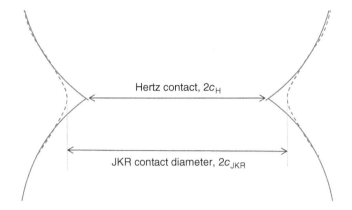

Figure 2.9   Hertz and JKR contact (exaggerated)

There are many types of forces that can cause cohesion and adhesion, including the presence of surface liquid due to being stored at high humidity, for example, and electrostatic interactions. (In the case of electrostatics these forces can also be repulsive.) The most general type of cohesive force arises from intermolecular attraction, commonly known as the *van der Waals* force. Surface forces and their origins are considered further in Chapter 5.

In this section we show the effect of cohesion on the contact between particles and how the magnitude of these effects can be calculated. As noted throughout this book, cohesive forces are of increasing importance as the particle size is reduced.

As we have seen, for perfectly elastic spheres in the absence of cohesion, the radius of the area of contact is given by Equation (2.2). Figure 2.9 shows how the shape of the contact is deformed by the action of the forces of cohesion, so enlarging the contact area, the radius of which is now given by:

$$c^3_{JKR} = \frac{3x(1-\nu^2)}{8E}\left(W + \frac{3\pi\gamma x}{4} + \sqrt{\frac{3\pi\gamma xW}{2} + \left(\frac{3\pi\gamma x}{4}\right)^2}\right) \qquad (2.8)$$

where the suffix JKR indicates that this is from the well-known analysis of Johnson, Kendall and Roberts (1971), who established this result by performing an energy balance between the interfacial or surface energy and the elastic strain energy stored in the deformed particles. Their approach has been found to be in good agreement with experiment. In this equation γ indicates the cohesive surface energy (J/m^2), as defined above if the two particles are of the same material or the adhesive surface energy if they are not. In practice, a rather simpler but more approximate approach to incorporation of cohesion and adhesion is usually used in computational models of particle systems, as discussed in Chapter 4.

There are two simplifying cases of Equation (2.8):

(1) When γ goes to zero, i.e. the effect of surface forces becomes negligible, Equation (2.8) reduces to Equation (2.2), the Hertz result.

(2) When W goes to zero, i.e. there is no applied load, the radius of the contact circle c, is given by:

$$c^3 = \frac{9}{16}\pi\gamma x^2 \frac{(1-v^2)}{E} \tag{2.9}$$

There is therefore a non-zero contact area even when the external load goes to zero. This is the same contact area as would be produced in the absence of surface forces by a load $W = 3\pi\gamma x/2$.

An important consequence of this analysis is that if cohesion is present, there is a force necessary to separate the particles, which follows from putting c_{JKR} to zero in Equation (2.8). This gives the 'pull-off force' F_C:

$$-F_C = 3\pi\gamma x/8 \tag{2.10}$$

where F_C is by definition negative because it acts in the opposite direction from the cohesive forces.

Figure 2.10 shows in exaggerated form the sequence of particle interaction under the influence of surface forces. The particles first come into contact at or near point A, where the attractive forces are able to act (point B) to pull them into equilibrium (no load) contact at point C, corresponding to Equation (2.9). If there is an applied force, the overlap between the particles can increase to point D, accompanied by an increase in the repulsive force. At that point the overlap goes into reverse and the force–overlap curve is followed in the opposite direction through C and B to point E where the pull-off force is reached and, if this is overcome, the particles separate at point F.

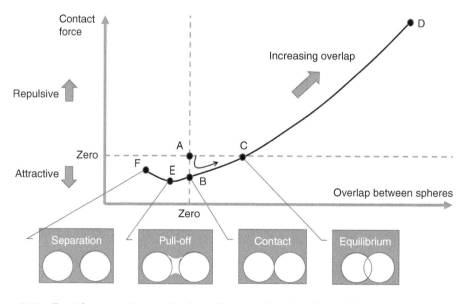

Figure 2.10   Particle interaction under the influence of surface forces (deformation exaggerated)

It is of interest to compare the pull-off force with the gravitational force on a particle F_g:

$$\frac{-F_C}{F_g} = \frac{3\pi\gamma x}{8} \bigg/ \frac{\pi x^3 \rho_p g}{6} = \frac{9}{4} \cdot \frac{\gamma}{x^2 \rho_p g} \tag{2.11}$$

where ρ_p is the particle density. In summary, the cohesive force increases with particle size but the particle weight increases much more strongly – with x^3 – so that as particle size increases, the gravitational force soon becomes dominant. The comparison between cohesive forces and particle weight is considered further below.

2.3.2 Surface Roughness and Contamination

The analysis of contact in Sections 2.2 and 2.3 assumes that the particles involved are perfectly smooth, clean spheres. In practice, of course, this is rarely the case for real particles, which may be of irregular non-spherical shape, rough and covered in surface contamination. Even if the particles are nearly spherical, they will most probably have surface roughnesses, or *asperities*, of much smaller size, as shown schematically in Figure 2.11. One way of taking account of this in calculations is to use the asperity radius in the contact equations in place of the particle radius.

The intermolecular forces which contribute to surface energy are of very short range so that unless the surfaces are very clean, the surface energy is likely to be that of the contaminant layer rather than the material itself. For example, metals will often be covered with a layer of oxide, which will determine the cohesion and adhesion to any other surface.

2.3.3 Comparison of Cohesion with Particle Weight

The effect of all of the above will be to reduce the cohesive forces between particles, as shown in Figure 2.12, which is in accordance with common experience. While Figure 2.12 suggests that a 1 mm particle, for example, will be sufficiently cohesive to support its own weight, in practice 1 mm particles are not observed to be cohesive unless they are very smooth and clean, whereas particles of 100 μm and below are often found to be cohesive.

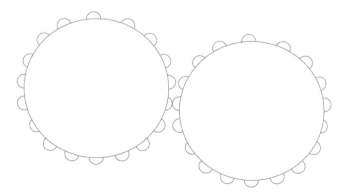

Figure 2.11 Asperity contact

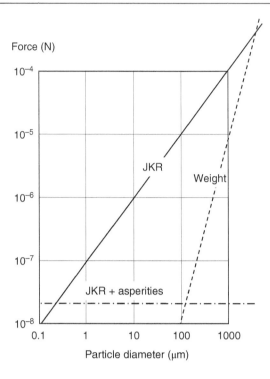

Figure 2.12   Predictions of JKR pull-off force compared with single particle weight, for properties corresponding to quartz particles (asperity size taken as 0.1 μm)

2.3.4   Measurement of Cohesion and Adhesion

Direct measurement of interparticle cohesive forces is difficult because both the physical dimensions and the magnitude of the forces acting are extremely small. The most common direct method is to attach a particle to the tip of an *atomic force microscope (AFM)* and to measure the interaction force against a flat surface such as silica. As in all such single-particle measurements, it is necessary to give careful thought to the statistical variation in results (i) between successive measurements using the same particle and (ii) between measurements for different particles. Such measurements are far from routine and are usually confined to research laboratories.

In practice, indirect measurements of cohesion are more frequently used. It has long been known that the angle of repose of a poured heap of particles is related to particle cohesion, but this measurement is not very reproducible and is influenced by factors such as preconditioning of the powder and operator skill. A more reproducible measurement is the dynamic angle of repose, as measured in a rotating drum. Figure 2.13(a) shows a set of images of the surface profile of the solids in a rotating drum, with the average interface position represented in red. A single dynamic angle of repose is obtained from the average position in Figure 2.13(b). As the drum rotates, the position of the interface will fluctuate about this average position within the bounds shown. The magnitude of this fluctuation is larger for more cohesive powders and can be used as an index of cohesion.

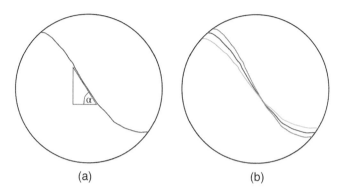

Figure 2.13 Schematic representation of the dynamic angle of repose measurement method in a rotating drum – (a) its definition and (b) how it might naturally vary during drum rotation. After Neveu *et al.* (2022)

2.4 FRICTION

Figure 2.14(a) shows a block of material 1 being pulled along an infinitely large plane of material 2, with a contact area A between the block and the plane.

According to the classical laws of friction, the force which must be applied in the direction parallel to the plane F_T, in order to cause the block to move is proportional to the normal force F_N, so that for sliding to occur

$$F_T = \mu F_N \tag{2.12}$$

where μ is known as the *friction coefficient*. This is known as the Coulomb condition[6]; the block will remain stationary for values of F_T less than μF_N.

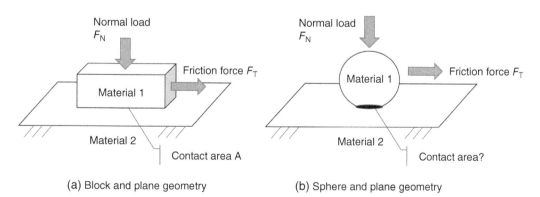

Figure 2.14 Frictional interaction (a) between a block and a plane and (b) between a sphere and a plane

---

[6] After Charles-Augustin de Coulomb (1736–1806), French engineer and physicist, also well known for his work on electricity.

The explanation above applies to the case where the area of contact between the sliding bodies remains constant. When the contact is between curved surfaces, however, such as the sphere–plane geometry shown in Figure 2.14(b) and the sphere–sphere geometries we have considered in the case of purely normal contacts, the situation becomes complicated for a number of reasons:

(1) As we have seen for normal contacts, the area of contact depends on the normal load.

(2) The deformations in the normal and tangential directions are not independent, so that changes in either will influence the area of contact.

(3) It is possible for relative movement or *slip* to occur over part of the contact area, without the entire contact area slipping (this is termed *micro-slip* and is considered further in Section 2.5 in connection with impacts).

(4) For all these reasons, the loading and unloading curves can be different, so that they show *hysteresis*.

The general behaviour in Figure 2.14(b) is therefore that the particle and the surface deform elastically under the influence of both the normal and tangential loads until slip occurs at the interface. A simplified way of dealing with this in computations is to treat the normal and tangential deformations separately and assume that the contact area depends only on the current normal load and is therefore constant during a computational time step. This is considered further in Chapter 4.

It should be noted that not all granular materials follow Equation (2.12), particularly at low normal loads. Very smooth particles, such as polymer and glass spheres, can show anomalously high friction coefficients at low loads where cohesive and adhesive forces are more noticeable. Figure 2.15 shows an example of the results from an experiment in

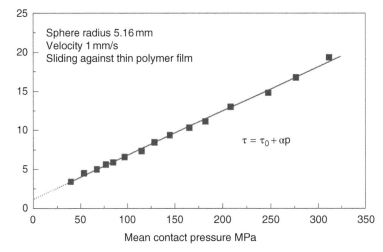

Figure 2.15 Shear stress versus contact pressure for a spherical particle sliding on a plane, showing adhesion (shear stress = shear force/area of contact; contact pressure = normal force/area of contact; Adams *et al.*, 1999)

which a single spherical particle is moved tangentially along a plane. Note that the line of best fit does not go through the origin, because of the adhesive effect, which is only noticeable at low loads. Hence, in general we can write:

$$\tau = \tau_0 + \alpha p \tag{2.13}$$

where $p = W/A$ and $\tau = F/A$

2.5   IMPACT AND BOUNCE

2.5.1   Impacts in the Normal Direction

Using the ideas of Sections 2.2 and 2.3 it is possible to describe what happens when a particle collides with another particle or a different kind of surface, as shown in Figure 2.16.

When a particle impacts on a surface, in this case in a normal direction, it first makes contact at point a. Elastic deformation of both particle and surface then proceeds and the opposing force builds up as the deformation increases to point c, the point of maximum deformation, where all of the particle's initial kinetic energy has been converted to stored elastic energy. The process is then reversed, exactly retracing the curve to point a, where separation occurs and the particle rebounds.

The comparable process for the case where cohesion is present is similar to Figure 2.10.

Collisions between particles and between a particle and a wall are often quantified by the coefficient of restitution e, defined as the ratio of the rebound velocity v_r, and the incident velocity v_i:

$$e = -v_r/v_i \tag{2.14}$$

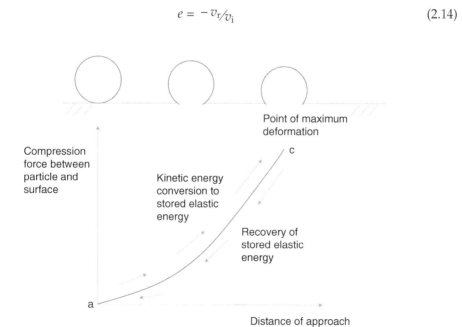

Figure 2.16   Fully elastic impact of a particle on a plane surface

where the minus sign takes account of the change of direction. Note that since the kinetic energy of a particle is $(1/2)mv^2$, the ratio of rebound to incident kinetic energies is e^2. The coefficient of restitution is not a material property like Young's modulus, for example, but depends on the details of the collision. In a normal collision, if both materials behave in a purely elastic way, e is equal to one[7]. However, at a higher incident velocity using the same materials, plastic deformation may occur and e may be greatly reduced because plastic deformation dissipates energy which cannot be recovered on rebound.

The velocity of impact required to produce yield v_y, can be derived from the analysis of the stress field at the contact (Figure 2.6) and is beyond the scope of this book. As might be expected, yield is predicted to occur when the kinetic energy of the collision exceeds a certain value, which depends on the geometry of contact and the material properties:

$$\frac{1}{2}mv_y^2 \approx f(Y, E^*, \text{radius of contact}) \tag{2.15}$$

where $1/m = 1/m_1 + 1/m_2$, m_1 and m_2 are the masses of the two colliding bodies, E^* is the composite Young's modulus of the two colliding bodies and Y is the yield stress of the softer body. For impact of a sphere on a large plane, this reduces to:

$$\rho_p v_y^2 = f(Y, E^*) \tag{2.16}$$

Figure 2.17 shows how experimental values of e for selected metals decrease with increasing velocity of impact, due to yielding.

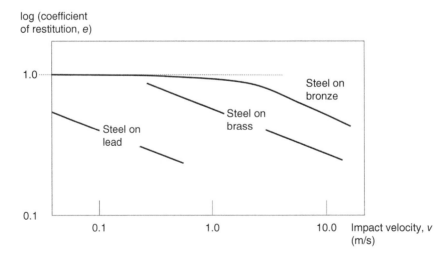

Figure 2.17 Example of the dependence of coefficient of restitution on impact velocity. At higher impact velocities, e is proportional to $v^{-1/4}$, due to yielding. After Goldsmith (1960); see also Seville and Wu (2016)

---

[7] In fact, for e to equal one all the elastic strain energy has to be returned to the departing particle(s), which may not happen if elastic waves can carry it away from the impact site.

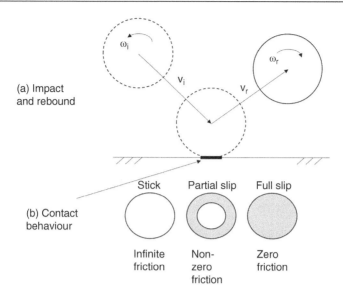

Figure 2.18 Oblique impact of a sphere with a flat surface. (a) Impact and rebound and (b) contact behaviour.

2.5.2 Oblique Impacts

Section 2.5 was concerned only with normal impacts. In general, the impact between particles and between particles and surfaces is oblique, as shown in Figure 2.18. This is a much more complex situation, where it becomes necessary to consider the angular rotation of the particles and the frictional behaviour of the surface. The interaction between the particle and the surface in Figure 2.18 can be of three types:

(1) Stick – if the friction is high, so that the particle cannot move tangentially against the surface.

(2) Partial slip – in which the outer annulus of the contact zone is slipping and the inner part is stuck.

(3) Full slip – in which the friction is sufficiently low that tangential movement is possible over the entire contact circle.

   Analysis of this situation demands a computational approach, considering each annular ring of the contact circle, and is beyond the scope of this book.

2.6 LIQUID BRIDGES

Particles are often found in multi-phase mixtures containing liquids, four possibilities being shown in Figure 2.19. If the liquid content is small it is common to find discrete *liquid bridges* at the contacts between particles, which are termed *pendular*. As the liquid content increases, some of the spaces between particles are filled, which is termed the

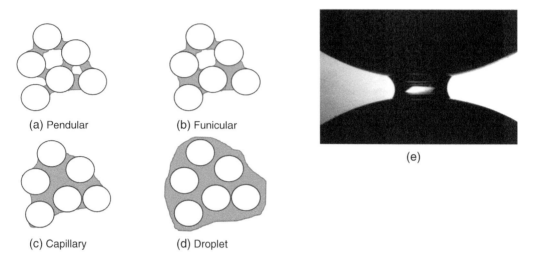

(a) Pendular   (b) Funicular

(c) Capillary   (d) Droplet

(e)

Figure 2.19 Distribution of liquids in an assembly of particles: (a) pendular; (b) funicular; (c) capillary; (d) droplet and (e) a pendular bridge between two spheres

funicular state. Increasing the liquid content to the point where all the pores are filled is called the *capillary* state.

Here we consider only the pendular region, which can give rise to surprisingly high forces between particles, as makers of sand castles and pastry will know.

Figure 2.20 shows a single pendular bridge between two spherical particles in contact. The liquid bridge has a half-angle β measured at the centre of each sphere and it is assumed that the liquid perfectly wets the sphere surface. A reasonable approximation is that the shape of the bridge is *toroidal*, with radii of curvature r_1 and r_2, as shown. In order to visualize the forces acting on each particle, imagine cutting the circular neck

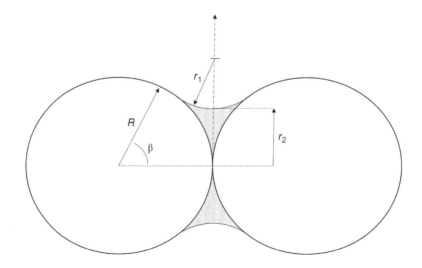

Figure 2.20 A pendular liquid bridge between two equal spheres in contact

of the bridge on the plane of symmetry. The force due to the liquid bridge F_L, which acts to pull the two particles together, is the sum of two components:

(1) a surface tension force F_1, which acts at the circular gas–liquid interface on the plane of symmetry;

(2) a 'pressure-deficit' force F_2, which arises from the fact that the pressure inside the bridge is less than outside it.

In general, whenever there is a curved interface as in Figure 2.20, a pressure difference will act over it; its magnitude depends on the curvature.

$$F_1 = 2\pi r_2 \gamma_L \tag{2.17}$$

$$F_2 = \pi r_2^2 \Delta P \tag{2.18}$$

where γ_L is the *surface tension* of the liquid (strictly, the interfacial surface energy of the liquid–gas interface) and ΔP is the difference between the pressure inside and outside the bridge, which, by the Laplace equation, is:

$$\Delta P = \gamma_L \left(\frac{1}{r_1} - \frac{1}{r_2} \right) \tag{2.19}$$

Combining these equations:

$$F_L = F_1 + F_2 = \pi r_2 \gamma_L \left(\frac{r_1 + r_2}{r_1} \right) \tag{2.20}$$

By trigonometry,

$$r_1 = R \left(\sec\beta - 1 \right) \tag{2.21}$$

$$r_2 = R \left(1 + \tan\beta - \sec\beta \right) \tag{2.22}$$

Substituting into Equation (2.20) gives:

$$F_L = \frac{2\pi R \gamma_L}{1 + \tan(\beta/2)} \tag{2.23}$$

Equation (2.23) indicates, not surprisingly, that the attractive force due to the liquid bridge increases with surface tension, but decreases with increasing size of the liquid bridge, indicated by the half-angle β. One reason for this unexpected result is that surface roughness has been neglected. Consider the effect of surface roughness, as shown in Figure 2.21, which acts to separate the two surfaces. When even a small separation between spheres, s, is included in the liquid bridge force calculation, the results in Figure 2.21 are obtained.

Note that the maximum liquid bridge force $2\pi R\gamma$ or $\pi x\gamma$, is of a similar form to the JKR pull-off force for cohesive spheres given in Equation (2.10); both are proportional to particle diameter.

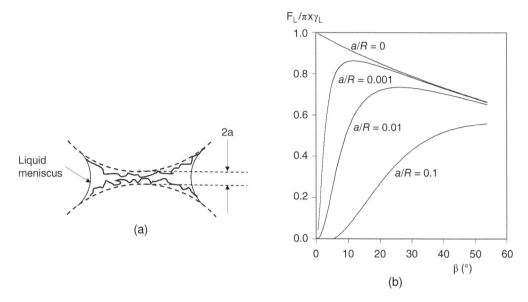

Figure 2.21   Liquid bridges – effect of roughness on attractive force (a) roughness acts to separate the spherical surfaces; (b) resulting dimensionless liquid bridge force as a function of bridge half-angle β and surface separation a/R (particle radius R; surface separation $2a$). Seville and Wu (2016) / with permission of Elsevier

Note also that the liquid bridge forces considered here are all for a static situation, i.e. no relative movement of the particles. If relative movement does occur, in collisions between wet particles, for example, viscous forces may also be important. These may be calculated by means of the Reynolds lubrication equation, which gives the force between two equal spheres of radius R and separation $2a$, being separated at a rate $2v$ (where $v = da/dt$) in a liquid of viscosity μ:

$$F_V = 3\pi\mu R^2 v/2a \qquad\qquad (2.24)$$

In Chapter 12 we consider the case of particles with wet surfaces impacting on each other in an agglomeration process. If their kinetic energy is dissipated in the collision, they will stick to each other and growth will occur.

This introduction to the forces due to liquid bridges between particles covers only a small part of a large subject. For example:

- Real particles are seldom spherical so that in addition to any effect of roughness, the contact geometry may be more like point–plane or edge–plane, which would modify the form of the expression for liquid bridge force.

- In the pendular state, the liquid bridges are isolated and liquid can only move between them due to migration of molecules over the particle surface, which is slow. As the quantity of liquid increases (causing angle β in Figure 2.20 to increase), there comes a point where neighbouring liquid bridges can make contact. At that point, liquid can flow easily between the bridges, the spaces between particles can fill up and a transition to the funicular region begins to take place.

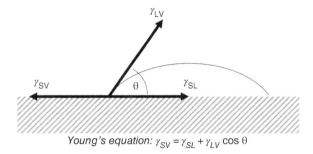

Young's equation: $\gamma_{SV} = \gamma_{SL} + \gamma_{LV} \cos \theta$

Figure 2.22 Equilibrium of a drop on a surface, showing the interfacial surface tensions for the solid–vapour interface (SV), the liquid–vapour interface (LV) and the solid–liquid interface (SL), and the relationship between them

- The liquid which is present must exist in equilibrium with the vapour pressure of the liquid in the surrounding gas; otherwise evaporation or condensation will occur.

- In the calculations presented so far, no mention has been made of *contact angle*; we have assumed that the liquid *perfectly wets* the solid, so that the contact angle is zero. The contact angle arises as a consequence of the balance between the different interfacial surface energies, as shown in Figure 2.22. This is known as Young's equation:

$$\gamma_{SV} = \gamma_{SL} + \gamma_{LV} \cos \theta \qquad (2.25)$$

where the subscripts refer to the solid–vapour interface (SV), the liquid–vapour interface (LV) and the solid–liquid interface (SL). If the contact angle is small, the forces between particles will be reduced but the same trends will be seen as for zero contact angle. In an extreme case of very high contact angle, or non-wetting, as shown in Figure 2.23, the force acting between the particles will be repulsive. This might be the case for an aqueous liquid and particles with hydrophobic surfaces, for example.

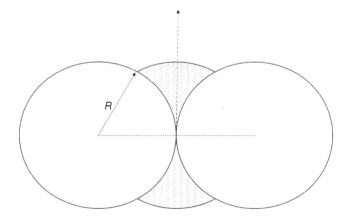

Figure 2.23   A pendular liquid bridge with a high contact angle (non-wetting solid)

These and other aspects of liquid bridges are covered in reviews by Willett *et al.* (2007) and Simons (2007). As we have seen, the presence of liquid between particles can have a big effect on system behaviour, even when the quantity is small. These effects can be a problem in some processes but are exploited in others, notably in size enlargement, which is considered in detail in Chapter 12.

Further reading on mechanical properties of particles: Johnson (1985); Kendall (2001); Seville *et al.* (1997) and Ward and Sweeney (2004).

2.7   WORKED EXAMPLES

WORKED EXAMPLE 2.1

What is the value of Poisson's ratio for an incompressible material?
Consider the bar geometry shown in Figure 2.5. The initial volume of the bar V_0, is:

$$V_0 = L_0 b_0^2$$

If the material is incompressible then the volume is conserved so that this must be equal to the new volume of the bar:

$$V_0 = (L_0 + \Delta L)(b_0 - \Delta b)^2$$

Equating these two equations and omitting second-order terms (since changes in length and width are assumed small) gives:

$$\frac{\Delta b}{b_0} = \frac{\Delta L}{2L_0}$$

Substituting in ε_{11} and ε_{22} ($=\varepsilon_{33}$), Poisson's ratio, ν, is given by:

$$\nu = \frac{-\varepsilon_{22}}{\varepsilon_{11}} = \frac{1}{2}$$

Thus, for an incompressible material, ν is equal to 0.5.

WORKED EXAMPLE 2.2

Two 100 μm diameter elastic spheres of polyethylene are brought together such that the radius of the contact area is 10% of their diameter. What force is necessary to achieve this? What is the corresponding change in the centre–centre distance?

Solution

Since the spheres are to be considered elastic, Hertz analysis applies and Equation (2.2) can be used:

$$c^3 = \frac{3}{8} \frac{(1-\nu^2)}{E} Wx$$

where Young's modulus and Poisson's ratio are taken from Table 2.1 as 1 GPa and 0.45, respectively. Putting values into this equation, we have:

$$c^3 = \frac{3}{8} \frac{(1-0.45^2)}{10^9} \cdot 100 \cdot 10^{-6} W = 3.10^{-14} W$$

Putting $c = x/10 = 10\,\mu m$ gives a force $W = 3.3 \times 10^{-2}$ N or 33 mN. The resulting linear deformation (centre to centre) is given by Equation (2.4) or may be obtained from the general relationship between δ and c in the Hertz region [compare Equations (2.2) and (2.4)]:

$$\delta = (2c)^2/x = (20 \cdot 10^{-6})^2/100 \cdot 10^{-6} = 4 \cdot 10^{-6}\,m \ \ or \ \ 4\,\mu m$$

WORKED EXAMPLE 2.3

For the same conditions as above, if cohesion applies, with an interfacial energy of cohesion $50\,mJ/m^2$, how much force does it take to pull the particles apart, and how does this compare with the single particle weight? Comment on the result.

Solution

From Equation (2.10), pulling the particles apart requires a force F_C, here shown as negative because it opposes contact:

$$-F_C = 3\pi\gamma x/8$$

Inserting values gives $-F_C = 5.9 \cdot 10^{-6}$ N.

The weight of a single particle is given by $(\pi x^3/6)\rho_S g$. Taking the density of polyethylene as $950\,kg/m^3$ (Table 2.1) gives a value of $4.9 \cdot 10^{-9}$ N, so that the ratio of JKR pull-off force to single particle weight is $5.9 \cdot 10^{-6}/4.9 \cdot 10^{-9}$ or of order 10^3. This result might be applicable to perfectly clean molecularly smooth surfaces. If, however, the surfaces are rough or contaminated with dust, the pull-off force will decrease in proportion to the radius of contact, so that at a roughness radius of around $0.1\,\mu m$, for example, the calculated cohesive contact force and the particle weight will be similar in magnitude, as shown (for different particle properties) in Figure 2.12.

WORKED EXAMPLE 2.4

Again taking the conditions of Example 2.2, what is the maximum force arising from the presence of a liquid bridge of water at the contact point? Comment on the result.

Solution

From Equation (2.23), the maximum liquid bridge force is $\pi x \gamma_L$ where γ_L is the surface tension of the liquid. For water this is $72\,mJ/m^2$, giving a value of $22.6 \cdot 10^{-6}$ N. This is larger than the cohesive pull-off force by a factor of about 4. Again, this value will be reduced by surface roughness, but as Figure 2.21(a) shows, if the liquid bridge is large enough it can immerse the roughness elements in such a way that it is still the radius of the parent particle which determines the liquid bridge shape. Because the bridge can extend to a certain distance without breaking, liquid bridge forces are important at greater particle separations than van der Waals forces.

TEST YOURSELF

2.1 Give examples of high-speed collision-dominated flow and low-speed friction-dominated flow.

2.2 What features distinguish elastic behaviour from elastoplastic or plastic behaviour?

2.3 Why is the force–distance relationship for an elastic material linear when the material is in a block and nonlinear when sphere–sphere contact is concerned?

2.4 What are the conditions for the Hertz analysis for particle contact to apply?

2.5 What difference does it make to the contact behaviour if the surfaces of particles are (a) rough and (b) contaminated with a different material?

2.6 What is the difference between cohesion and adhesion?

2.7 What is the relationship between the shear force required to move a particle along a surface and the normal load on that particle?

2.8 Why does the presence of a liquid bridge at the contact point between particles tend to pull them together? Can it push them apart?

2.9 What happens to the liquid bridge force as the amount of liquid in the bridge decreases? Why? Does this agree with your everyday experience of slightly wet particles?

2.10 Explain the sequence of impact of a sphere on a plane if the interaction is (a) elastic and (b) elastoplastic. What difference does it make if there is a layer of liquid on the surface?

2.11 What are the conditions for a particle to be 'captured' by an impact on a surface?

EXERCISES

2.1 Worked Example 2.1 shows that if a material is incompressible and formed into a bar of square cross-section, its Poisson's ratio is 0.5. Prove that this is also true for a different shape of bar. (Hint: try a cylinder.)

2.2 Look at the ceiling of your room. Is it dusty? Why? Estimate the largest particle size that can stick to it, stating any assumptions made.

2.3 What are the perfect conditions for building sandcastles?

2.4 Two spherical particles of diameter 1 mm are joined by a liquid bridge of water at 25°C and atmospheric pressure. (a) What is the maximum surface tension force between them? (b) If they are separated by a distance of 10 μm, at what relative velocity of separation does the viscous force equal the surface tension force? (c) What happens as the separation goes to zero and why is this physically unrealistic? (d) Comment on the conditions needed for impacting particles to be 'captured' by a wet surface.

2.5 In frictional movement between a sphere and a plane, part of the contact area can be deforming while some of the area may be slipping. Why do you think this is and how could you use the laws of friction to help you model this behaviour?

3

Motion of Particles in a Fluid

This chapter deals with the motion of solid particles in fluids. First, the motion of single isolated particles is studied. The objective here is to develop an understanding of the forces resisting the motion of any such particle and provide methods for the estimation of the steady velocity of the particle relative to the fluid. Next, the motion of many particles in close proximity with each other in a fluid is covered, with a focus on the gravity settling of suspensions. The subject matter of the chapter will be used in subsequent chapters on fluidization, gas cyclones and pneumatic transport.

3.1 SINGLE PARTICLES IN A FLUID

3.1.1 Motion of Single Solid Particles in a Fluid

For a sphere moving relative to a large body of fluid (i.e. well away from any surfaces) in the absence of gravity and other effects, one might expect the drag force F_D, to depend on the relative velocity U, the sphere diameter x, and the density ρ_f, and viscosity μ, of the fluid:

$$F_D = f(U, \rho_f, \mu, x) \qquad (3.1)$$

It is advantageous to write this relationship in dimensionless form.

The dimensions of the variables in question are as follows:

Drag force F_D: $\dfrac{ML}{T^2}$

Relative velocity U: $\dfrac{L}{T}$

Fluid density ρ_f: $\dfrac{M}{L^3}$

Introduction to Particle Technology, Third Edition. Martin Rhodes and Jonathan Seville.
© 2024 John Wiley & Sons Ltd. Published 2024 by John Wiley & Sons Ltd.
Website: www.wiley.com/go/rhodes/particle3e

Fluid viscosity μ: $\dfrac{M}{LT}$

Dimensional analysis gives the dimensionless form of the drag force as: $\dfrac{F_D}{\rho_f U^2 x^2}$

The simplest dimensionless arrangement of U, ρ_f, μ and x in the function term is: $\dfrac{U\rho_f x}{\mu}$,

known as the single particle Reynolds[1] number Re_p.

And so, the dimensionless relationship becomes:

$$\frac{F_D}{\rho U^2 x^2} = f\left(\frac{U\rho_f x}{\mu}\right) \tag{3.2}$$

Now we have a relationship involving only two variables instead of the original five. We would therefore expect, for a sphere, a single curve relating the dimensionless drag force to the particle Reynolds number.

In practice, the dimensionless drag force, known as the drag coefficient C_D, is defined as:

$$C_D = \frac{R'}{\frac{1}{2}\rho_f U^2} \tag{3.3}$$

where R' is the drag force per unit projected area and $\frac{1}{2}\rho_f U^2$ is the velocity head.

And so, from dimensional analysis, we have:

$$C_D = f\left(Re_p\right) \tag{3.4}$$

Theoretical analysis shows that the drag force resisting very slow steady relative motion (creeping motion) between a rigid sphere of diameter x and a fluid of infinite extent and of viscosity μ, is composed of two components:

a pressure drag force, which is the integral of the normal force acting over the surface of the particle, otherwise known as *form drag*, because of its sensitivity to the particle shape:

$$F_p = \pi x \mu U \tag{3.5}$$

a shear stress drag force, which is the integral of the shear stress over the surface of the particle, otherwise known as *skin friction drag*:

$$F_s = 2\pi x \mu U \tag{3.6}$$

---

[1] After Osborne Reynolds, Irish scientist and engineer (1842–1912). The particle Reynolds number is analogous to the more familiar number used to describe the regimes of flow in a pipe, which for a pipe of diameter D takes the form $Re = D U \rho_f / \mu$.

The total drag force is then the sum of the two:

$$F_D = 3\pi x\mu \tag{3.7}$$

This is known as Stokes' law.[2] Experimentally, Stokes' law is found to hold almost exactly for single particle Reynolds numbers $Re_p \leq 0.1$, within 3% for $Re_p \leq 0.5$ and within 9% for $Re_p \leq 1.0$,

Thus, for a sphere the projected area is $\pi x^2/4$ and so, from Equation (3.3),

Stokes' law, in terms of the drag coefficient C_D, becomes:

$$C_D = 24/Re_p \tag{3.8}$$

At higher relative velocities (higher Reynolds numbers), the inertia of the fluid begins to dominate (the fluid must accelerate out of the way of the particle). Exact mathematical expressions for the drag coefficient do not exist at these higher Reynolds numbers. However, experimental findings provide the relationship between the drag coefficient and the particle Reynolds number in the form of the so-called standard drag curve (Figure 3.1). Four regions are identified: the Stokes' law region; the Newton's law[3] region in which the drag coefficient is independent of Reynolds number; an intermediate region between the Stokes' law and Newton's law regions; and the boundary layer separation region.

Several correlations have been proposed for C_D over the entire range of particle Reynolds number; the one presented in Equation (3.8) is that of Turton and

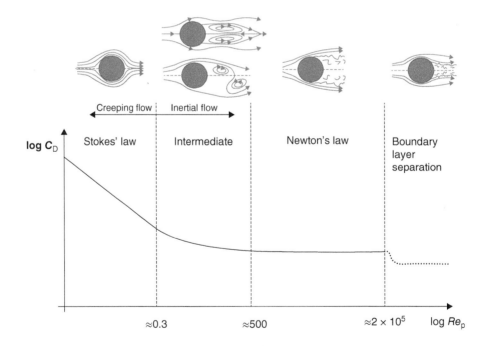

Figure 3.1   Standard drag curve for the motion of a sphere in a fluid

---

[2] After George Stokes, Irish physicist and mathematician (1819–1903).
[3] After Isaac Newton, English mathematician, physicist, astronomer, alchemist and theologian (1643–1727).

Levenspiel (1986), which is claimed to fit the data with a root mean square deviation of 0.025 for $Re_p < 2.6 \times 10^5$.

$$C_D = \frac{24}{Re}\left(1 + 0.173\,Re_p^{0.657}\right) + \frac{0.413}{1 + 16300\,Re_p^{-1.09}} \qquad (3.9)$$

Note that as Re_p tends to zero, C_D in Equation (3.9) tends to $24/Re_p$, the Stokes' law expression.

Non-spherical particles

The effect of the shape of non-spherical particles on their drag coefficient is difficult to define due to the difficulty in describing particle shape for irregular particles and the problem of orientation. Shape affects drag coefficient far more in the intermediate and Newton's law regions than in the Stokes' law region. It is difficult to characterize its effects in a simple way, but there are some general observations. For example, in the Stokes' law region, particles tend to fall with their longest dimension roughly parallel to the direction of motion, whereas in the Newton's law region particles tend to present their maximum area to the oncoming fluid.

As noted in Chapter 1, it is useful to define a single number to describe the shape of a particle. One simple approach is to describe the shape in terms of its sphericity ϕ, the ratio of the surface area of a sphere of volume equal to that of the particle to the surface area of the particle. For example, a cube of side 1 unit has a volume of 1 cubic unit and a surface area of 6 square units. A sphere of the same volume has a diameter x_v of 1.24 units. The surface area of a sphere of diameter 1.24 units is 4.836 units. The sphericity of a cube is therefore 0.806 (= 4.836/6). Other methods of describing particle shape are given in Chapter 1.

For non-spherical particles the particle Reynolds number is based on the equivalent volume sphere diameter, i.e. the diameter of the sphere having the same volume as that of the particle. Figure 3.2 shows drag curves for particles of different sphericities. This covers regular and irregular particles. The plot should be used with caution since sphericity on its own may not be sufficient to describe the shape for all types of particles.

Haider and Levenspiel (1989) recommend the following expression for the drag curve for isometric non-spherical particles:

$$C_D = \frac{24}{Re_p}\left[1 + (8.1716\,\exp(-4.0655\phi))\,Re_p^{(0.0964 + 0.5565\phi)}\right] + \frac{73.69\,Re_p\,\exp(-5.0748\phi)}{Re_p + 5.378\,\exp(6.2122\phi)} \qquad (3.10)$$

3.1.2   Particles Falling Under Gravity Through a Fluid

The relative motion under gravity of particles in a fluid is of particular interest. In general, the forces of buoyancy, drag and gravity act on the particle, such that:

$$\text{Acceleration force = Gravity force} - \text{Buoyancy force} - \text{Drag force} \qquad (3.11)$$

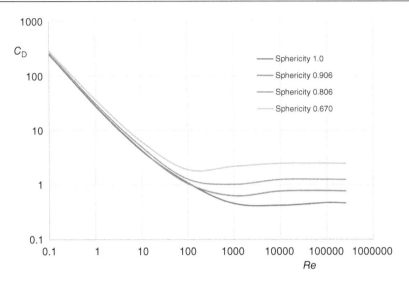

Figure 3.2   Drag coefficient C_D versus Reynolds number Re_p for particles of sphericity ϕ ranging from 0.67 to 1.0. (Note Re_p and C_D are based on the equivalent volume diameter)

A particle falling from rest in a fluid will initially accelerate as the shear stress drag, which increases with relative velocity, will be small. As the particle accelerates, the drag force increases, causing the acceleration to reduce. Eventually a force balance is achieved when the acceleration is zero and a maximum relative velocity is reached. This is known as the single particle *terminal velocity*.

For a spherical particle, Equation (3.11) becomes:

$$\frac{\pi x^3}{6}\rho_p g - \frac{\pi x^3}{6}\rho_f g - R'\frac{\pi x^2}{4} = 0 \tag{3.12}$$

Combining Equation (3.12) with Equation (3.3):

$$\frac{\pi x^3}{6}\left(\rho_p - \rho_f\right)g - C_D \frac{1}{2}\rho_f U_T^2 \frac{\pi x^2}{4} = 0 \tag{3.13}$$

where U_T is the single-particle terminal velocity. Equation (3.13) gives the following expression for the drag coefficient under terminal velocity conditions:

$$C_D = \frac{4}{3}\frac{gx}{U_T^2}\left[\frac{\left(\rho_p - \rho_f\right)}{\rho_f}\right] \tag{3.14}$$

Thus, in the Stokes' law region, with $C_D = 24/Re_p$, the single particle terminal velocity is given by:

$$U_T = \frac{x^2\left(\rho_p - \rho_f\right)g}{18\mu} \tag{3.15}$$

Note that in the Stokes' law region the terminal velocity is proportional to the square of the particle diameter.

In the Newton's law region, with $C_D = 0.44$, the terminal velocity is given by:

$$U_T = 1.74 \left[\frac{x \left(\rho_p - \rho_f \right) g}{\rho_f} \right]^{1/2}$$ (3.16)

Note that in this region the terminal velocity is independent of the fluid viscosity and proportional to the square root of the particle diameter.

Generally, when calculating the terminal velocity for a given particle or the particle diameter for a given velocity, it is not known which region of operation is relevant. One way around this is to formulate the relationship between dimensionless terminal velocity U_T^* and dimensionless particle size x^* as defined below:

$$U_T^* = U_T \left[\frac{\rho_f^2}{g \mu \left(\rho_p - \rho_f \right)} \right]^{1/3}$$ (3.17)

$$x^* = x \left[\frac{\left(\rho_p - \rho_f \right) g \rho_f}{\mu^2} \right]^{1/3}$$ (3.18)

From our expression for the drag coefficient under terminal velocity conditions (3.14): we may formulate the following dimensionless groups:

$$\frac{C_D}{Re_p} = \frac{4}{3} \frac{g \mu \left(\rho_p - \rho_f \right)}{U_T^3 \rho_f^2}$$ (3.19)

and

$$C_D Re_p^2 = \frac{4}{3} \frac{x^3 \rho_f \left(\rho_p - \rho_f \right) g}{\mu^2}$$ (3.20)

And so, from Equations (3.17) and (3.19):

$$U_T^* = \left[\frac{4 Re_p}{3 C_D} \right]^{1/3}$$ (3.21)

and from Equations (3.18) and (3.20),

$$x^* = \left[\frac{3 C_D Re_p^2}{4} \right]^{1/3}$$ (3.22)

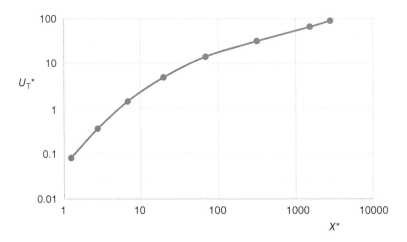

Figure 3.3 Chart of dimensionless terminal velocity U_T^* versus dimensionless particle diameter x^* for spherical particles

Hence, based on the standard drag relationship between C_D and Re_p, we can develop a relationship between dimensionless terminal velocity and dimensionless particle size. The form of this relationship for spherical particles is shown in Figure 3.3. Such plots are available for spherical and non-spherical particles in the literature (e.g. Haider and Levenspiel, 1989).

However, for ease of use, it is convenient to have these plots in equation form. Haider and Levenspiel (1989) recommend the following expression for spherical particles:

$$U_T^* = \left[\frac{1}{\left(\dfrac{18}{[x^*]^2}\right)^{K_1} + \left(\dfrac{3K_1}{4[x^*]^{0.5}}\right)^{K_2}} \right]^{1/K_2} \tag{3.23}$$

with $K_1 = 0.7554$ and $K_2 = 0.8243$ and for non-spherical particles with sphericity ≥ 0.67, they recommend the following expression:

$$U_T^* = \left[\frac{18}{\left(x_v^*\right)^2} + \frac{(2.3348 - 1.7439\phi)}{\left(x_v^*\right)^{0.5}} \right]^{-1} \tag{3.24}$$

where ϕ is the particle sphericity and x_v^* is the dimensionless diameter of a sphere with the same volume as the non-spherical particle.

The procedures for estimating U_T are then as follows.
To estimate U_T for a given x for spherical particles:
Calculate x^* from Equation (3.18).

Calculate U_T^* using Equation (3.23).
Determine terminal velocity from Equation (3.17).

To estimate x for a given U_T for spherical particles:
Calculate U_T^* from Equation (3.17).
From Equation (3.23), find x^* (use function *Goal Seek* in *Excel* or calculator).
From Equation (3.18) determine particle size x.

To estimate U_T for a given x for isometric non-spherical particles:
Calculate x^* from Equation (3.18) (using x_v).
Calculate U_T^* using Equation (3.24).
Determine terminal velocity from Equation (3.17).

To estimate x for a given U_T for isometric non-spherical particles:
Calculate U_T^* from Equation (3.17).
From Equation (3.24), find x^* (use function *Goal Seek* in *Excel* or calculator).
From Equation (3.18) determine particle size x_v.

For non-isometric particles and particles with sphericity < 0.67 consult Haider and Levenspiel (1989).

Small particles in gases and all common particles in liquids quickly accelerate to their terminal velocity. As an example, a 100 μm particle falling from rest in water requires 1.5 ms to reach its terminal velocity of 2 mm/s. Table 3.1 gives some interesting comparisons of terminal velocities, acceleration times and distances for sand particles falling from rest in air.

3.1.3 Effect of Boundaries on Terminal Velocity

When a particle is falling through a fluid in the presence of a solid boundary the terminal velocity reached by the particle is less than that for an infinite fluid. In practice, this is really only relevant to the falling sphere method of measuring liquid viscosity, which is restricted to the Stokes' law region. In the case of a particle falling along the axis of a vertical pipe this is described by a wall factor f_w, the ratio of the velocity in the pipe U_D to the velocity in an infinite fluid U_∞. The correlation of Francis (1933) for f_w is given in Equation (3.25).

$$f_w = \left(1 - \frac{x}{D}\right)^{2.25} \quad Re_p \leq 0.3; \quad x/D \leq 0.97 \qquad (3.25)$$

Table 3.1 Sand particles falling from rest in air (particle density 2600 kg/m^3)

Size	Time to each 99% of U_T (s)	U_T (m/s)	Distance travelled in this time (m)
30 μm	0.033	0.07	0.00185
3 mm	3.5	14	35
3 cm	11.9	44	453

3.1.4   Unsteady Motion

So far, we have considered only steady motion of a particle in a fluid at a constant velocity. Unsteady motion is generally much more complex because the inertia of the particle is important as well as its immersed weight. A detailed discussion is given by Clift *et al.* (1978) from which some general points can be derived. In the low Reynolds number creeping flow range, the drag on a sphere in unsteady motion through a stagnant fluid can be written as:

$$F_D = 3\pi\mu x U + \frac{1}{2}\rho_f\left[\frac{\pi x^3}{6}\right]\frac{dU}{dt} + \frac{3}{2}x^2\sqrt{\pi\mu\rho_f}\int_{-\infty}^{t}\left[\frac{dU}{dt}\right]_{t=s}\frac{ds}{\sqrt{t-s}} \tag{3.26}$$

where U is the instantaneous particle velocity. The total instantaneous drag force is composed of three terms. The first is the steady Stokes' drag at the instantaneous velocity. The second term on the right is known as the *added mass* or *virtual mass* term, which arises from the acceleration of the fluid entrained with the particle. For a sphere, this added mass of fluid corresponds to a volume equal to half that of the sphere. The last term on the right is the *Basset*[4] or *History term* and involves an integral over all past accelerations of the sphere, which in practice is very difficult to calculate. From Equation (3.26) an equation of motion of a sphere in low Reynolds number creeping flow can be developed. The equation can be extended semi-empirically to describe accelerated motion at higher Reynolds numbers (see Clift *et al.* 1978). This development is beyond the scope of this chapter, but some general conclusions are given below.

For particles in gases, where particle density is much greater than fluid density, the added mass and History terms are often small, so the drag force in the Stokes' law region can be approximated by the steady-state drag [Equation (3.7)]. However, caution is advised since this is not always true, even for particles in gases. It is therefore advisable to estimate U by making the quasi-steady approximation and then evaluate the two other terms to verify that they are small enough to confirm the validity of the approximation. For particles in liquids, the quasi-steady assumption is almost never justified. Therefore, the kind of calculations made for particles in gases, for example, in inertial devices for separating particles from gases (see Chapter 8), cannot be applied to particles in liquids.

Given the above, the drag force on a particle accelerating in a gas may be approximated by Stokes' law, using the instantaneous velocity, so for acceleration from rest:

Acceleration force = Gravity force − Buoyancy force − Drag force

$$\frac{\pi x^3 \rho_p}{6}\frac{dU}{dt} = \frac{\pi x^3 \rho_p g}{6} - \frac{\pi x^3 \rho_f g}{6} - 3\pi\mu x U \tag{3.27}$$

$$\frac{dU}{dt} = k_1 - k_2 U \tag{3.28}$$

where $k_1 = g\left(1 - \dfrac{\rho_f}{\rho_p}\right)$ $k_2 = \dfrac{18\mu}{x^2 \rho_p}$ and $\dfrac{k_2}{k_1} = \dfrac{18\mu}{x^2 g\left(\rho_p - \rho_f\right)}$

Note k_1/k_2 is U_T for the Stokes' law region.

---

[4] After Alfred Barnard Basset, English mathematician and physicist (1854–1930).

Integrating:

$$\int_0^U \frac{dU}{(k_1 - k_2 U)} = \int_0^t dt \tag{3.29}$$

$$-\frac{1}{k_2} \ln\left(1 - \frac{k_2}{k_1} U\right) = t \tag{3.30}$$

$$1 - \frac{k_2}{k_1} U = e^{-k_2 t} \tag{3.31}$$

$$U = U_T\left(1 - e^{-k_2 t}\right) \tag{3.32}$$

Since $\rho_f \ll \rho_p$ for most systems involving gases unless at very high pressure
Then,

$$\frac{1}{U_T} = \frac{18\mu}{x^2 g \rho_p} = \frac{k_2}{g} \tag{3.33}$$

So,

$$U = U_T\left(1 - e^{-gt/U_T}\right) \tag{3.34}$$

The figure below compares the acceleration of sand particles of diameters 15, 25 and 40 μm in air.

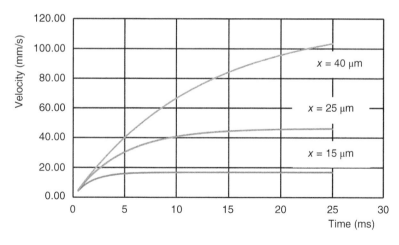

3.1.5 Further Reading

For further details on the motion of single particles in fluids (accelerating motion, added mass, bubbles and drops, non-Newtonian fluids) the reader is referred to Coulson and Richardson's Chemical Engineering (2019), Clift *et al.* (1978), Haider and Levenspiel (1989), and Kay and Nedderman (1985).

3.1.6   Worked Examples on the Motion of Single Particles in a fluid

WORKED EXAMPLE 3.1

Calculate the upper limit of particle diameter x_{max} as a function of particle density ρ_p for gravity sedimentation in the Stokes' law region. Plot the results as x_{max} versus ρ_p over the range $0 \leq \rho_p \leq 8000\,\text{kg/m}^3$ for settling in water and in air at ambient conditions. Assume that the particles are spherical and that Stokes' law holds for $Re_p \leq 0.3$.

Solution

The upper limit of particle diameter in the Stokes' law region is governed by the upper limit of single particle Reynolds number:

$$Re_p = \frac{\rho_f x_{max} U}{\mu} = 0.3$$

In gravity sedimentation in the Stokes' law region, particles accelerate rapidly to their terminal velocity. In the Stokes' law region the terminal velocity is given by Equation (3.15):

$$U_T = \frac{x^2 \left(\rho_p - \rho_f\right) g}{18\mu} \tag{3.15}$$

Solving these two equations for x_{max} we have:

$$
\begin{aligned}
x_{max} &= \left[0.3 \times \frac{18\mu^2}{g\left(\rho_p - \rho_f\right)\rho_f}\right]^{1/3} \\
&= 0.82 \times \left[\frac{\mu^2}{\left(\rho_p - \rho_f\right)\rho_f}\right]^{1/3}
\end{aligned}
$$

Thus, for air (density $1.2\,\text{kg/m}^3$ and viscosity $1.84 \times 10^{-5}\,\text{Pa s}$):

$$x_{max} = 5.37 \times 10^{-4} \left[\frac{1}{\left(\rho_p - 1.2\right)}\right]^{1/3}$$

and for water (density $1000\,\text{kg/m}^3$ and viscosity $0.001\,\text{Pa s}$):

$$x_{max} = 8.19 \times 10^{-4} \left[\frac{1}{\left(\rho_p - 1000\right)}\right]^{1/3}$$

These equations for x_{max} as a function are plotted in Figure 3.W1.1 for particle densities greater than and less than the fluid densities.

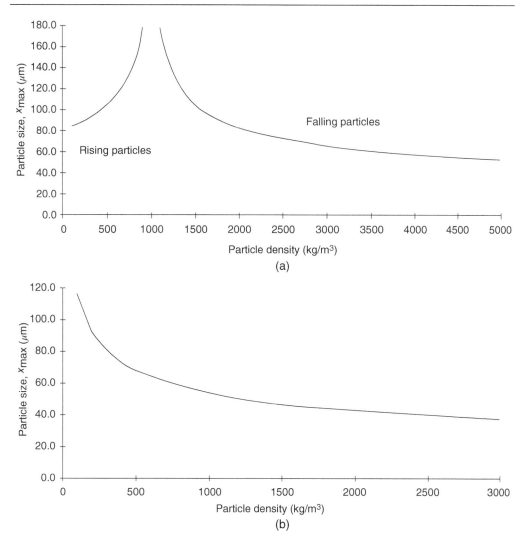

Figure 3.W1.1 (a) Limiting particle size for Stokes' law in water. (b) Limiting particle size for Stokes' law in air

WORKED EXAMPLE 3.2

A gravity separator for the removal of oil droplets (assumed to behave as rigid spheres) from water consists of a rectangular chamber containing inclined baffles as shown schematically in Figure 3.W2.1.

(a) Derive an expression for the ideal collection efficiency of this separator as a function of droplet size and properties, separator dimensions, fluid properties and fluid velocity (assumed uniform).

(b) Hence, calculate the percentage change in collection efficiency when the throughput of water is increased by a factor of 1.2 and the density of the oil droplets changes from 750 to $800\,\mathrm{kg/m^3}$.

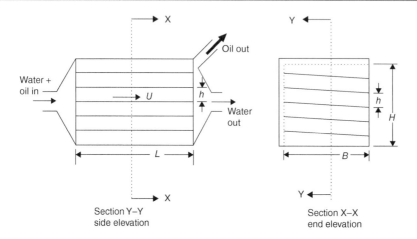

Figure 3.W2.1 Schematic diagram of oil–water separator

Solution

(a) Referring to Figure 3.W2.1, we will assume that all particles rising to the undersurface of a baffle will be collected. Therefore, any particle which can rise a distance h or greater in the time required for it to travel the length of the separator will be collected. Let the corresponding minimum vertical droplet velocity be $U_{T_{min}}$.

Assuming uniform fluid velocity and negligible relative velocity between drops and fluid in the horizontal direction,

$$\text{Drop residence time,} \, t = L/U$$
$$\text{Then } U_{T_{min}} = hU/L$$

Assuming that the droplets are small enough for Stokes' law to apply and that the time and distance for acceleration to terminal velocity is negligible, then droplet velocity will be given by Equation (3.15):

$$U_T = \frac{x^2\left(\rho_p - \rho_f\right)g}{18\mu}$$

This is the minimum velocity for drops to be collected whatever their original position between the baffles. Thus:

$$U_{T_{min}} = \frac{x^2\left(\rho_p - \rho_f\right)g}{18\mu} = \frac{hU}{L}$$

Assuming that drops of all sizes are uniformly distributed over the vertical height of the separator, then drops rising a distance less than h in the time required for them to travel the length of the separator will be fractionally collected depending on their original vertical position between two baffles. Thus, for droplets rising at a velocity of $0.5U_{T_{min}}$ only 50% will be collected: i.e. only those drops originally in the upper half of the space between adjacent baffles. For drops rising at a velocity of $0.25\, U_{T_{min}}$ only 25% will be collected.

$$\text{Thus, efficiency of collection,} \, \eta = \frac{\text{actual } U_T \text{ for droplet}}{U_{T_{min}}}$$

And so,

$$\eta = \left[\frac{x^2 \left(\rho_p - \rho_f \right) g}{18\mu} \right] / \frac{hU}{L}$$

(b) Comparing collection efficiencies when the throughput of water is increased by a factor of 1.2 and the density of the oil droplets changes from 750 to 850 kg/m^3.

Let original and new conditions be denoted by subscripts 1 and 2, respectively.

Increasing the throughput of water by a factor of 1.2 means that $U_2/U_1 = 1.2$.

Therefore, from the expression for collection efficiency derived above:

$$\frac{\eta_2}{\eta_1} = \left(\frac{\rho_{p2} - \rho_f}{U_2} \right) / \left(\frac{\rho_{p1} - \rho_f}{U_1} \right)$$

$$\frac{\eta_2}{\eta_1} = \left(\frac{850 - 1000}{750 - 1000} \right) \times \frac{1}{1.2} = 0.5$$

The decrease in collection efficiency is therefore 50%.

WORKED EXAMPLE 3.3

A sphere of diameter 10 mm and density 7700 kg/m^3 falls under gravity at terminal conditions through a liquid of density 900 kg/m^3 in a tube of diameter 12 mm. The measured terminal velocity of the particle is 1.6 mm/s. Calculate the viscosity of the fluid. Verify that Stokes' law applies.

Solution

To solve this problem, we first convert the measured terminal velocity to the equivalent velocity which would be achieved by the sphere in a fluid of infinite extent. Assuming Stokes' law we can determine the fluid viscosity. Finally, we check the validity of Stokes' law.

Using the Francis wall factor expression [Equation (3.25)]:

$$\frac{U_{T_\infty}}{U_{T_D}} = \frac{1}{\left(1 - x/D \right)^{2.25}} = 56.34$$

Thus, terminal velocity for the particle in a fluid of infinite extent is:

$$U_{T_\infty} = U_{T_D} \times 56.34 = 0.0901 \text{ m/s}$$

Equating this value to the expression for U_{T_∞} in the Stokes' law region [Equation (3.15)]:

$$U_{T_\infty} = \frac{\left(10 \times 10^{-3} \right)^2 \times \left(7700 - 900 \right) \times 9.81}{18\mu}$$

Hence, fluid viscosity $\mu = 4.11$ Pa s.

Checking the validity of Stokes' law:

$$\text{Single particle Reynolds number } Re_p = \frac{x\rho_f U}{\mu} = 0.197$$

Re_p is less than 0.3 and so the assumption that Stokes' law holds is valid.

WORKED EXAMPLE 3.4

A mixture of spherical particles of two materials A and B is to be separated using a rising stream of liquid. The size range of both materials is 15–40 μm. (a) Show that a complete separation is not possible using water as the liquid. The particle densities for materials A and B are 7700 and 2400 kg/m³, respectively. (b) Which fluid property must be changed to achieve complete separation? Assume Stokes' law applies.

Solution

(a) First, consider what happens to a single particle introduced into the centre of a pipe in which a fluid is flowing upwards at a velocity U which is uniform across the pipe cross section. We will assume that the particle is small enough to consider the time and distance for its acceleration to terminal velocity to be negligible. Referring to Figure 3.W4.1(a), if the fluid

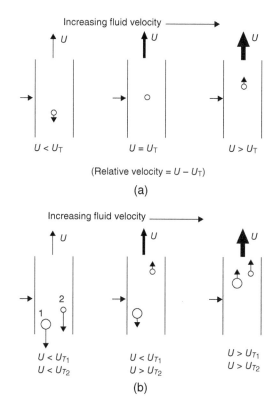

Figure 3.W4.1 Relative motion of particles in a moving fluid (a) single particle, (b) two different particles

velocity is greater than the terminal velocity of the particle U_T, then the particle will move upwards; if the fluid velocity is less than U_T then the particle will fall; and if the fluid velocity is equal to U_T then the particle will remain at the same vertical position. In each case the velocity of the particle relative to the pipe wall is $(U - U_T)$. Now consider introducing two particles of different sizes and density having terminal velocities U_{T_1} and U_{T_2}. Referring to Figure 3.W4.1(b), at low fluid velocities $(U < U_{T_2} < U_{T_1})$, both particles will fall. At high fluid velocities $(U > U_{T_1} > U_{T_2})$, both particles will be carried upwards. At intermediate fluid velocities $(U_{T_1} > U > U_{T_2})$, particle 1 will fall and particle 2 will rise. Thus, we have the basis of a separator according to particle size and density. From the analysis above we see that to be able to completely separate particles A and B, there must be no overlap between the ranges of terminal velocity for the particles: i.e. all sizes of the denser material A must have terminal velocities which are greater than all sizes of the less dense material B.

Assuming Stokes' law applies, Equation (3.15), with fluid density and viscosity $1000 \, kg/m^3$ and $0.001 \, Pa \, s$, respectively, gives

$$U_T = 545x^2 \left(\rho_p - 1000\right)$$

Based on this equation, the terminal velocities of the extreme sizes of particles A and B are:

Size (μm) →	15	40
U_{T_A} (mm/s)	0.82	5.84
U_{T_B} (mm/s)	0.17	1.22

We see that there is an overlap of the ranges of terminal velocities. We can therefore select no fluid velocity which would completely separate particles A and B.

(b) Inspecting the expression for terminal velocity in the Stokes' law region [Equation (3.15)] we see that changing the fluid viscosity will have no effect on our ability to separate the particles, since change in viscosity will change the terminal velocities of all particles in the same proportion. However, changing the fluid density will have a different effect on particles of different densities and this is the effect we are looking for. The critical condition for the separation of particles A and B is when the terminal velocity of the smallest A particle is equal to the terminal velocity of the largest B particle.

$$U_{T_{B40}} = U_{T_{A15}}$$

Hence,

$$545 \times x_{40}^2 \times (2400 - \rho_f) = 545 \times x_{15}^2 \times (7700 - \rho_f)$$

From which, critical minimum fluid density $\rho_f = 1533 \, kg/m^3$.

WORKED EXAMPLE 3.5

A sphere of density $2500 \, kg/m^3$ falls freely under gravity in a fluid of density $700 \, kg/m^3$ and viscosity $0.5 \times 10^{-3} \, Pa \, s$. Given that the terminal velocity of the sphere is $0.15 \, m/s$, calculate its diameter. What would be the edge length of a cube of the same material falling in the same fluid at the same terminal velocity?

Solution

In this case we know the terminal velocity U_T, and need to find the particle size x. Since we do not know which region is appropriate, we must first calculate the dimensionless terminal velocity U_T^* from Equation (3.17):

$$U_T^* = U_T \left[\frac{\rho_f^2}{g\mu\left(\rho_p - \rho_f\right)} \right]^{1/3} \tag{3.17}$$

$$U_T^* = 0.15 \left[\frac{700^2}{9.81 \times 0.5 \times 10^{-3}(2500 - 700)} \right]^{1/3} = 5.72$$

From Equation (3.23), find x^* (use function *Goal Seek* in *Excel* or calculator):

$$U_T^* = \left[\frac{1}{\left(\dfrac{18}{[x^*]^2}\right)^{K_1} + \left(\dfrac{3K_1}{4[x^*]^{0.5}}\right)^{K_2}} \right]^{1/K_2} \tag{3.23}$$

Solving for x^* with $K_1 = 0.7554$ and $K_2 = 0.8243$ and $U_T^* = 5.72$, gives:

$$x^* = 670\,\mu m$$

If the particle is a cube, U_T^* has the same value (5.72) and x^* is calculated from Equation (3.24):

$$U_T^* = \left[\frac{18}{\left(x_v^*\right)^2} + \frac{(2.3348 - 1.7439\phi)}{\left(x_v^*\right)^{0.5}} \right]^{-1} \tag{3.24}$$

where ϕ is the particle sphericity (0.806 for a cube. See Section 3.1.1)

Solving with $U_T^* = 5.72$ and $\phi = 0.806$, gives $x_v^* = 927\,\mu m$.

And so, the volume of the particle is $\dfrac{\pi x_v^3}{6} = 4.171 \times 10^{-10}\,m^3$

Giving a cube side length of $(4.171 \times 10^{-10})^{1/3} = 7.5 \times 10^{-4}\,m$ (0.75 mm).

WORKED EXAMPLE 3.6

An isometric particle of equivalent volume diameter 0.5 mm, density 2000 kg/m^3 and sphericity 0.67 falls freely under gravity in a fluid of density 1.6 kg/m^3 and viscosity 2×10^{-5} Pa s. Estimate the terminal velocity reached by the particle.

Solution

In this case we know the particle size and we are required to determine its terminal velocity without knowing which region is appropriate. The first step is, therefore, to calculate the dimensionless particle size x^* from Equation (3.18)

$$x^* = x \left[\frac{\left(\rho_p - \rho_f \right) g \rho_f}{\mu^2} \right]^{1/3} \tag{3.18}$$

with $x = x_v$ the equivalent volume particle diameter.

$$x^* = 0.5 \times 10^{-3} \left[\frac{(2000 - 1.6)9.81 \times 1.6}{\left(2 \times 10^{-5} \right)^2} \right]^{1/3} = 21.4$$

With $x^* = 21.4$, we use Equation (3.24) to calculate U_T^*

$$U_T^* = \left[\frac{18}{\left(x_v^* \right)^2} + \frac{(2.3348 - 1.7439\phi)}{\left(x_v^* \right)^{0.5}} \right]^{-1} \tag{3.24}$$

$$U_T^* = \left[\frac{18}{21.4^2} + \frac{(2.3348 - 1.7439 \times 0.67)}{21.4^{0.5}} \right]^{-1} = 3.43$$

From Equation (3.17) we now calculate U_T

$$U_T^* = U_T \left[\frac{\rho_f^2}{g\mu \left(\rho_p - \rho_f \right)} \right]^{1/3} \tag{3.17}$$

$$U_T = \frac{3.43}{\left[\frac{1.6^2}{9.81 \times \left(2 \times 10^{-5} \right) \times (2000 - 1.6)} \right]^{1/3}} = 1.84 \, \text{m/s}$$

Hence, terminal velocity $U_T = 1.84$ m/s.

WORKED EXAMPLE 3.7

Determine the time taken for particles of density 2500 kg/m^3 of sizes 10, 20 and 30 μm to reach 95% of terminal velocity falling from rest under gravity in air of density 1.2 kg/m^3 and viscosity 18.4×10^{-6} Pa s.

Assuming Stokes' law applies (we will check this later), from Equation (3.15).

$$U_T = \frac{x^2 g \left(\rho_p - \rho_f \right)}{18\mu}$$

Under the specified conditions, $U_T = \dfrac{x^2 (2500 - 1.2)}{18 \times 18.4 \times 10^{-6}} = x^2 \times 7.545 \times 10^6$

Terminal velocities are therefore:

Size (μm)	10	20	30
U_T (mm/s)	7.4	29.6	66.6

To check that Stokes' law applies we calculate the Reynolds number at U_T.

Size (μm)	10	20	30
Re_p	0.0048	0.0386	0.130

Reynolds numbers are less than 0.3 and so Stokes' law applies here.

Equation (3.34) relates time and velocity for a particle falling from rest under gravity in the Stokes' region assuming fluid density is negligible compared to particle density (which it is in this case):

$$U = U_T\left(1 - e^{-gt/U_T}\right) \tag{3.34}$$

When U is 95% of U_T

$$0.95 = 1 - e^{-gt/U_T}$$

And so

$$e^{-gt/U_T} = 0.05$$

Solving for the three particle sizes:

Size (μm)	10	20	30
Time to $0.95U_T$ (ms)	2.3	9.0	20.3

The figure below shows the change in velocity with time for the three particles.

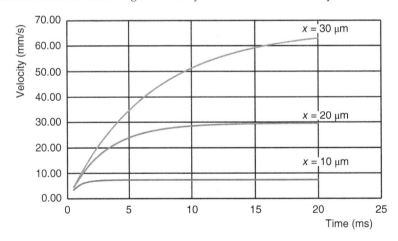

TEST YOURSELF – SINGLE PARTICLES IN A FLUID

3.1 The drag force resisting very slow steady relative motion (creeping motion) between a rigid sphere is composed of which two components and what are their relative contributions to the total drag force?

3.2 Under Stokes' law conditions the terminal velocity is independent of which quantity?

3.3 The particle Reynolds number is a function of which quantities?

3.4 Explain why a particle falling from rest in a fluid reaches a maximum or terminal velocity.

3.5 Write down the word equation for the force balance on a solid particle falling through a fluid under gravity.

3.6 In the Newton's law region of drag, the terminal velocity of a particle is independent of which quantity?

3.2 SETTLING OF A SUSPENSION OF PARTICLES

3.2.1 Introduction

When many particles flow in a fluid in close proximity to each other the motion of each particle is influenced by the presence of the others. The simple analysis for the fluid–particle interaction for a single particle is no longer valid but can be adapted to model the multiple particle system.

For a suspension of particles in a fluid, Stokes' law is assumed to apply if the Reynolds number is sufficiently small, but an effective suspension viscosity and effective average suspension density are used:

$$\text{Effective viscosity } \mu_e = \mu/f(\varepsilon) \tag{3.35}$$

$$\text{Average suspension density } \rho_{ave} = \varepsilon\rho_f + (1-\varepsilon)\rho_p \tag{3.36}$$

where ε is the void fraction or volume fraction occupied by the fluid. The effective viscosity of the suspension is seen to be equal to the fluid viscosity, μ modified by a function $f(\varepsilon)$ of the fluid volume fraction.

The drag coefficient for a single particle in the Stokes' law region (Section 3.1.1) is given by $C_D = 24/Re_p$. Substituting the effective viscosity and average density for the suspension, Stokes' law becomes

$$C_D = \frac{24}{Re_p} = \frac{24\mu_e}{U_{rel}\rho_{ave}x} \tag{3.37}$$

where $C_D = R'/(\frac{1}{2}\rho_{ave}U_{rel}^2)$ and U_{rel} is the relative velocity of the particle to the fluid.

Under terminal velocity conditions for a particle falling under gravity in a suspension, the force balance,

Drag force = weight – buoyancy force

becomes

$$\left(\frac{\pi x^2}{4}\right)\frac{1}{2}\rho_{ave}U_{rel}^2C_D = \left(\rho_p - \rho_{ave}\right)\left(\frac{\pi x^3}{6}\right)g \tag{3.38}$$

giving

$$U_{rel} = \left(\rho_p - \rho_{ave}\right)\frac{x^2g}{18\mu_e} \tag{3.39}$$

Substituting for average density ρ_{ave} and effective viscosity μ_{e} of the suspension, we obtain the following expression for the terminal falling velocity for a particle in a suspension:

$$U_{\mathrm{rel_T}} = \left(\rho_{\mathrm{p}} - \rho_{\mathrm{f}}\right) \frac{x^2 g}{18\mu} \varepsilon f(\varepsilon) \tag{3.40}$$

Comparing this with the expression for the terminal free fall velocity of a single particle in a fluid in Stokes' law region [Equation (3.15)], we find that:

$$U_{\mathrm{rel_T}} = U_{\mathrm{T}} \varepsilon f(\varepsilon) \tag{3.41}$$

$U_{\mathrm{rel_T}}$ is known as the particle settling velocity in the presence of other particles or the *hindered settling velocity*.

In the following analysis, it is assumed that the fluid and the particles are incompressible and that the volume flow rates, Q_{f} and Q_{p}, of the fluid and the particles are constant. We define U_{fs} and U_{ps} as the superficial velocities of the fluid and particles, respectively:

$$\text{Superficial fluid velocity } U_{\mathrm{fs}} = \frac{Q_{\mathrm{f}}}{A} \tag{3.42}$$

$$\text{Superficial particle velocity } U_{\mathrm{ps}} = \frac{Q_{\mathrm{p}}}{A} \tag{3.43}$$

where A is the vessel cross-sectional area. The superficial particle velocity U_{ps} is also known as volumetric particle flux.

For a uniform suspension, the flow areas occupied by the fluid and the particles are as follows:

$$\text{Flow area occupied by the fluid, } A_{\mathrm{f}} = \varepsilon A \tag{3.44}$$

$$\text{Flow area occupied by the particles, } A_{\mathrm{p}} = (1 - \varepsilon)A \tag{3.45}$$

And so, continuity gives:

$$\text{For the fluid}: Q_{\mathrm{f}} = U_{\mathrm{fs}}A = U_{\mathrm{f}}A\varepsilon \tag{3.46}$$

$$\text{For the particles}: Q_{\mathrm{p}} = U_{\mathrm{ps}}A = U_{\mathrm{f}}A(1 - \varepsilon) \tag{3.47}$$

Hence the actual[5] velocities of the fluid and the particles U_{f} and U_{p} are given by:

$$\text{Actual velocity of the fluid } U_{\mathrm{f}} = U_{\mathrm{fs}}/\varepsilon \tag{3.48}$$

$$\text{Actual velocity of the particles } U_{\mathrm{p}} = U_{\mathrm{ps}}/(1 - \varepsilon) \tag{3.49}$$

---

[5] Actual fluid and particle velocities are spatial and temporal averages.

3.2.2   Settling Flux as a Function of Suspension Concentration

When particles in suspension are allowed to settle, say in a measuring cylinder in the laboratory, there is no net flow through the vessel so

$$Q_p + Q_f = 0 \tag{3.50}$$

Hence

$$U_p(1 - \varepsilon) + U_f\varepsilon = 0 \tag{3.51}$$

and

$$U_f = -U_p \frac{(1 - \varepsilon)}{\varepsilon} \tag{3.52}$$

In hindered settling under gravity the relative velocity between the particles and the fluid $(U_p - U_f)$ is U_{rel_T}. Thus, using the expression for U_{rel_T} [Equation (3.41)], we have:

$$U_p - U_f = U_{rel_T} = U_T \varepsilon f(\varepsilon) \tag{3.53}$$

Combining Equation (3.53) with Equation (3.52) gives the following expression for U_p, the hindered settling velocity of particles relative to the vessel wall in batch settling:

$$U_p = U_T \varepsilon^2 f(\varepsilon) \tag{3.54}$$

The effective viscosity function $f(\varepsilon)$, has been shown theoretically to be

$$f(\varepsilon) = \varepsilon^{2.5} \tag{3.55}$$

for uniform spheres forming a suspension of solid volume fraction less than 0.1 [$(1 - \varepsilon) \le 0.1$].

Richardson and Zaki (1954) showed by experiment that for $Re_p < 0.3$ (under Stokes' law conditions where drag is independent of fluid density),

$$U_p = U_T \varepsilon^{4.65} \left[\text{giving} f(\varepsilon) = \varepsilon^{2.65} \right] \tag{3.56}$$

and for $Re_p > 500$ (under Newton's law conditions where drag is independent of fluid viscosity)

$$U_p = U_T \varepsilon^{2.4} \left[\text{giving} f(\varepsilon) = \varepsilon^{0.4} \right] \tag{3.57}$$

In general, the Richardson–Zaki relationship is given as:

$$U_p = U_T \varepsilon^n \tag{3.58}$$

Khan and Richardson (1989) recommend the use of the following correlation for the value of exponent n over the entire range of Reynolds numbers:

$$\frac{4.8 - n}{n - 2.4} = 0.043 Ar^{0.57} \left[1 - 2.4 \left(\frac{x}{D} \right)^{0.27} \right] \tag{3.59}$$

where Ar is the Archimedes number $[x^3 \rho_f (\rho_p - \rho_f) g / \mu^2]$ and x is the particle diameter and D is the vessel diameter. The most appropriate particle diameter to use here is the surface-volume mean (see Section 1.5).

Expressed as a volumetric particle settling *flux* U_{ps}, Equation (3.58) becomes:

$$U_{ps} = U_p (1 - \varepsilon) = U_T (1 - \varepsilon) \varepsilon^n \tag{3.60}$$

where *flux* refers to a flow of particles per unit area; for example, volumetric flux is defined as the volume of particles moving across a defined unit area perpendicular to the flow direction and has units of volume/(area × time) ($m^3/m^2 s$). Mass flux is defined similarly and has units of mass/(area × time) ($kg/m^2 s$).

Expressed as a dimensionless volumetric particle settling flux,

$$\frac{U_{ps}}{U_T} = (1 - \varepsilon) \varepsilon^n \tag{3.61}$$

First and second derivatives of Equation (3.61) demonstrate that a plot of dimensionless particle settling flux versus suspension volumetric concentration, $(1 - \varepsilon)$, has a maximum at $\varepsilon = n/(n + 1)$ and an inflection point at $\varepsilon = (n - 1)/(n + 1)$. The theoretical form of such a plot is therefore that shown in Figure 3.4.

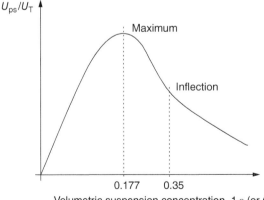

Figure 3.4 Variation of dimensionless settling flux with suspension concentration, based on Equation (3.61) (for $Re_p < 0.3$, i.e. $n = 4.65$)

3.2.3 Sharp Interfaces in Sedimentation

Interfaces or discontinuities in concentration occur in the sedimentation or settling of particle suspensions.

In the remainder of this chapter, for convenience, the symbol C will be used to represent the particle volume fraction $(1 - \varepsilon)$. Also, for convenience the particle volume fraction will be called the concentration of the suspension.

Consider Figure 3.5, which shows the interface between a suspension of concentration C_1 containing particles settling at a velocity U_{p1} and a suspension of concentration C_2 containing particles settling at a velocity U_{p_2}.

The interface is falling at a velocity U_{int}. All velocities are measured relative to the vessel walls. Assuming incompressible fluid and particles, the mass balance over the interface gives:

$$\left(U_{p_1} - U_{int}\right)C_1 = \left(U_{p_2} - U_{int}\right)C_2$$

Hence

$$U_{int} = \frac{U_{p_1}C_1 - U_{p_2}C_2}{C_1 - C_2} \tag{3.62}$$

Or, since $U_p C$ is volumetric particle flux, $U_{ps,}$:

$$U_{int} = \frac{U_{ps_1} - U_{ps_2}}{C_1 - C_2} \tag{3.63}$$

where U_{p1} and U_{p_2} are the volumetric particle fluxes in suspensions of concentration C_1 and C_2, respectively. Thus,

$$U_{int} = \frac{\Delta U_{ps}}{\Delta C} \tag{3.64}$$

$$\text{and, in the limit as } \Delta C \to 0, U_{int} = \frac{dU_{ps}}{dC} \tag{3.65}$$

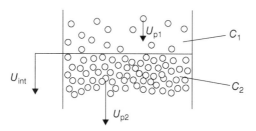

Figure 3.5 Concentration interface in sedimentation

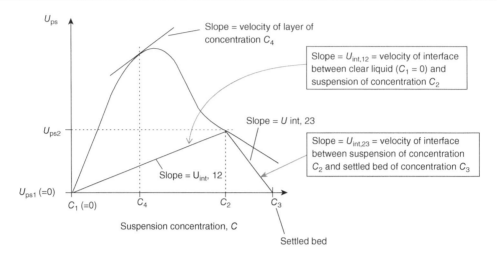

Figure 3.6 Determination of interface and layer velocities from a batch flux plot

Hence, on a flux plot (a plot of U_{ps} versus concentration):

(a) The gradient of the curve at concentration C is the velocity of a layer of suspension of this concentration.

(b) The slope of a chord joining two points at concentrations C_1 and C_2 is the velocity of a discontinuity or interface between suspensions of these concentrations.

This is illustrated in Figure 3.6.

3.2.4 The Batch Settling Test

The simple batch settling test can supply all the information for the design of a *thickener* for the separation of particles from a fluid. In this test a suspension of particles of known concentration is prepared in a measuring cylinder. The cylinder is shaken to thoroughly mix the suspension and then placed upright to allow the suspension to settle. The positions of the interfaces which form are monitored in time. Two types of settling occur depending on the initial concentration of the suspension. The first type of settling is depicted in Figure 3.7 (Type 1 settling). Three zones of constant concentration are formed. These are: zone A, clear liquid ($C = 0$); zone B, a suspension of concentration equal to the initial suspension concentration (C_B); and zone S, the sediment concentration (C_s). Figure 3.8 is a typical plot of the height of the interfaces AB, BS and AS with time for this type of settling. On this plot the slopes of the lines give the velocities of the interfaces. For example, interface AB descends at constant velocity; interface BS rises at constant velocity. The test ends when the descending AB meets the rising BS forming an interface between clear liquid and sediment (AS), which is stationary.

In the second type of settling (Type 2 settling), shown in Figure 3.9, a zone of variable concentration, zone E, is formed in addition to the zones of constant concentration (A, B and S). The suspension concentration within zone E varies with position. However,

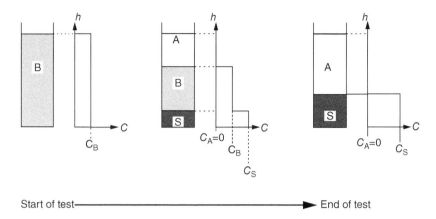

Figure 3.7 Type 1 batch settling. Zones A, B and S are zones of constant concentration. Zone A is a clear liquid; zone B is a suspension of concentration equal to the initial suspension concentration; and zone S is a suspension of settled bed or sediment concentration

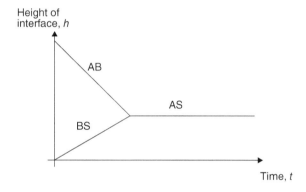

Figure 3.8 Change in positions of interface AB, BS and AS with time in Type 1 batch settling (e.g. AB is the interface between zone A and zone B; see Figure 3.7)

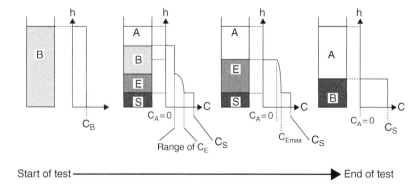

Figure 3.9 Type 2 batch settling. Zones A, B and S are zones of constant concentration. Zone A is clear liquid; zone B is a suspension of concentration equal to the initial suspension concentration; and zone S is a suspension of settled bed concentration. Zone E is a zone of variable concentration

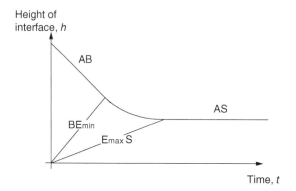

Figure 3.10   Change in positions of interface AB, BE_{min}, $E_{max}S$ and AS with time in Type 2 batch settling (e.g. AB is the interface between zone A and zone B. BE_{min} is the interface between zone B and the lowest suspension concentration in the variable zone E; see Figure 3.9)

the minimum and maximum concentrations within this zone $C_{E_{min}}$ and $C_{E_{max}}$, are constant. Figure 3.10 is a typical plot of the height of the interfaces AB, BE_{min}, $E_{max}S$ and AS with time for this type of settling.

The occurrence of Type 1 or Type 2 settling depends on the initial concentration of the suspension, C_B. In simple terms, if an interface between zone B and a suspension of concentration greater than C_B but less that C_S rises faster than the interface between zones B and S then a zone of variable concentration will form. Examination of the particle flux plot enables us to determine which type of settling is occurring. Referring to Figure 3.11, a tangent to the curve is drawn through the point ($C = C_s$, $U_{ps} = 0$). The concentration at the point of tangent is C_{B2}. The concentration at the point of intersection of the projected tangent with the curve is C_{B1}. Type 1 settling occurs when the initial suspension

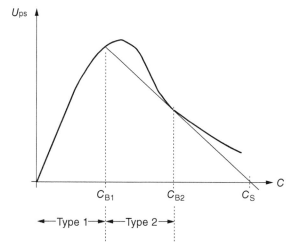

Figure 3.11   Determining if settling will be Type 1 or Type 2. A line through C_S tangent to the flux curve gives C_{B1} and C_{B2}. Type 2 settling occurs when initial suspension concentration is between C_{B1} and C_{B2}

concentration is less than C_{B1}. Type 2 settling occurs when the initial suspension concentration lies between C_{B1} and C_{B2}. Strictly, beyond C_{B2}, Type 1 settling will again occur, but this is of little practical significance.

3.2.5 Relationship Between the Height–Time Curve and the Flux Plot

Following the AB interface in the simple batch settling test gives rise to the height–time curve shown in Figure 3.12 (Type 2 settling). In fact, there will be a family of such curves for different initial concentrations. The following analysis permits the derivation of the particle flux plot from the height–time curve.

Referring to Figure 3.12, at time t the interface between clear liquid and suspension of concentration C is at a height h from the base of the vessel and the velocity of the interface is the slope of the curve at this time:

$$\text{Velocity of interface} = \frac{dh}{dt} = \frac{h_1 - h}{t} \qquad (3.66)$$

This is also equal to U_p, the velocity of the particles at the interface relative to the vessel wall. Hence,

$$U_p = \frac{h_1 - h}{t} \qquad (3.67)$$

Now consider planes or waves of higher concentration which rise from the base of the vessel. At time t a plane of concentration C has risen a distance h from the base. Thus, the velocity at which a plane of concentration C rises from the base is h/t. This plane or wave of concentration passes up through the suspension. The velocity of the particles relative to the plane is therefore:

$$\text{Velocity of particles relative to plane} = U_p + \frac{h}{t} \qquad (3.68)$$

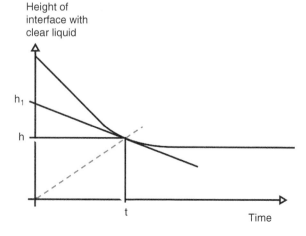

Figure 3.12 Analysis of batch settling test

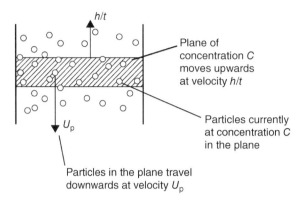

Figure 3.13 Analysis of batch settling; relative velocities of a plane of concentration C and the particles in the plane

As the particles pass through the plane they have a concentration, C (refer to Figure 3.13). Therefore, the volume of particles which have passed through this plane in time t is

$$= \text{area} \times \text{velocity of particles} \times \text{concentration} \times \text{time}$$

$$= A\left(U_p + \frac{h}{t}\right)Ct \tag{3.69}$$

But, at time t this plane is interfacing with the clear liquid, and so at this time all the particles in the test have passed through the plane.

$$\text{The total volume of particles in the test} = C_B h_0 A \tag{3.70}$$

where h_0 is the initial suspension height.
Therefore,

$$C_B h_0 A = A\left(U_P + \frac{h}{t}\right)Ct \tag{3.71}$$

Hence, substituting for U_p from Equation (3.67), we have

$$C = \frac{C_B h_0}{h_1} \tag{3.72}$$

Thus, for each point on the height–time plot we have the particle velocity U_p [from Equation (3.67)] and the corresponding suspension concentration C [from Equation (3.72)]. From these pairs of values, we can create the flux plot (U_{ps} or $U_p C$ versus C).

3.2.6 Worked Examples on Settling of a Suspension of Particles

WORKED EXAMPLE 3.8

A height–time curve for the sedimentation of a suspension, of initial suspension concentration 0.1, in vertical cylindrical vessel is shown in Figure 3.W8.1. Determine:

(a) The velocity of the interface between clear liquid and suspension of concentration 0.1.

(b) The velocity of the interface between clear liquid and a suspension of concentration 0.175.

(c) The velocity at which a layer of concentration 0.175 propagates upwards from the base of the vessel.

(d) The final sediment concentration.

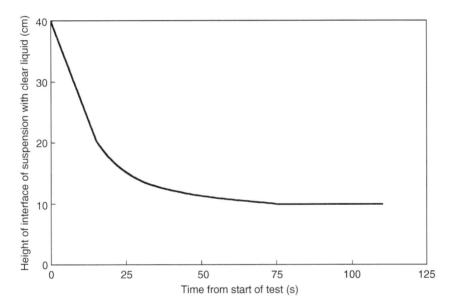

Figure 3.W8.1 Batch settling test; height–time curve

Solution

(a) Since the initial suspension concentration is 0.1, the velocity required in this question is the velocity of the AB interface. This is given by the slope of the straight portion of the height–time curve.

$$\text{Slope} = \frac{20 - 40}{15 - 0} = 1.333 \, \text{cm/s}$$

(b) We must first find the point on the curve corresponding to the point at which a suspension of concentration 0.175 interfaces with the clear suspension. From Equation (3.72) with $C = 0.175$, $C_B = 0.1$ and $h_0 = 40$ cm, we find:

$$h_1 = \frac{0.1 \times 40}{0.175} = 22.85 \, \text{cm}$$

A line drawn through the point $t = 0, h = h_1$ tangential to the curve locates the point on the curve corresponding to the time at which a suspension of concentration 0.175 interfaces with the clear suspension (Figure 3.W8.2). The coordinates of this point are $t = 26$ s, $h = 15$ cm. The velocity of this interface is the slope of the curve at this point:

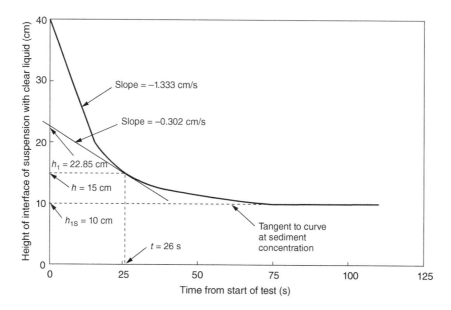

Figure 3.W8.2   Batch settling test

$$\text{Slope of curve at 26 s, 15 cm} = \frac{15 - 22.85}{26 - 0} = -0.302 \text{ cm/s}$$

downward velocity of interface= 0.30 cm/s.

(c) From the consideration above, after 26 s the layer of concentration 0.175 has just reached the clear liquid interface and has travelled a distance of 15 cm from the base of the vessel in this time.

Therefore, upward propagation velocity of this layer $= \dfrac{h}{t} = \dfrac{15}{16} = 0.577 \text{ cm/s}$.

(d) To find the concentration of the final sediment we again use Equation (3.72). The value of h_1 corresponding to the final sediment (h_{1s}) is found by drawing a tangent to the part of the curve corresponding to the final sediment and projecting it to the h axis.

In this case $h_{1S} = 10$ cm and so from Equation (3.72),

$$\text{Final sediment concentration, } C = \frac{C_0 h_0}{h_{1S}} = \frac{0.1 \times 4.0}{10} = 0.4$$

WORKED EXAMPLE 3.9

A suspension in water of uniformly sized spheres (diameter 150 μm, density 1140 kg/m^3) has a solids concentration of 25% by volume. The suspension settles to a bed of solids concentration of 55% by volume. Calculate:

(a) The rate at which the water–suspension interface settles.

(b) The rate at which the sediment–suspension interface rises (assume water properties: density, 1000 kg/m^3; viscosity, 0.001 Pa s).

Solution

(a) Solids concentration of initial suspension $C_B = 0.25$.

Equation (3.62) allows us to calculate the velocity of interfaces between suspensions of different concentrations.

The velocity of the interface between initial suspension (B) and clear liquid (A) is therefore:

$$U_{int,AB} = \frac{U_{pA}C_A - U_{pB}C_B}{C_A - C_B}$$

Since $C_A = 0$, the equation reduces to

$$U_{int,AB} = U_{pB}$$

where U_{pB} is the hindered settling velocity of particles relative to the vessel wall in batch settling and is given by Equation (3.58):

$$U_p = U_T \varepsilon^n$$

Assuming Stokes' law applies, then $n = 4.65$ and the single particle terminal velocity is given by Equation (3.15):

$$U_T = \frac{x^2 \left(\rho_p - \rho_f \right) g}{18\mu}$$

$$U_T = \frac{9.81 \times \left(150 \times 10^{-6} \right)^2 \times (1140 - 1000)}{18 \times 0.001}$$

$$= 1.717 \times 10^{-3} \text{ m/s}$$

To check that the assumption of Stokes' law is valid, we calculate the single-particle Reynolds number:

$$Re_p = \frac{\left(150 \times 10^{-6} \right) \times 1.717 \times 10^{-3} \times 1000}{0.001}$$

= 0.258, which is less than the limiting value for Stokes' law (0.3) and so the assumption is valid.

The void fraction of the initial suspension $\varepsilon_B = 1 - C_B = 0.75$

$$\text{Hence, } U_{pB} = 1.717 \times 10^{-3} \times 0.75^{4.65}$$

$$= 0.45 \times 10^{-3} \text{ m/s}$$

Hence, the velocity of the interface between the initial suspension and the clear liquid is 0.45 mm/s. The fact that the velocity is positive indicates that the interface is moving downwards.

(b) Here again we apply Equation (3.62) to calculate the velocity of interfaces between suspensions of different concentrations.

The velocity of the interface between initial suspension (B) and sediment (S) is therefore

$$U_{int,BS} = \frac{U_{pB}C_B - U_{pS}C_S}{C_B - C_S}$$

With $C_B = 0.25$ and $C_s = 0.55$ and since the velocity of the sediment U_{pS} is zero, we have:

$$U_{int,BS} = \frac{U_{pB}0.25 - 0}{0.25 - 0.55} = -0.833U_{pB}$$

And from part (a), we know that $U_{pB} = 0.45$ mm/s, and so $U_{int,BS} = -0.375$ mm/s.

The negative sign signifies that the interface is moving upwards. So, the interface between initial suspension and sediment is moving upwards at a velocity of 0.375 mm/s.

WORKED EXAMPLE 3.10

For the batch flux plot shown in Figure 3.W10.1, the sediment has a solids concentration of 0.4 volume fraction of solids.

(a) Determine the range of initial suspension concentrations over which a zone of variable concentration is formed under batch settling conditions.

(b) Calculate and plot the concentration profile after 50 min in a batch settling test of a suspension with an initial concentration 0.1 volume fraction of solids, and initial suspension height of 100 cm.

(c) At what time will the settling test be complete?

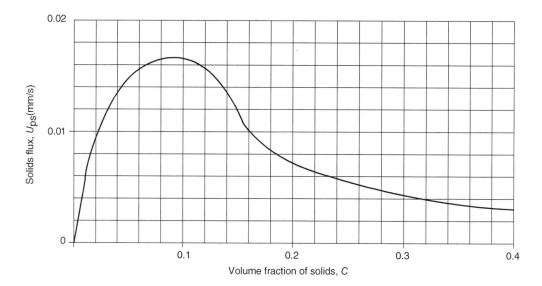

Figure 3.W10.1   Batch flux plot

Solution

(a) Determine the range of initial suspension concentrations by drawing a line through the point $C = C_s = 0.4$, $U_{ps} = 0$ tangential to the batch flux curve. This is shown as line XC_s in Figure 3.W10.2. The range of initial suspension concentrations for which a zone of variable concentration is formed in batch settling (Type 2 settling) is defined by $C_{B_{min}}$ and $C_{B_{max}}$. $C_{B_{min}}$ is the value of C at which the line XC_s intersects the settling curve and $C_{B_{max}}$ is the value of C at the tangent. From Figure 3.W10.2, we see that $C_{B_{min}} = 0.036$ and $C_{B_{max}} = 0.21$.

(b) To calculate the concentration profile, we must first determine the velocities of the interfaces between the zones A, B, E and S and hence find their positions after 50 min.

The line AB in Figure 3.W10.2 joins point A representing the clear liquid (0, 0) and point B representing the initial suspension (0.1, U_{ps}). The slope of line AB is equal to the velocity of the interface between zones A and B. From Figure 3.W10.2, $U_{int, AB} = +0.166$ mm/s or $+1.00$ cm/min.

The slope of the line from point B tangential to the curve is equal to the velocity of the interface between the initial suspension B and the minimum value of the variable concentration zone $C_{E_{min}}$.

From Figure 3.W10.2,

$$U_{int,BE_{min}} = -0.111 \, \text{mm/s or} -0.66 \, \text{cm/min}$$

The slope of the line tangential to the curve and passing through the point representing the sediment (point $C = C_S = 0.4$, $U_{ps} = 0$) is equal to the velocity of the interface between the maximum value of the variable concentration zone $C_{E_{max}}$ and the sediment.

From Figure 3.W10.2,

$$U_{int,E_{max}S} = -0.0355 \, \text{mm/s or} -0.213 \, \text{cm/min}$$

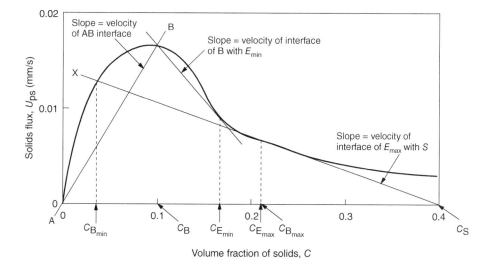

Figure 3.W10.2 Graphical solution to batch settling problem in Worked Example 3.10

Therefore, after 50 min the distances travelled by the interfaces will be:

AB interface   50.0 cm (1.00×50) downwards

BE_{min} interface   33.2 cm upwards

$E_{max}S$ interface   10.6cm upwards

Therefore, the positions of the interfaces (distance from the base of the test vessel) after 50 min will be:

AB interface   50.0 cm

BE_{min} interface   33.2 cm

$E_{max}S$ interface   10.6cm

From Figure 3.W10.2 we determine the minimum and maximum values of suspension concentration in the variable zone:

$$C_{E_{min}} = 0.16$$
$$C_{E_{max}} = 0.21$$

Using this information, we can plot the concentration profile in the test vessel 50 min after the start of the test. A sketch of the profile is shown in Figure 3.W10.3. The shape of the concentration profile within the variable concentration zone may be determined by the following method. Recalling that the slope of the batch flux plot (Figure 3.W10.1) at a value of suspension concentration C is the velocity of a layer of suspension of that concentration, we find the slope at two or more values of concentration and then determine the positions of these layers after 50 min:

• Slope of batch flux plot at $C = 0.18$ is 0.44 cm/min upwards.

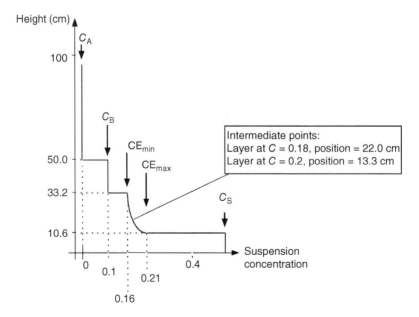

Figure 3.W10.3   Sketch of concentration profile in batch settling test vessel after 50 min

Hence, the position of a layer of concentration 0.18 after 50 min is 22.0 cm from the base.

- Slope of batch flux plot at $C = 0.20$, is 0.27 cm/min upwards.

Hence, the position of a layer of concentration 0.20 after 50 min is 13.3 cm from the base.

These two points are plotted on the concentration profile in order to determine the shape of the profile within the zone of variable concentration.

Figure 3.W10.4 is a sketched plot of the height–time curve for this test constructed from the information above. The shape of the curved portion of the curve can again be determined by plotting the positions of two or more layers of suspension of different concentration. The initial suspension concentration zone (B) ends when the AB line intersects the BE_{min} line, both of which are plotted from a knowledge of their slopes.

The time for the end of the test is found in the following way. The end of the test is when the position of the $E_{max}S$ interface coincides with the height of the final sediment. The height of the final sediment may be found using Equation (3.72) [see part (d) of Worked Example 3.8]:

$$C_S h_S = C_B h_0$$

where h_S is the height of the final sediment and h_0 is the initial height of the suspension (at the start of the test). With $C_S = 0.4$, $C_B = 0.1$ and $h_0 = 100$ cm, we find that $h_S = 25$ cm. Plotting h_S on Figure 3.W10.4, we find that the $E_{max}S$ line intersects the final sediment line at about 120 min and so the test ends at this time.

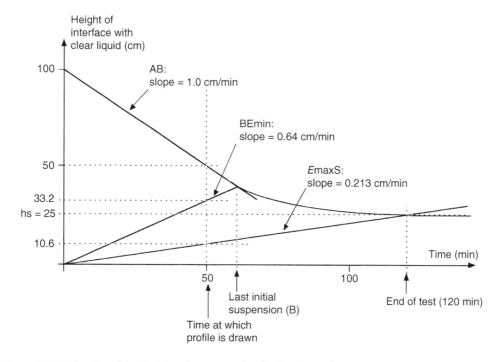

Figure 3.W10.4 Sketch of height–time curve for the batch settling test in Worked Example 3.10

TEST YOURSELF – SETTLING OF A SUSPENSION PARTICLES

3.7 What is the Richardson–Zaki relationship and what does it describe?

3.8 On a plot of volumetric particle flux versus suspension concentration (flux plot), what does the gradient of the curve at concentration C represent?

3.9 On a plot of volumetric particle flux versus suspension concentration (flux plot), what does the slope of a chord joining two points at concentrations C_1 and C_2 represent?

3.10 What is the main difference between Types 1 and 2 settling?

3.11 Make a sketch showing the zones of constant concentration formed during a batch settling test on a suspension displaying Type 1 settling.

3.12 Make a sketch showing the zones of constant concentration formed during a batch settling test on a suspension displaying Type 2 settling.

EXERCISES – SINGLE PARTICLES IN A FLUID

3.1 The settling chamber, shown schematically in Figure 3.E1.1, is used as a primary separation device in the removal of dust particles of density $1500 \, \text{kg}/\text{m}^3$ from a gas of density $0.7 \, \text{kg}/\text{m}^3$ and viscosity $1.90 \times 10^{-5} \, \text{Pa s}$.

   (a) Assuming Stokes' law applies, show that the efficiency of collection of particles of size x is given by the expression:

$$\text{Collection efficiency,} \, \eta_x = \frac{x^2 g \left(\rho_p - \rho_f \right) L}{18 \mu H U}$$

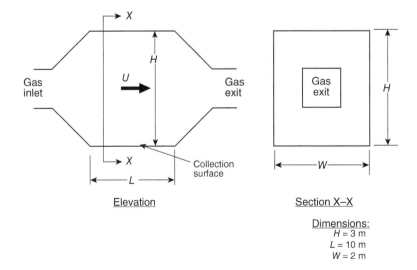

Figure 3.E1.1 Schematic diagram of the settling chamber

where U is the uniform gas velocity through the parallel-sided section of the chamber. State any other assumptions made.

(b) What is the upper limit of particle size for which Stokes' law applies?

(c) When the volumetric flow rate of gas is 0.9 m³/s, and the dimensions of the chamber are those shown in Figure 3.E1.1, determine the collection efficiency for spherical particles of diameter 30 μm.

[Answer: (b) 57 μm; (c) (86%).]

3.2 A particle of equivalent sphere volume diameter 0.2 mm, density 2500 kg/m³ and sphericity 0.67 falls freely under gravity in a fluid of density 1.0 kg/m³ and viscosity 2×10^{-5} Pa s. Estimate the terminal velocity reached by the particle.

[Answer: 1.12 m/s.]

3.3 Spherical particles of density 2500 kg/m³ and in the size range 20–100 μm are fed continuously into a stream of water (density, 1000 kg/m³ and viscosity, 0.001 Pa s) flowing upwards in a vertical, large diameter pipe. What maximum water velocity is required to ensure that no particles of diameter greater than 60 μm are carried upwards with the water?

[Answer: 3.1 mm/s.]

3.4 Spherical particles of density 2000 kg/m³ and in the size range 20–100 μm are fed continuously into a stream of water (density, 1000 kg/m³ and viscosity, 0.001 Pa s) flowing upwards in a vertical, large diameter pipe. What maximum water velocity is required to ensure that no particles of diameter greater than 50 μm are carried upwards with the water?

[Answer: 1.57 mm/s.]

3.5 A particle of equivalent volume diameter 0.3 mm, density 2000 kg/m³ and sphericity 0.8 falls freely under gravity in a fluid of density 1.2 kg/m³ and viscosity 2×10^{-5} Pa s. Estimate the terminal velocity reached by the particle.

[Answer: 1.59 m/s.]

3.6 Assuming that a car is equivalent to a flat plate 1.5 m square, moving normal to the airstream, and with a drag coefficient, $C_D = 1.1$, calculate the power required for steady motion at 100 km/h on level ground. What is the Reynolds number? For air assume a density of 1.2 kg/m³ and a viscosity of 1.71×10^{-5} Pa s.

[Answer: 32.9 kW; 2.95×10^6.]

3.7 A cricket ball is thrown with a Reynolds number such that the drag coefficient is 0.4 ($Re \approx 10^5$).

(a) Find the percentage change in velocity of the ball after 100 m horizontal flight in air.

(b) With a higher Reynolds number and a new ball, the drag coefficient falls to 0.1. What is now the percentage change in velocity over 100 m horizontal flight?

(In both cases take the mass and diameter of the ball as 0.15 kg and 6.7 cm, respectively, and the density of air as 1.2 kg/m³.) Readers unfamiliar with the game of cricket may substitute a baseball.

[Answer: (a) 43.1%; (b) 13.1%.]

3.8 The resistance F of a sphere of diameter x, due to its motion with velocity u through a fluid of density ρ and viscosity μ, varies with Reynolds number $(Re = \rho ux/\mu)$ as given below:

$\log_{10}Re$	2.0	2.5	3.0	3.5	4.0
$C_D = \dfrac{F}{\frac{1}{2}\rho u^2(\pi x^2/4)}$	1.05	0.63	0.441	0.385	0.39

Find the mass of a sphere of 0.013 m diameter which falls with a steady velocity of 0.6 m/s in a large deep tank of water of density 1000 kg/m^3 and viscosity 0.0015 Pa s.

[Answer: 0.0021 kg.]

3.9 A particle of 2 mm in diameter and density of 2500 kg/m^3 is settling in a stagnant fluid in the Stokes' law region.

(a) Calculate the viscosity of the fluid if the fluid density is 1000 kg/m^3 and the particle falls at a terminal velocity of 4 mm/s.

(b) What is the drag force on the particle at these conditions?

(c) What is the particle drag coefficient at these conditions?

(d) What is the particle acceleration at these conditions?

(e) What is the apparent weight of the particle?

3.10 Starting with the force balance on a single particle at terminal velocity, show that:

$$C_D = \frac{4}{3}\frac{gx}{U_T^2}\left[\frac{\rho_p - \rho_f}{\rho_f}\right]$$

where the symbols have their usual meaning.

3.11 A spherical particle of density 1500 kg/m^3 has a terminal velocity of 1 cm/s in a fluid of density 800 kg/m^3 and viscosity 0.001 Pa s. Estimate the diameter of the particle.

[Answer: 181 μm]

3.12 Estimate the largest diameter of spherical particle of density 2000 kg/m^3 which would be expected to obey Stokes' law in air of density 1.2 kg/m^3 and viscosity 18×10^{-6} Pa s.

3.13 Determine the time taken for particles of density 1200 kg/m^3 and sizes 15, 25 and 50 μm to reach 99% of terminal velocity falling from rest under gravity in air of density 1.0 kg/m^3 and viscosity 18×10^{-6} Pa s. Validate any assumptions.

[Answer: 3.8 ms, 10.8 ms and 42.6 ms for 15, 25 and 50 μm particles respectively]

EXERCISES – SETTLING OF A SUSPENSION OF PARTICLES

3.14 A suspension in water of uniformly sized spheres of diameter 100 μm and density 1200 kg/m^3 has a solids volume fraction of 0.2. The suspension settles to a bed of solids volume fraction 0.5. (For water, density is 1000 kg/m^3 and viscosity is 0.001 Pa s.)

The single particle terminal velocity of the spheres in water may be taken as 1.1 mm/s.

Calculate:

(a) The velocity at which the clear water–suspension interface settles.

(b) The velocity at which the sediment–suspension interface rises.

[Answer: (a) 0.39 mm/s; (b) 0.26 mm/s.]

3.15 A height–time curve for the sedimentation of a suspension in a vertical cylindrical vessel is shown in Figure 3.E15.1. The initial solids concentration of the suspension is 150 kg/m^3.

Determine:

(a) The velocity of the interface between clear liquid and suspension of concentration 150 kg/m^3.

(b) The time from the start of the test at which the suspension of concentration 240 kg/m^3 is in contact with the clear liquid.

(c) The velocity of the interface between the clear liquid and suspension of concentration 240 kg/m^3.

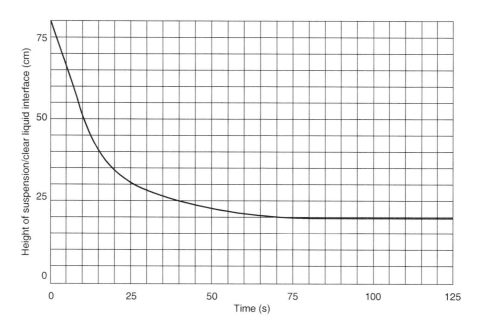

Figure 3.E15.1 Batch settling test results. Height–time curve for use in Exercises 3.15 and 3.17

(d) The velocity at which a layer of concentration $240 \, \text{kg/m}^3$ propagates upwards from the base of the vessel.

(e) The concentration of the final sediment.

[Answer: (a) 2.91 cm/s; (b) 22 s; (c) 0.77 cm/s downwards; (d) 1.50 cm/s upwards; (e) $600 \, \text{kg/m}^3$.]

3.16 A suspension in water of uniformly sized spheres of diameter 90 μm and density $1100 \, \text{kg/m}^3$ has a solids volume fraction of 0.2. The suspension settles to a bed of solids volume fraction 0.5. (For water, density is $1000 \, \text{kg/m}^3$ and viscosity is 0.001 Pa s.)

The single particle terminal velocity of the spheres in water may be taken as 0.44 mm/s.

Calculate:

(a) The velocity at which the clear water–suspension interface settles.

(b) The velocity at which the sediment–suspension interface rises.

[Answer: (a) 0.156 mm/s; (b) 0.104 mm/s.]

3.17 A height–time curve for the sedimentation of a suspension in a vertical cylindrical vessel is shown in Figure 3.E15.1. The initial solids concentration of the suspension is $200 \, \text{kg/m}^3$.

Determine:

(a) The velocity of the interface between clear liquid and suspension of concentration 200 kg/m^3.

(b) The time from the start of the test at which the suspension of concentration $400 \, \text{kg/m}^3$ is in contact with the clear liquid.

(c) The velocity of the interface between the clear liquid and suspension of concentration $400 \, \text{kg/m}^3$.

(d) The velocity at which a layer of concentration $400 \, \text{kg/m}^3$ propagates upwards from the base of the vessel.

(e) The concentration of the final sediment.

[Answers: (a) 2.9 cm/s downwards; (b) 32.5 s; (c) 0.40 cm/s downwards; (d) 0.846 cm/s upwards; (e) $800 \, \text{kg/m}^3$.]

3.18

(a) Spherical particles of uniform diameter 40 μm and particle density $2000 \, \text{kg/m}^3$ form a suspension of solids volume fraction 0.32 in a liquid of density $880 \, \text{kg/m}^3$ and viscosity 0.0008 Pa s. Assuming Stokes' law applies, calculate (i) the sedimentation velocity and (ii) the sedimentation volumetric flux for this suspension.

(b) A height–time curve for the sedimentation of a suspension in a cylindrical vessel is shown in Figure 3.E18.1. The initial concentration of the suspension for this test is $0.12 \, \text{m}^3/\text{m}^3$.

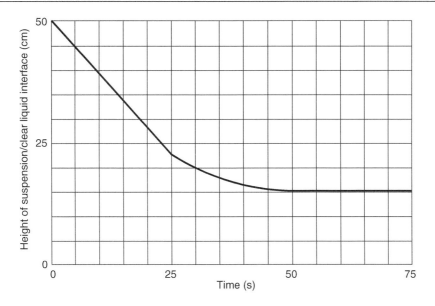

Figure 3.E18.1 Batch settling test results. Height–time curve for use in Exercise 3.18

Calculate:

    (i) The velocity of the interface between clear liquid and a suspension of concentration, 0.12 m^3/m^3.

    (ii) The velocity of the interface between clear liquid and a suspension of concentration 0.2 m^3/m^3.

    (iii) The velocity at which a layer of concentration, 0.2 m^3/m^3 propagates upwards from the base of the vessel.

    (iv) The concentration of the final sediment.

    (v) The velocity at which the sediment propagates upwards from the base.

[Answer: (a) (i) 0.203 mm/s, (ii) 0.065 mm/s, (b) (i) 1.11 cm/s downwards, (ii) 0.345 cm/s downwards, (iii) 0.514 cm/s upwards, (iv) 0.4, (v) 0.30 cm/s upwards.]

3.19 A height–time curve for the sedimentation of a suspension in a vertical cylindrical vessel is shown in Figure 3.E19.1. The initial solids concentration of the suspension is 100 kg/m^3.

Determine:

(a) The velocity of the interface between clear liquid and suspension of concentration 100 kg/m^3.

(b) The time from the start of the test at which the suspension of concentration 200 kg/m^3 is in contact with the clear liquid.

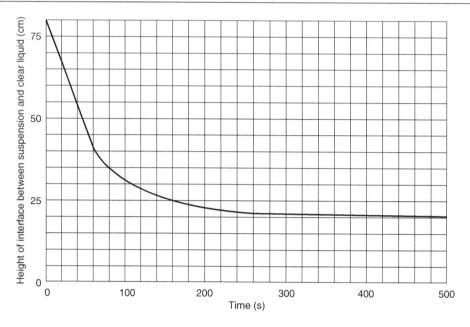

Figure 3.E19.1   Batch settling test results. Height–time curve for use in Exercises 3.19 and 3.21

(c) The velocity of the interface between the clear liquid and suspension of concentration 200 kg/m^3.

(d) The velocity at which a layer of concentration 200 kg/m^3 propagates upwards from the base of the vessel;

(e) The concentration of the final sediment.

[Answer: (a) 0.667 cm/s downwards; (b) 140 s; (c) 0.0976 cm/s downwards; (d) 0.189 cm/s upwards; (e) 400 kg/m^3.]

3.20   A suspension in water of uniformly sized spheres of diameter 80 μm and density 1300 kg/m^3 has a solids volume fraction of 0.10. The suspension settles to a bed of solids volume fraction 0.4. (For water, density is 1000 kg/m^3 and viscosity is 0.001 Pa s.)

The single particle terminal velocity of the spheres under these conditions is 1.0 mm/s.

Calculate:

(a) The velocity at which the clear water–suspension interface settles.

(b) The velocity at which the sediment–suspension interface rises.

[Answer: (a) 0.613 mm/s; (b) 0.204 mm/s.]

3.21   A height–time curve for the sedimentation of a suspension in a vertical cylindrical vessel is shown in Figure 3.E19.1. The initial solids concentration of the suspension is 125 kg/m^3.

Determine:

(a) The velocity of the interface between clear liquid and suspension of concentration 125 kg/m^3.

(b) The time from the start of the test at which the suspension of concentration 200 kg/m^3 is in contact with the clear liquid.

(c) The velocity of the interface between the clear liquid and suspension of concentration 200 kg/m^3.

(d) The velocity at which a layer of concentration 200 kg/m^3 propagates upwards from the base of the vessel.

(e) The concentration of the final sediment.

[Answer: (a) 0.667 cm/s downwards; (b) 80 s; (c) 0.192 cm/s downwards; (d) 0.438 cm/s upwards; (e) 500 kg/m^3.]

3.22 Use the batch flux plot in Figure 3.E22.1 to answer the following questions. (Note that the sediment concentration is 0.44 volume fraction.)

(a) Determine the range of initial suspension concentration over which a variable concentration zone is formed under batch settling conditions.

(b) For a batch settling test using a suspension with an initial concentration 0.18 volume fraction and initial height 50 cm, determine the settling velocity of the interface between clear liquid and suspension of concentration 0.18 volume fraction.

(c) Determine the position of this interface 20 min after the start of this test.

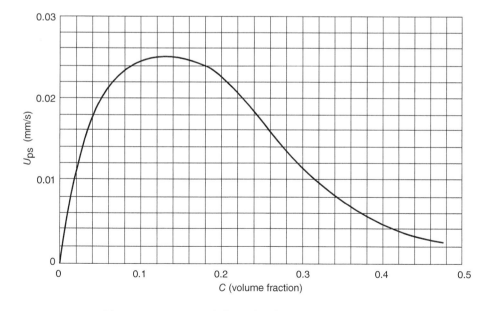

Figure 3.E22.1 Batch flux plot for use in Exercise 3.22

(d) Produce a sketch showing the concentration zones in the settling test 20 min after the start of this test.

[Answer: (a) 0.135 to 0.318; (b) 0.80 cm/min; (c) 34 cm from base; (d) BE interface is 12.5 cm from base.]

3.23 Consider the batch flux plot shown in Figure 3.W8.1. Given that the final sediment concentration is 0.36 volume fraction:

(a) Determine the range of initial suspension concentration over which a variable concentration zone is formed under batch settling conditions.

(b) Calculate and sketch the concentration profile after 40 min of the batch settling test with an initial suspension concentration of 0.08 and an initial height of 100 cm.

(c) Estimate the height of the final sediment and the time at which the test is complete.

[Answers: (a) 0.045 to 0.20; (c) 22.2 cm; 83 min.]

3.24 Uniformly sized spheres of diameter 50 μm and density 1500 kg/m^3 are uniformly suspended in a liquid of density 1000 kg/m^3 and viscosity 0.002 Pa s. The resulting suspension has a solids volume fraction of 0.30.

The single particle terminal velocity of the spheres in this liquid may be taken as 0.00034 m/s ($Re_p < 0.3$). Calculate the velocity at which the clear water-suspension interface settles.

3.25 Calculate the settling velocity of glass spheres having a diameter of 155 μm in water at 293 K. The slurry contains 60 wt% solids. The density of the glass spheres is 2467 kg/m^3.

How does the settling velocity change if the particles have a sphericity of 0.3 and an equivalent diameter of 155 μm?

3.26 A suspension in water of uniformly sized spheres (diameter 150 μm and density 1140 kg/m^3) has a solids concentration of 25% by volume. The suspension settles to a bed of solids concentration 62% by volume. Calculate the rate at which the spheres settle in the suspension. Calculate the rate at which the settled bed height rises.

3.27 If 20 μm particles with a density of 2000 kg/m^3 are suspended in a liquid with a density of 900 kg/m^3 at a concentration of 50 kg/m^3, what is the solids volume fraction of the suspension? What is the bulk density of the suspension?

3.28 Given Figure 3.E28.1 for the height–time curve for the sedimentation of a suspension in a vertical cylindrical vessel with an initial uniform solids concentration of 100 kg/m^3:

(a) What is the velocity at which the sediment–suspension interface rises?

(b) What is the velocity of the interface between the clear liquid and suspension of concentration 133 kg/m^3?

(c) What is the velocity at which a layer of concentration 133 kg/m^3 propagates upwards from the base of the vessel?

(d) At what time does the sediment–suspension interface start rising?

(e) At what time is the concentration of the suspension in contact with the clear liquid no longer 100 kg/m^3?

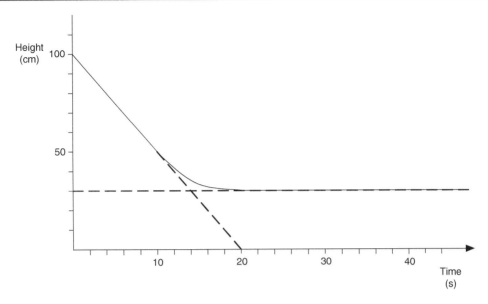

Figure 3.E28.1   Plot of height of clear liquid interface versus time during settling test for use in Exercise 3.28

4

Discrete Element Method Modelling

One of the most significant changes in the practice of particle technology has been the development and now widespread use of computational modelling techniques. This chapter introduces the most common of such methods: the *discrete element method (DEM)*. At the core of the method are the simplified physical models of particle–particle contact, which have already been introduced in Chapter 2. This chapter shows how those models can be implemented in a simulation involving many particles. Such simulations also require reliable detection of contacts and systematic updating of particle positions (timestepping); methods for both are briefly described. This chapter introduces the principles behind the coupling of DEM with computational fluid dynamics (CFD) in order to model both the solids and the fluid, and approaches to the modelling of large systems, where the number of particles makes it impractical to treat each one individually. Finally, we summarize the principles of model calibration and validation, giving some practical examples of how this can be done.

4.1 INTRODUCTION

The design of processes involving particles requires prediction of their behaviour. As in other areas of science and engineering, the behaviour of particles can be predicted to some extent using analytical models. This is more difficult for particles than for fluids because there is no overall mathematical description that covers the complexity of their behaviour and so their bulk properties cannot be simply derived. To take a simple example, the bulk density of a quantity of particles does not take a single value: it depends on whether the particles are static or flowing, how they have been placed in a container, whether they have been compacted and many other factors. Their bulk density will also change from point to point within a process (or even a bag), according to local conditions.

This chapter is contributed by Kit Windows-Yule, University of Birmingham, UK

Introduction to Particle Technology, Third Edition. Martin Rhodes and Jonathan Seville.
© 2024 John Wiley & Sons Ltd. Published 2024 by John Wiley & Sons Ltd.
Website: www.wiley.com/go/rhodes/particle3e

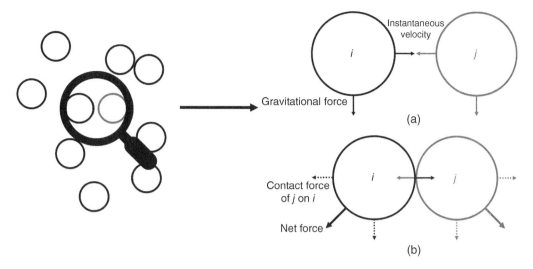

Figure 4.1 A simple schematic representation of a DEM simulation. Panel (a) represents two particles approaching one another preceding a collision. Panel (b) represents a point in time immediately following the collision of the particles

The dynamics of particulate and particle–fluid systems may be modelled, to varying degrees of realism, using a variety of methods including (but not limited to) Monte Carlo simulations, cellular automata, the multiphase particle-in-cell method (MP-PIC), or even CFD-like continuum simulations implementing the 'kinetic theory of granular flow'(KTGF)[1] (Andrews and O'Rourke, 1996; Campbell, 2006; Rosato and Windows-Yule, 2020). However, by far the most common method for simulating granular flows is the discrete element method (DEM).[2] In DEM, each individual particle within a given system of interest is modelled as a computational object, possessing a number of properties – size, density, stiffness, coefficient of friction etc. – directly corresponding to those of the real material(s) to be modelled. The motion of each individual particle within the system of interest is then simulated simply through the application of Newton's laws, with interactions between particles and other objects simulated through the application of suitable 'contact models' (see Section 4.5).

Figure 4.1 shows a schematic representation of a simple DEM model. Each particle is individually identified within the model. Each one can have different properties, so it is possible to model distributions of size, for example, and non-spherical shapes (see Section 4.9). Particles can interact with one another according to simple (or more complex) rules, which are usually selected to mimic real physical interactions such as collisions. In some versions of DEM, interstitial fluid can be included (see Section 4.6), so that particles are subject to the forces of fluid drag. DEM is therefore a computational framework within which user-defined variations are possible.[3] The output from the model computation can be extremely detailed, including the motion of, and forces acting

---

[1] An accessible introduction to Monte Carlo methods and cellular automata in the context of particle technology can be found in Chapter 4 of Rosato and Windows-Yule. Introductions to MP-PIC and the kinetic theory of granular flow can be found in Andrews and O'Rourke, and Campbell, respectively.

[2] Sometimes known by other names, such as the discrete particle method (DPM).

[3] In addition to commercial DEM codes, there are a number of open-access DEM codes.

upon, every particle within a system, at any arbitrary point in time, from which many other measures of particle behaviour can be derived. However, as with any model, the result is an approximation to reality and depends on the skill of the user in setting up the model and making the best choices for the various model parameters. (see Section 4.7).

This chapter is an introduction to the fundamental principles of DEM, including the way in which particles, the forces acting upon them, and their dynamics are modelled, the key assumptions of these models, and the issues which these assumptions may potentially cause. In further sections, the practicalities of constructing and implementing a DEM model are introduced, including how to choose suitable parameters to describe the particles and their interactions, and how to model scientifically and industrially relevant systems in a manner which is both accurate and practically achievable. In Section 4.10, we show how the rich data produced by these simulations can be post-processed and analyzed to provide meaningful information regarding the dynamics and mechanics of the system being modelled.

Through the use of interactive examples, which can be found at www.wiley.com/go/rhodes/particle3e, the reader is guided step-by-step through the development, from the ground up, of a simple DEM model, and shown how detailed, physical information may be extracted from DEM data (Section 4.10).

4.2   PRINCIPLES – THE "HARD-SPHERE" AND "SOFT-SPHERE" APPROACHES

Particles in processes make frequent contact or collisions with each other and with surfaces such as walls and blades. There are two distinct approaches to modelling such contacts, known as the *'hard-sphere'* and the *'soft-sphere'* models. In the hard-sphere model (HSM), particles are represented as rigid spheres, and interactions (i.e. collisions between particles and with other objects) are assumed to be instantaneous (of zero time duration) and binary (only two particles collide at any one time), with point-like contacts. Under these assumptions, the post-collisional velocities of two interacting particles can be simply derived using the conservation of momentum.

Figure 4.2 shows the simplest case: two spherical particles of equal size and density colliding along the same line; in this case the velocity of particle i immediately after contact is $-e_n v_i$, where v_i is the initial velocity before collision, e_n is the coefficient of restitution and the negative sign indicates that the direction of motion has been reversed.

A restitution coefficient $e_n = 1$ indicates a perfectly elastic collision, while in a collision between two particles with $e_n = 0$, the particles' kinetic energy would be entirely dissipated. The subscript n indicates that the coefficient of restitution is in the normal direction, the same direction as the collision. If the collision is not in a normal direction, and in general it will not be, there is a *tangential* part of the interaction to be considered, which can be calculated using a *tangential* coefficient of restitution.

The hard-sphere model treats particles a bit like billiard balls,[4] with 'hard' interactions, the results of which are relatively simple to calculate. Real particles seldom behave like this. As shown in Chapter 2, real particles deform when they come into contact and it is this deformation which results in a repulsive force which acts to separate them. Real collisions also take a finite time to occur. All of these features are incorporated in the

---

[4] Although – obviously – billiards is a 2D game and DEM is (usually) performed in 3D, and even billiard balls show a small deformation on contact.

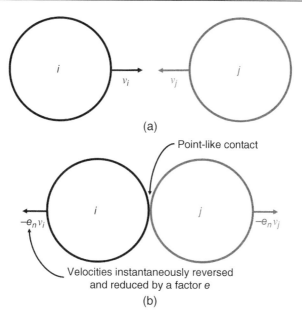

(a)

Point-like contact

Velocities instantaneously reversed
and reduced by a factor e

(b)

Figure 4.2 Schematic diagram of a collision between two equally massive particles possessing equal and opposite initial velocities, as simulated with the hard-sphere model. Panel (a) shows the particles' pre-collisional velocities and panel (b) their velocities immediately post-collision

soft-sphere model, which includes as much of the real physics of interaction as the user wants to consider. Since this approach is the one more commonly used, and is in principle a better approximation to reality, it is the only one considered in the rest of this chapter.

As shown in Chapter 2, contact between particles can be elastic, up to a certain limit, beyond which plastic deformation occurs. In the simplest soft-sphere approach (Figure 4.3), when particles collide their deformation is represented as an 'overlap', δ_{ij}. While in contact, the colliding particles are modelled as two masses connected by a 'Hookean spring' of stiffness k_n and a 'dashpot' with dissipation coefficient γ. A Hookean[5] spring produces a force f which is proportional to the overlap, and represents their elastic response. (An ideal spring dissipates no energy. When it is compressed, it stores energy which it then gives back as it recovers its shape.)

$$f = k_n \delta_{ij} \qquad (4.1)$$

A dashpot is a mechanical device rather like a piston filled with a viscous fluid, which resists motion in either direction, producing a force which is proportional to the relative *velocity* of motion v_n:

$$f = \gamma_n v_n \qquad (4.2)$$

This is often termed a 'damping force' and the device is sometimes known as a 'damper'. A dashpot dissipates energy, which cannot be recovered. This element can therefore account for the plastic/dissipative part of the collision.

---

[5] After Robert Hooke (1635–1703), English scientist who worked on elasticity and after whom Hooke's Law of linear elasticity is named. (See Chapter 2).

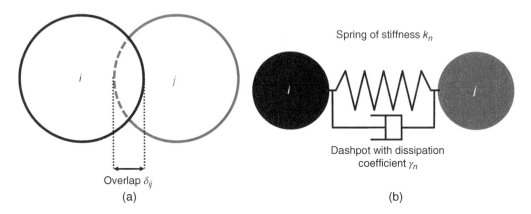

Figure 4.3 (a) Schematic diagram of two colliding particles as modelled using the soft-sphere approach and (b) representation of the system from (a) as a pair of masses joined by an elastic spring with stiffness k and a dashpot with dissipation coefficient γ, as in the simple, linear spring-dashpot model

When combined, these two elements – the spring and the dashpot – are capable of representing a range of quite complex interactions.[6] In a simple collision in the normal direction the combined force is then:

$$f_\mathrm{n} = k_\mathrm{n}\delta_\mathrm{ij} + \gamma_\mathrm{n}v_\mathrm{n} \qquad (4.3)$$

So far, only collisions in the normal direction have been considered and the friction between the particles has not been included. The spring–dashpot model can be relatively simply extended to model tangential forces between particles, as well as normal ones, through an additional spring and dashpot, alongside a frictional 'slider' between the two masses representing the particles (Figure 4.4).

The tangential force component can be expressed mathematically in a way equivalent to the normal force, i.e.

$$f_\mathrm{t} = -k_\mathrm{t}\xi - \gamma_\mathrm{t}v_\mathrm{t} \qquad (4.4)$$

where ξ is an analogue to the overlap δ for tangential motion. DEM particles are typically subject to Coulomb friction, introduced in Section 2.4 of Chapter 2, thus imposing the inequality

$$f_\mathrm{t} \leq \mu f_\mathrm{n} \qquad (4.5)$$

on the frictional forces modelled. In the case of friction, as long as the normal force f_n, between particles remains below a yield threshold $f_\mathrm{n}^\mathrm{max}$, the tangential force increases linearly with the normal force, the two being related by a constant μ_s, the coefficient of sliding friction. For normal forces exceeding $f_\mathrm{n}^\mathrm{max}$, the particles are assumed to 'slip', and the tangential force between them remains constant at a value $f_\mathrm{t} = \mu_\mathrm{s} f_\mathrm{n}^\mathrm{max}$.

---

[6] The spring-dashpot combination is very common in mechanical engineering, a good example being found in the suspension of road vehicles.

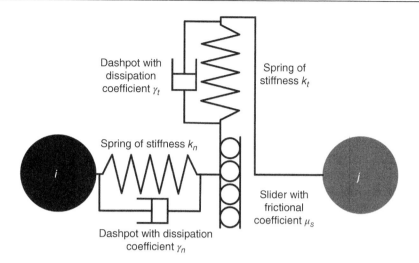

Figure 4.4 Spring–dashpot representation of a pair of colliding spheres with both normal and tangential forces

Equation (4.5) approximately describes sliding friction. Other forms of friction can occur, such as rolling friction – a parameter often used to capture the effects of the non-spherical nature of some (indeed most) 'real' particles whilst still simulating spherical particles. Such interactions depend very much on particle shape and surface roughness, which lie beyond the scope of this introduction to DEM. Since particle-particle contact is in reality very complex, this type of modelling always requires that simplifying assumptions are made.

An important additional factor is that the particles may be either cohesive (attracted to each other) or adhesive (attracted to other surfaces) or both, as introduced in Chapter 2. Methods of incorporating this into DEM are considered in Section 4.5.4.

4.3   UPDATING PARTICLE POSITIONS – 'TIME-STEPPING'

Between collisions, particles move through space according to Newton's laws. According to the second law ($f = ma$), the acceleration acting on a particle at a given point in time, t_n is given by:

$$a_n = \frac{f_{net,n}}{m} \tag{4.6}$$

where f_{net} is the sum of all the forces acting, which can include 'body forces' such as gravity and electric fields as well as fluid drag and contact forces. It is then possible to calculate the new velocity and position of the particle after a short time, provided that the time step dt ($= t_{n+1} - t_n$) in between the two points in time t_n and t_{n+1} is sufficiently small that any changes in velocity and acceleration during that step can be reasonably assumed infinitesimal. The velocity at time t_{n+1} can then be estimated as:

$$v_{n+1} = v_n + a_n dt \tag{4.7}$$

and the particle's new position at t_{n+1} as

$$r_{n+1} = r_n + v_n dt \tag{4.8}$$

The process is then repeated.

The methodology described above represents a simple numerical integration scheme known as the Euler method (Butcher, 2016), which is one of the simpler and more computationally efficient schemes available. For any integration scheme, it is important to choose an appropriate time step: too large and the simulation may be unstable; too small and it will be computationally expensive. The Euler method also has the unfortunate property of being non-conservative – that is to say that a particle may, somewhat unphysically, *gain energy* during free flight, the problem being exacerbated by larger time steps. This particular issue can be circumvented by using the slightly more complex (though still very efficient) velocity-Verlet integrator. While a detailed discussion of this scheme is beyond the scope of this book, the interested reader may find an example of the integrator in use in the interactive examples available at www.wiley.com/go/rhodes/particle3e.

A common mistake made by inexperienced DEM users is the implementation of overly large time steps in an effort to speed up simulation, inadvertently rendering their simulations inaccurate – or in extreme cases entirely unphysical. One criterion which is often used to avoid such problems is to maintain a time step, known as the *Rayleigh Time step*, which is smaller than the time taken for elastic vibrations[7] at the surface of a particle to pass around it. The Rayleigh time step is given by:

$$\tau_R = \frac{\pi R \sqrt{\frac{\rho_p}{G}}}{0.1631\nu + 0.8766} \tag{4.9}$$

where ρ_p is the particle density, R the particle radius, G the particle shear stiffness and ν the Poisson's ratio. The shear stiffness can be defined as $G = \frac{E}{2(1+\nu)}$ where E is the Young's modulus.

It is important to note that Equation (4.9) is only an estimate of the critical time step, based on several unproven assumptions, and thus is not guaranteed to produce stable behaviour for all simulation setups. As such, many authors take dt as some percentage of the Rayleigh time, commonly 20% thereof, so as to allow for a margin of error. Another common method for the estimation of a suitable combination of particle stiffness and time step relies on a more 'trial and error' approach. The underlying principle of this approach is to adjust the two relevant parameters until a 'realistic' maximum overlap between interacting particles is achieved throughout the range of stresses and velocities relevant to the simulation at hand. The maximum overlap might typically be either one-fiftieth or one-hundredth of a particle diameter.[8]

---

[7] These are known as Rayleigh waves and can be visualized as the waves which travel at the surface of a body when it is struck. For example, a point impact on a sphere, such as the Earth, causes waves to spread out from the impact point and travel around the body, often several times, interacting as they do so. The computation time therefore needs to be short compared with the time for a complete circuit of waves to develop.

[8] These are small overlaps. Readers should recall that the equations for particle contact interactions developed in Chapter 2 only apply for deformations which are small compared with the particle diameter. More importantly for practical implementation, large overlaps quickly result in very large repulsive forces, which can cause simulations to fail as a result of apparent 'explosions' of strongly repelled particles.

The approach to time-stepping described above may seem wasteful if the system being modelled is dilute so that 'events' such as particle collisions are infrequent. An alternative approach can then be adopted, called *event-driven simulation* in which complete particle trajectories are calculated between events. However, the advantages of this approach reduce strongly as the volume fraction of particles increases and it is not suitable for use where fluid drag needs to be included in the simulation, so it will not be considered further here. The interested reader may find an introduction to event-driven simulations in Luding (2004).

4.4 CONTACT DETECTION

Preceding sections have shown how to model the motion of particles through space and the contacts between particles when they collide. Another crucial part of DEM, however, is contact detection: how to determine when two particles are in contact. This might seem easy: two particles must touch if their separation is less than the sum of their radii. However, contact detection can easily become very computationally intensive, as explained below.

Consider a system of N_p particles. If nothing is known about the system, it will be necessary to test every particle against every other particle in order to find out if they are in contact, giving a number of contact evaluations as:

$$N_{eval} = N_p \left(N_p - 1 \right) / 2 \tag{4.10}$$

Since N_p is usually large compared with 1, the computational expense of contact detection is proportional to N_p^2. A system comprising two individual particles will require only a single evaluation. A system of 100 particles, however, will require almost 5000 evaluations, while a bed comprising 1 000 000 particles will require almost 5×10^{11} evaluations. To put this into context, assume (somewhat generously) that a PC's CPU can test for a contact 10^{11} times per second. For a system of 1 000 000 particles using a time step of 10^{-6} s, even discounting the time required to perform all other operations (time-stepping, contact modelling, etc.) would require more than 11 days to produce a single second of simulation. For 10 000 000 particles, this would require several years. Considering that even a small, laboratory-scale fluidized bed can easily contain of the order of 10 000 000 particles, the above timescales are clearly unacceptable, meaning that a more computationally efficient alternative is needed. This is typically achieved by using a 'neighbourhood search' method, as illustrated for the simple case of monosized particles in Figure 4.5, which takes advantage of the fact that, for any given particle, contacts will most obviously occur with neighbouring particles.

The simplest version of such a neighbourhood search, known as the *linked cell method*, works by dividing the computational volume into a series of individual 'cells', each with a side length greater than or equal to the diameter of a particle. Tests for particle contacts are then only performed for particles in the same or neighbouring cells, as illustrated (for two dimensions) in Figure 4.5. For each particle, possible contact with only those in the nine cells closest to it needs be considered (or 27 in 3D). Instead of comparing the position of each particle against potentially millions of others, at most only a few tens of other particles typically need be considered. This number can in fact be reduced further: taking

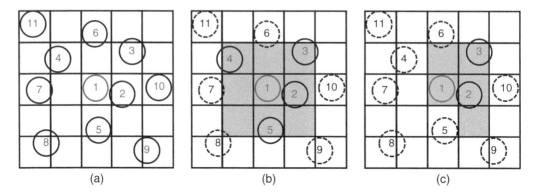

(a)   (b)   (c)

Figure 4.5 Simple schematic illustrating the advantages of neighbourhood search algorithms. Panel (a) provides a very simplified image of a two-dimensional DEM domain. Using a simple neighbourhood search, one need only check each particle against others in the 9 computational cells nearest to said particle (panel (b)). In reality, as each particle in the system will be checked, it is actually only necessary to check half of the nearest neighbour cells (panel (c))

account of all nearest-neighbour cells [Figure 4.5(b)] results in each particle being checked twice. That is to say particle 1 will be checked against particle 2, and later particle 2 will be checked against particle 1, and so on. This redundancy may be avoided, however, by considering only half of the neighbouring cells in each case [Figure 4.5(c)], meaning only one check is required per particle. By limiting the number of interactions to only those within neighbouring cells, the computational cost of the linked-cell algorithm scales with N_p rather than N_p^2.

The explanation above considers only mono-sized spherical particles. Real systems will contain a particle size distribution and usually non-spherical particles. For a size distribution, the obvious problem is that a cell size which is bigger than the largest particle diameter may contain many of the smaller particles. Consider, as an example, a binary system with a size ratio of 10:1. Even for the minimum cell size ($d_{cell} = d_{p,max}$), a single cell might contain more than 700 of the smaller particles, and a three-dimensional neighbourhood more than 19 000, thus again requiring an undesirable number of contact checks. In such systems, a multi-level grid approach may be used. The interested reader is referred to Ogarko and Luding (2012).

4.5 CONTACT MODELLING

Table 4.1 lists the most commonly used parameters in DEM modelling. There are two main types of parameters – those which are properties of a given particle (mass, diameter, density, etc.), and 'pair-wise' properties corresponding to interactions between particles (frictional coefficients, restitution coefficients, etc.). Note that all of the parameters in Table 4.1 are in principle experimentally measurable, although the first four – diameter, mass, density and to some extent shape – are much more easily measured in practice than the others listed, which generally require specialist research techniques, although there are indirect ways of obtaining these values, as described later.

Table 4.1 Summary of the most common parameters implemented in DEM simulations

Property	Symbol
Diameter	x
Mass	m
Density	ρ_P
Shape[a]	—
Sliding friction coefficient[b]	μ_s
Rolling friction coefficient[b]	μ_r
Torsion friction coefficient[b]	μ_t
Normal restitution coefficient[b]	e_n
Tangential restitution coefficient[b]	e_t
Young's modulus	E
Poisson ratio	ν
Cohesive energy density[c]	Ω

[a] 'Shape' is a generic term which may incorporate a range of properties.
[b] Parameters corresponding to pair-wise interactions (see main text).
[c] Cohesive energy density is one of several representations of cohesion, and is discussed further in Section 4.5.4.

4.5.1   Linear Spring–Dashpot Model

This is the simplest of all DEM contact models, used to describe simple non-cohesive interactions between particles showing a combination of elastic and plastic deformation.

The general form of the interaction was introduced as Equation (4.3):

$$f_n = k_n \delta_{ij} + \gamma_n v_n$$

The viscous damping coefficient γ_n can be related to the experimentally measurable restitution coefficient e in a variety of ways. The most common [though not necessarily the most accurate (Thornton *et al.*, 2013)] formulation is given below:

$$\gamma_n = \sqrt{\frac{4 m_{ij} k_n}{1 + \left(\dfrac{\pi}{\ln e}\right)^2}} \tag{4.11}$$

where m_{ij} is the reduced mass of the two particles, given by:

$$\frac{1}{m_{ij}} = \frac{1}{m_i} + \frac{1}{m_j}, m_{ij} = \frac{m_i m_j}{m_i + m_j} \tag{4.12}$$

See Worked Example 4.2, where we derive Equation (4.11).

The spring coefficient k_n can also be related to measurable parameters, though in practice it is commonly used as a fitting parameter, being set high enough to ensure the overlap between particles remains realistically small during collisions, but low enough that the computational expense of simulations does not become unfeasibly high.

This approach is therefore a compromise because the spring coefficient should be directly related to the Young's modulus of the material(s) of the particle(s), as described in Chapter 2. In practice, the use of a spring coefficient (Young's modulus) which is smaller than the physically measurable value in order to speed up the computation often results in a good simulation of particle motion (Chen *et al.*, 2017), but potentially a poor representation of the transmission of stresses through the system.

4.5.2 Hertzian Model

The simplicity of the linear spring–dashpot model means that it is easy both to implement and analyse, as well as being highly computationally efficient. However, the model is oversimplified, most notably because it incorporates a linear spring, whereas in reality the stiffness of a contact increases with the contact area, as shown in Chapter 2. This assumption can be corrected by using a contact model in which the spring coefficient k_n is no longer Hookean (i.e. linear) but Hertzian (with force increasing as deformation$^{3/2}$), being defined as:

$$k_n = \frac{4}{3} E_{ij} \sqrt{R_{ij} \delta_{ij}} \tag{4.13}$$

where $R_{ij} = \dfrac{R_i R_j}{R_i + R_j}$ is the effective radius of colliding particles and E_{ij} is the effective elastic modulus, which can be computed from the colliding particles' individual Young's moduli and Poisson ratios via the relation:

$$\frac{1}{E_{ij}} = \frac{1 - \nu_i^2}{E_i} + \frac{1 - \nu_j^2}{E} \tag{4.14}$$

Substituting Equation (4.13) into Equation (4.3), we obtain:

$$f_n = \frac{4}{3} E_{ij} \sqrt{a_{ij} \delta_{ij}} \delta_{ij} + \gamma_n \upsilon_n = \frac{4}{3} E_{ij} \sqrt{a_{ij}} \delta_{ij}^{\frac{3}{2}} + \gamma_n \upsilon_n \tag{4.15}$$

which is clearly no longer linear in δ_{ij}, meaning that the more physically accurate model comes at the cost of increased computational time.

For the Hertzian model, the dissipation coefficient is commonly chosen as:

$$\gamma \propto \sqrt{m_{ij} k_n}, \tag{4.16}$$

yielding a constant restitution coefficient, as in the linear spring–dashpot case.

4.5.3 Elastoplastic Models

Though the Hertzian model is physically more accurate than the point-like interaction assumed in the linear spring–dashpot model, its validity is still limited to the case of relatively small overlap between particles, as it cannot model plastic deformation – that is, it

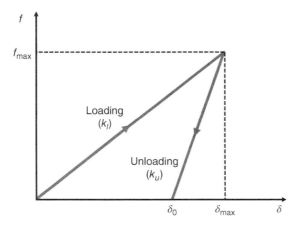

Figure 4.6 Schematic representation showing force versus overlap distance (loading–unloading) curve for the Walton–Braun model

is only technically valid in the Hertzian elastic regime. Modelling of larger, plastic deformations, requires a different model. A simple one [due to Walton and Braun (1986)] is shown conceptually in Figure 4.6, in which there is no dashpot but the dissipation is modelled by giving the 'loading' and 'unloading' phases of a particle collision two separate stiffness constants, k_l and k_u, with $k_u > k_l$, which results in a restitution coefficient which can be determined as:

$$e = \sqrt{\frac{k_l}{k_u}} \tag{4.17}$$

4.5.4   Cohesive Models

Real particles are often cohesive. As discussed in Chapter 2, naturally occurring van der Waals forces are always present and strongly influence the behaviour of smaller particles. A widely used contact model for van der Waals forces is the 'JKR' model described in Chapter 2, which results from an energy balance between the interfacial or surface energy and the elastic strain energy stored in the deformed particles.

   The complexity of the full JKR model means that it is not easy to incorporate directly into a DEM model, nor as efficient to run as more concise models, meaning that in practice a simpler version is often used. In one such simple method, commonly referred to as the simplifed JKR (sJKR) model, the interaction force due to cohesion $f_{n,jkr}$ is given by:

$$f_{n,jkr} = \Omega A \tag{4.18}$$

where Ω is called the *cohesive energy density* and A is the contact area between interacting objects:

$$A = 4\pi c^2 \tag{4.19}$$

where c is the contact radius. By simple geometry, assuming that the overlap between particles is small:

$$c = \sqrt{a_{ij}\delta_{ij}} \tag{4.20}$$

so that:

$$A = 4\,\pi\delta_{ij}a_{ij}. \tag{4.21}$$

In this form, the cohesion force can be simply added to Equation (4.3) as an additional term, i.e.

$$f_n = k_n\delta_{ij} + \gamma_n v_n + \Omega A \tag{4.22}$$

In this approach the measure of cohesion which is entered into the calculation is the cohesive energy density[9] with units J/m^3, which when multiplied by the contact area becomes J/m or N. The cohesive energy density is again commonly used as a fitting parameter because it is difficult to relate to physical measurements. Other models for cohesion are available, including several for forces due to liquid bridges, as discussed in Chapter 2, but implementation of these into DEM is beyond the scope of this book. Interested readers are referred to Guo and Curtis (2015).

4.6   COUPLING WITH COMPUTATIONAL FLUID DYNAMICS (CFD–DEM)

Computational fluid dynamics (CFD) is a widely used method for computing flow fields in fluid systems. It may be interfaced with DEM in order to model the behaviour of fluid–particle systems where both phases are important and they influence each other. CFD solves the Navier–Stokes equations for fluid flow in a user-defined geometry so as to provide a series of three-dimensional fields (velocity, temperature, pressure etc.) which together fully describe the flow within the system of interest. The flow fields are typically discretized across a three-dimensional 'grid' or 'mesh', with each 'cell' within the mesh housing a set of values representing the local flow velocity, pressure, etc. In order to simulate a two-phase fluid/solid system, this information must be 'overlaid' onto a DEM simulation, i.e. the DEM and CFD solvers must be *coupled*. The term 'coupling' in the current context can be thought of as passing the relevant forces from one simulation to another: in the simplest case, knowledge of the local fluid velocity, pressure gradient and viscous force at each point in the system is passed from the CFD part of the simulation to the DEM part, where it is used to determine the relevant drag forces acting on particles. This is known as 'one-way' coupling (i.e. the fluid influences the particles, but the particles do not influence the fluid), and is a reasonable assumption for systems of

---

[9] A word of warning to those using simplified JKR models pre-implemented in existing commercial or open-source codes: the approach described here is one example of multiple similar but differing models, which may calculate A in a variety of non-equivalent ways, and may interchangeably use the term 'cohesion energy density' to mean a parameter with units J/m^3 (cohesive energy density) or J/m^2 (interfacial energy of cohesion as considered in Chapter 2). It is strongly recommended that users read the documentation and/or source code to ensure they are aware of the precise implementation being used, and the physical meaning of the terms included.

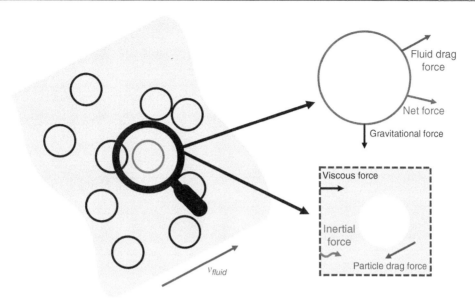

Figure 4.7 A simple schematic representation of a coupled CFD–DEM simulation

relatively low volume fraction of particles possessing relatively low Stokes numbers (see Chapter 3). For denser systems, such as bubbling fluidized beds or wet mills, where this assumption does not hold, the influence of the particles on the fluid motion must also be considered (two-way coupling). CFD-DEM models may also be *'resolved'* (the resolution of the mesh grid is smaller than the particle size, meaning that the fluid feels the effects of each particle individually) or *'unresolved'* (the mesh cells are larger than the particle size, meaning that only bulk properties such as porosity are fed to the CFD model). Figure 4.7 provides a simple visual representation of CFD-DEM coupling. Figure 4.8 outlines the coupling process for an unresolved simulation.

4.7 CALIBRATION

In principle, the input parameters for a DEM model can be real measured values, as listed in Table 4.1. In practice, as discussed earlier, some of the parameters, such as the coefficient of restitution, are difficult to measure at the particle level and will in any case have a distribution of values. A different approach is to calibrate the model using well-known macroscopic powder test results from, for example, flow or shear measurements, some of which are discussed in more detail in Chapter 9.

In calibration, a DEM model is used to produce a simulation of a simple test device – a rotating drum, for example, as shown in Figure 4.9. The model is then run for a number of different values of the parameter(s) of interest.

An important principle of any calibration campaign is the 'N-parameter N-test' rule, i.e. for each unknown parameter there must be (at least) one set of data to compare with from a characterization test. To take a simple example, assume that all the necessary parameters for a powder sample have been chosen except for its coefficient of sliding friction. In this case, there is one unknown parameter and one test series may therefore be used to

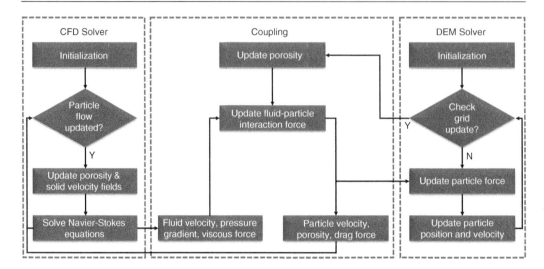

Figure 4.8 Diagram illustrating a common method for the coupling of CFD and DEM in order to simulate a multiphase (particle–fluid) system. Adapted from Che *et al.* (2023)

determine it. A number of DEM simulations can be carried out using different values of the unknown parameter and the best fit between the output of the model and experimental 'reality' will then indicate the best value of the unknown parameter. In this instance, a dynamic angle of repose test (see Chapter 2) such as the rotating drum shown in Figure 4.9 might be chosen to find the 'true' or 'experimental' angle of repose. A DEM simulation or 'digital model' of the test can then be constructed and run for different values of the friction coefficient until the simulated angle of repose matches the experimental angle, as illustrated in Figure 4.10. The same principle could equally be applied to

Figure 4.9 A 'digital-twin' of a rotating drum powder tester Image reused with permission from Herald et al. (2022). See also Lumay et al., 2012.

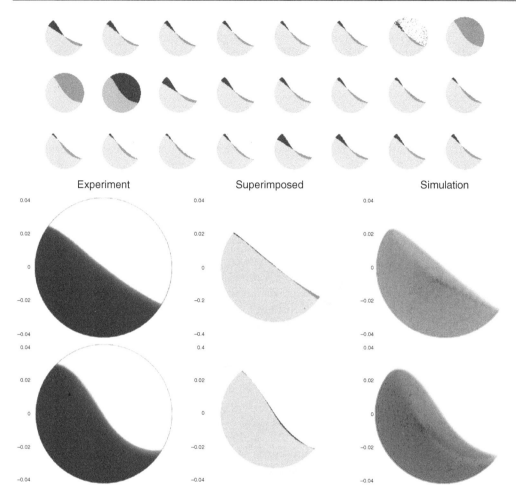

Figure 4.10 Example of the use of a powder characterization device to calibrate a DEM simulation, in this case by tuning parameters until they reproduce the free surface profile of a dynamic angle of repose tester. The above images represent simulations with randomly chosen parameters. The bottom images show a fully calibrated simulation. With acknowledgements to A. Leonard Niçusan.

any powder characterization tester, e.g. comparing an experimental and simulated yield stress or internal angle of friction for a shear cell (Chapter 9), a Hausner ratio for a tapped density tester (Chapter 6), or a flow rate or Beverloo plot for a flowmeter (Chapter 9)

The process described above is a simple example of an 'inverse problem'. As in any inverse problem, however, the calibration task can be 'ill-posed'. For example, if there are two unknown parameters (say the sliding friction coefficient and the rolling friction coefficient) and only one test method (in this case, the angle of repose test) – it cannot be guaranteed that there is a unique solution to the problem. That is to say that there may be multiple different combinations of the two unknown parameters which may reproduce the same angle of friction. In this case, another test method is required. In general it is better for the conditions in the test method to correspond more closely to the application which is being investigated. For example, the rotating drum in Figure 4.9 is a relatively low-stress environment while a shear cell will generally impose higher stresses on the particles. The former may thus be more relevant to small-scale pharmaceutical processes whereas the latter may be more relevant to large-scale particle-dense processes. A detailed discussion of the difficulties and most effective approaches to the calibration of DEM simulations can be found in the review by Windows-Yule and Neveu (2022).

4.8   MODELLING LARGE SYSTEMS

DEM is a computationally intensive technique and simulations taking days or weeks to perform are normally only possible in a research environment. Current hardware is, at the time of the writing of this book, typically limited to the simulation of millions of particles on a typical desktop computer, or billions with access to suitable high-performance clusters or supercomputers. In a real industrial system, however, the number of particles may easily reach trillions (10^{12}) or quadrillions (10^{15}), as illustrated in Worked Example 4.4, at the end of the chapter.

In order to bridge the gap between the current capability of DEM and an industrial system size, a number of methods have been developed to reduce the computational load. There are two main approaches: (1) reducing the scale of the system by only modelling a section of it, or (2) reducing the number of particles to be simulated by representing *groups* of multiple, actual-size particles by larger 'meso-particles' which behave in the same way. Both of these techniques carry their own individual limitations and modelling assumptions, yet both have also been shown to produce quantitatively accurate results at scales which would be impossible using full-scale simulations.

Figures 4.11 and 4.12 show some options for modelling sections of a larger system. Simple solid boundaries are shown in Figure 4.11(a). In order to model larger systems which have a degree of symmetry, 'periodic boundaries' can be used, as in Figure 4.11(b), in which particles crossing (in this case) the right-hand boundary are reintroduced with the same properties and velocity and at the same vertical position on the left-hand boundary. This type of boundary is suitable for modelling, for example, the central part of a batch-type rotating kiln mixer operating at steady state, as shown in Figure 4.12. (This approach would not be suitable for modelling flow near the ends of the kiln, where the flow is not lengthwise symmetrical.) In an alternative type – the 'deletion'

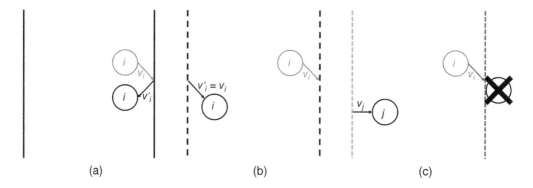

(a)   (b)   (c)

Figure 4.11   The most common types of boundaries used in DEM. Panels (a), (b), and (c) provide schematic representations of, respectively, solid boundaries, periodic boundaries, and deletion boundaries

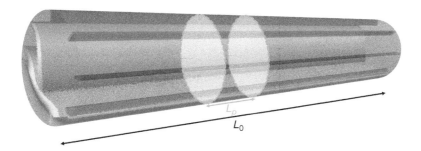

Figure 4.12 Modelling the central section of a rotating drum. With acknowledgements to Dominik Werner.

boundary – shown in Figure 4.11(c), particles passing through the boundary may be deleted. Deletion boundaries are often paired with insertion boundaries, at which new particles may be introduced to the system at specified positions and with specified geometries, so as to maintain a continuous flux of particles through the system. These are features which are useful in modelling parts of continuous flow processes.

As mentioned above, the number of particles to be simulated can be an issue, because, for a given type of problem, the computational time increases as N_p^2, or at best N_p. In most types of processes, particles do not move entirely independently of each other but in groups, 'clusters' or 'parcels'. This has led some engineers to adopt a compromise modelling technique which is often called *coarse-graining (CG)*, in which a group of particles is represented by a single larger one, as shown in Figure 4.13.

CG can be highly effective in reducing computational cost, as even increasing the particle diameter by a factor of two will reduce the number of particles in the system by a factor of $2^3 = 8$ (see Figure 4.13), and thus – dependent on the efficiency of a given DEM implementation – can potentially increase computation speed from anywhere between a factor of $\sim$8 and a factor of $\sim$64. Greater reductions may be possible in some cases. CG is, however, an approach to be used with extreme caution because, as shown throughout this book, so many properties of particles depend strongly on particle size and size distribution and it is not clear to what extent CG can reproduce this wide range of behaviour. A minimum requirement, of course, is that the coarse grain size must remain at least an order of magnitude below the scale of the process being modelled. A more complete discussion of coarse graining may be found in the review of Di Renzo *et al.* (2021).

4.9 MODELLING ASPHERICAL PARTICLES

Alongside the modelling of industrial-scale systems, another major challenge in the DEM community has, historically, been the modelling of aspherical particles – though, as we will see, there now exist many methods through which this may be achieved. A significant majority of DEM simulations consider only spherical particles, for the simple reason that such particles are easier and more computationally efficient to model. However, most industrial powders and grains are at least mildly aspherical, meaning that the use of spherical particles in the simulation is often questionable. As such, the field has developed a number of ways to model the behaviour of non-spherical particles. The simplest and most computationally efficient method does not involve the direct

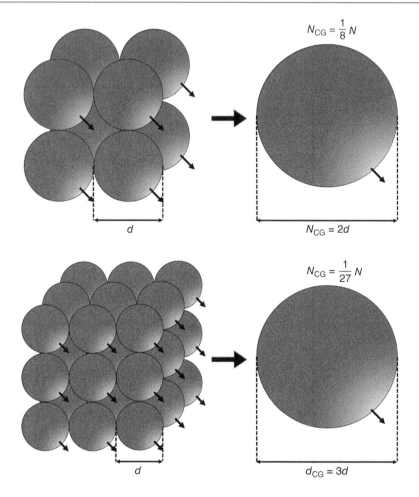

Figure 4.13 Schematic illustration of coarse-graining with a scaling factor $\lambda = 2$ (above) and $\lambda = 3$ (below)

modelling of particle geometry (i.e. particles are still simulated as spheres), but rather the use of an increased rolling friction parameter to provide a crude – though in many cases very effective – approximation of the influence of a particle's shape on its ability to rotate (Wensrich and Katterfeld, 2012).

The *multi-sphere method*, often referred to also as the glued-sphere method or the agglomerated sphere method, involves (as its various names imply) computationally 'gluing together' multiple spherical particles to form an agglomerate particle whose shape imitates that of the 'real' particles being modelled (Kodam *et al.*, 2009). As each constituent particle is still a sphere, this method avoids the computational expense involved with contact detection for truly aspherical particles; a notable disadvantage of this technique, however, is that it is very difficult to model sharp edges using only spherical particles.

Particles may also be modelled as *superquadrics* (Podlozhnyuk *et al.*, 2017) or as *polyhedra* (Govender *et al.*, 2015). A major advantage of the superquadric method is that the particle geometries can be described by a comparatively simple mathematical formula, thus aiding the detection and modelling of particle contacts. However, the model is limited to a finite range of shapes, and cannot create entirely asymmetrical geometries; sharp edges can also be computationally expensive to model. The polyhedral method, meanwhile, allows the modelling of truly arbitrary shapes, but contact detection and modelling are comparatively complex and (thus) computationally expensive.

4.10 DATA ANALYSIS AND VISUALIZATION

DEM produces a wide variety of output information on each particle in the simulation, including its position, rotation, instantaneous velocity and angular velocity, and forces acting, at each time step of the simulation. Since even a moderately sized simulation can easily contain of the order 10^6 particles over of the order 10^6 time steps, the total amount of information may become difficult to handle, and certainly for the human brain to interpret meaningfully. If the user is to be able to extract an understanding of the process being studied then further analysis must take place; this process can also be helped by suitable visualization of the data analysed. Figure 4.14 shows the results of some of the possible analyses for the DEM simulation of the rotating drum shown in Figure 4.9.

The first task in the analysis is to divide the volume up into volume elements or '*voxels*', allowing the particle data shown in panels (a) and (b) of Figure 4.14 to be averaged into a quasi-continuous distribution as shown in panels (c)–(f). Information can then be extracted for each particle which has a centre within the voxel of interest at that moment in time. When choosing the size of the voxels into which the data are 'binned', it is important to choose a suitable size – large enough that one may obtain adequate statistics, but small enough that the data are not unduly 'coarsened', such that local behaviours can be reliably observed.

Figure 4.14(c) shows the time-averaged velocity field for the rotating drum, averaged in the depth direction, giving a very clear indication of the way that particles move up collectively in 'solid body rotation' in the bottom half of the drum before moving down more independently and more rapidly in a flowing region in the upper part of the drum. Time-averaged patterns of this kind can be very useful in understanding a process but can also sometimes be misleading because they can mask the differences in instantaneous motion. Taking one volume element of the flow, for example, it is possible to obtain the distribution of velocity over time.

Figure 4.14(d) shows the time-averaged density field for the rotating drum, again averaged in the depth direction, showing a lower density (fewer particles per voxel) in the fast-flowing upper region. Again, it is possible to obtain the distribution of densities over time within each voxel.

Figure 4.14(e) and (f) provide two-dimensional fields showing the time-averaged *relative concentration* (the ratio of the local solids fraction of a given species divided by the total solids fraction) of smaller and larger particles in the binary system explored. Such plots can provide insight into any segregation occurring within the system – both qualitatively in their own right, or quantitatively through further analysis, for example by using the concentration values across all cells to determine a mixing index.

Figure 4.14(c)–(f) are averaged over the depth of the drum in order to provide a simple two-dimensional plot of results. The data from DEM are usually fully three-dimensional but such information is difficult to represent graphically. An alternative analysis technique which is often used is to extract information from a slice through the field of interest or that relating to a surface cut through the field.

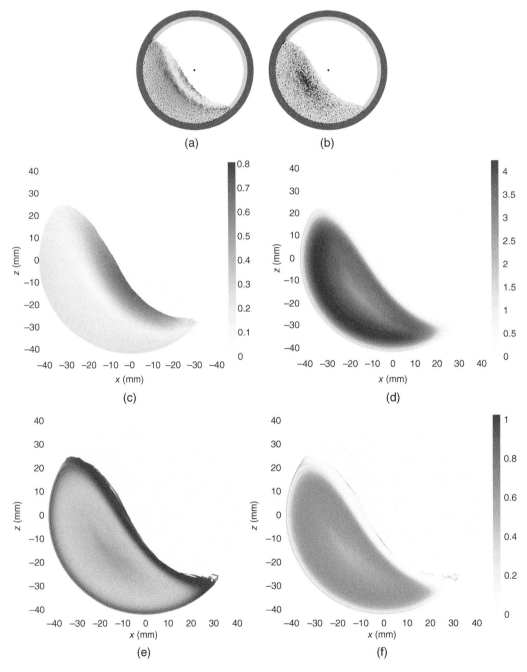

Figure 4.14 Examples of DEM analysis outputs for a rotating drum containing a binary mixture of small and large particles: (a) particle data, coloured by velocity; (b) particle data, coloured by particle size; (c) two-dimensional velocity distribution; (d) two dimensional solids fraction distribution; and (e, f) the relative concentration of large (left) and small (right) particles. With acknowledgements to A. Leonard Niçusan.

The examples above represent only a small sample of the information which may be extracted from DEM simulations, but cover some of the most general and widely used methodologies for presenting DEM data. Many more types of information can be extracted, such as the coordination number (the number of contacts each particle has with its neighbours), stress distributions, contact networks and force chains, collision rates, segregation rates and segregation patterns for binary, ternary and polydisperse systems.

4.11   VALIDATION

As discussed earlier in this chapter, DEM gives an approximation to reality, based on a large number of individual approximations and compromises. It is advisable to compare the results obtained from it with the outputs from real experiments wherever possible – a process typically referred to as 'validation'.[10] While, hypothetically, a perfectly calibrated model should not require validation, in reality one can rarely be 100% certain that the systems and conditions used for calibration will necessarily provide a perfect model for the full range of systems and conditions explored in the final simulations. As such, for rigour, even a well-calibrated simulation should still undergo validation.

There are many ways in which experimental data may be used to validate numerical simulations, including through diverse sensor measurements, or by applying suitable experimental imaging techniques to the system of interest. A general rule of thumb, however, is that richer, 'higher-dimensional' data, in general, facilitate more rigorous validation. For example, when simulating a system such as a high-shear mixer-granulator (see Chapter 12), it is not uncommon for authors to compare simulated and experimentally measured impeller torque or power-draw values as a means of validation. However, as such a measurement only provides a single-point comparison it is not, in reality, a suitable means of validating such a simulation: it is possible (indeed likely) that many distinct combinations of particle friction, restitution and cohesion may produce the same scalar value of power draw, meaning that it is impossible to tell if the combination you have chosen represents the 'real' parameter set.

When simulating systems such as fluidized beds, it is common practice to measure pressure drops at multiple points along the system's height, thus essentially providing a one-dimensional 'pressure drop profile'. Such a multi-point comparison clearly provides a more rigorous validation data set, but as it provides only one-dimensional information regarding one parameter, there is still a small but finite possibility that more than one combination of DEM parameters may yield a matching pressure profile. A still greater degree of rigour may be achieved by (a) comparing two- or three-dimensional fields as opposed to one-dimensional profiles and/or (b) comparing multiple different quantities. Figure 4.13 provides an example of a validation procedure which does both. In this case, two-dimensional occupancy fields *and* velocity fields produced from a CFD–DEM simulation of a spouted-bed coffee roaster are compared to equivalent experimental

---

[10] To clarify the terminology used, in DEM *calibration* is the initial process used to determine the correct model parameters to use in a simulation, and is often carried out using test systems (e.g. powder characterization devices) as opposed to the actual system to be modelled. *Validation* is the process of comparing the outputs of the modelled system itself to suitable experimental data from the 'real' system. For rigour, even if the same system is used for calibration and validation, the validation data set should not be included in the calibration data set in the same manner that training data should not be used for the validation of machine learning models, for example.

Occupancy:

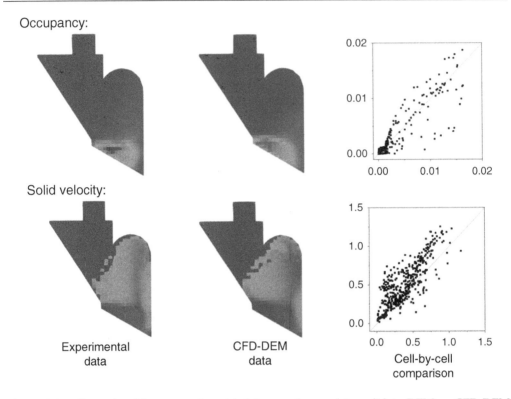

Solid velocity:

| Experimental | CFD-DEM | Cell-by-cell |
| data | data | comparison |

Figure 4.15 Example of how experimental data may be used to validate DEM or CFD-DEM models. Che *et al.* (2023)/Reproduced with permission of Elsevier

data acquired using positron emission particle tracking (Windows-Yule *et al.*, 2020). Both experimental and numerical data are discretized onto identical voxel grids (as discussed in Section 4.10), and the values for each non-empty cell compared to one another, providing 100s of comparison data points. The quality of the simulation can then be directly and quantitatively assessed through a suitable statistical measure such as Pearson's r or the reduced χ^2. The validation method described here is discussed in greater detail by Che et al., 2023 (Figure 4.15).

4.12   WORKED EXAMPLES

WORKED EXAMPLE 4.1

(a) A head-on collision between two spherical, steel particles is being simulated in DEM using the Hertz model (see Section 4.5.2). Both colliding particles have a diameter of 10 mm and a restitution coefficient of 0.9, and are travelling towards one another with a relative velocity of 1 m/s. If the simulated particles are overlapping by 1% of their radius, what is the repulsive force acting on them?

(b) What if the particles were instead overlapping by 20% of their radius? What does this tell you about DEM's modelling assumptions, and how we should choose our time steps?

Solution

(a) As it is specified that the collision is head-on, we need to consider only the normal force acting between the particles. This can be determined from Equation (4.3):

$$f_n = k_n \delta_{ij} + \gamma_n v_n$$

For the Hertzian model, the spring constant can be defined according to Equation (4.13):

$$k_n = \frac{4}{3} E_{ij} \sqrt{R_{ij} \delta_{ij}}$$

The effective radius can be calculated as:

$$R_{ij} = \frac{R_1 R_2}{R + R_2} = \frac{5 \times 5}{5 + 5} = 2.5 \, \text{mm} = 0.0025 \, \text{m}$$

As our particles are steel, a reasonable estimate for the Young's modulus and Poisson ratio could be, respectively, 200 GPa and 0.3. The effective elastic modulus can then be calculated from Equation (4.14):

$$\frac{1}{E_{ij}} = \frac{1 - \nu_i^2}{E_i} + \frac{1 - \nu_j^2}{E_j} = 2\frac{1 - 0.3^2}{2 \times 10^{11}} \rightarrow E_{ij} \tilde{\ } 110 \, \text{GPa}$$

Substituting into Equation (4.13) we obtain:

$$k_n = \frac{4}{3} \times 1.1 \times 10^{11} \times \sqrt{0.0025 \times 0.005 \times 0.01} = 5.2 \times 10^7 \text{N/m}$$

As per Equation (4.11), the damping coefficient can be defined as

$$\gamma_n = \sqrt{\frac{4 m_{ij} k_n}{1 + \left(\frac{\pi}{\ln \varepsilon}\right)^2}}$$

As our particles are spherical, the volume of each particle can be calculated as:

$$V = \frac{4}{3} \pi R^3 = 5.24 \times 10^{-7} \text{m}^3$$

Taking the density of steel as 7850 kg/m^3, the mass of each particle can be taken as:

$$m = \rho V = 0.0041 \, \text{kg}$$

The reduced mass m_{ij} can then be calculated from Equation (4.12) as:

$$\frac{1}{m_{ij}} = \frac{1}{m_i} + \frac{1}{m_j} = \frac{2}{m} \rightarrow m_{ij} = \frac{m}{2} = 0.0021$$

The damping coefficient can thus be determined as:

$$\gamma_n = \sqrt{\frac{4 \times 0.0021 \times 5.2 \times 10^7}{1 + \left(\frac{\pi}{\ln 0.9}\right)^2}} = 21.9$$

Finally, putting all values into Equation (4.3):

$$f_n = 2{,}568 \, \text{N}$$

(b) Repeating the above steps with $\delta_{ij} = 0.2R$ gives:

$$f_n = 231{,}622 \, \text{N}$$

The Hertz model is only valid for small overlaps between particles, and as this assumption is broken we see extremely high forces acting between our particles. Such an overlap would likely accelerate our particle at such a rate that it breaks our simulation!

The best (indeed only) way to avoid such large overlaps is to ensure a suitably small time step such that particles do not move far enough in a single time step to cause such unrealistic overlaps, but rather are pushed away from one another whilst only overlapping slightly.

WORKED EXAMPLE 4.2

Derive the expression for the damping coefficient shown in Equation (4.11) of the main text. *Hint: As we are assuming our interacting particles to represent a spring and a dashpot, we can consider the system as a damped harmonic oscillator (see, as an example, this useful open-source resource).*

Solution

Recalling Equation (4.3):

$$f_n = k_n \delta_{ij} + \gamma_n v_n$$

Considering our system as a simple harmonic oscillator, this equation may be simply re-written as:

$$m_{ij} a + \gamma_n v_n + k_n x = 0$$

Or, as a differential equation:

$$m_{ij} \frac{\mathrm{d}^2 x}{\mathrm{d}t^2} + \gamma_n \frac{\mathrm{d}x}{\mathrm{d}t} + k_n x = 0$$

For a simple, elastic harmonic oscillator, the natural frequency can be defined as:

$$\omega_0 = \sqrt{\frac{k_n}{m_{ij}}}$$

By defining also an effective viscosity η, as:

$$\eta = \frac{\gamma_n}{2 m_{ij}}$$

The above equation can be re-written as:

$$\frac{\mathrm{d}^2 x}{\mathrm{d}t^2} + 2\eta \frac{\mathrm{d}x}{\mathrm{d}t} + \omega_0^2 x$$

The solution to this equation is:

$$x(t) = \frac{v_0}{\omega} e^{-\eta t} \sin(\omega t)$$

Differentiating the above gives:

$$v(t) = \frac{dx}{dt} = \frac{v_0}{\omega} e^{-\eta t} [\omega \cos(\omega t) - \eta \sin(\omega t)]$$

So long as $\eta < \omega_0$, the duration t_c, of a collision can be taken as a half-oscillation, i.e.:

$$t_c = \frac{\pi}{\omega}$$

Substituting t_c into the above equation, we obtain:

$$
\begin{aligned}
v(t_c) &= \frac{v_0}{\omega} e^{-\eta t_c} \left[\omega \cos\left(\omega \cdot \frac{\pi}{\omega}\right) - \eta \sin\left(\omega \cdot \frac{\pi}{\omega}\right) \right] \\
&= \frac{v_0}{\omega} e^{-\eta t_c} \left[\omega \cos\left(\omega \cdot \frac{\pi}{\omega}\right) - \eta \sin\left(\omega \cdot \frac{\pi}{\omega}\right) \right] \\
&= \frac{v_0}{\omega} e^{-\eta t_c} [\omega \cos(\pi) - \eta \sin(\pi)] \\
&= -v_0 e^{-\eta t_c}
\end{aligned}
$$

As discussed in the main text, the restitution coefficient can be defined as the ratio of velocities before and after collision, i.e.

$$\varepsilon = -\frac{v(t_c)}{v_0}$$

$$\rightarrow \varepsilon = e^{-\eta t_c} = e^{-\frac{\pi \eta}{\omega}}$$

Rearranging the above, we obtain:

$$\ln \varepsilon = -\frac{\pi \eta}{\omega} = \frac{\pi}{\omega} \frac{\gamma_n}{2m_{ij}}$$

Since we are still considering our system as a damped harmonic oscillator, we can also substitute in:

$$\omega = \sqrt{\frac{k}{m_{ij}} - \left(\frac{\gamma_n}{2m_{ij}}\right)^2}$$

To give:

$$\ln \varepsilon = \frac{\pi}{\sqrt{\dfrac{k}{m_{ij}} - \left(\dfrac{\gamma_n}{2m_{ij}}\right)^2}} \frac{\gamma_n}{2m_{ij}}$$

which can be rearranged to yield:

$$\gamma_n = \sqrt{\frac{4m_{ij}k_n}{1 + \left(\dfrac{\pi}{\ln \varepsilon}\right)^2}}$$

as required.

WORKED EXAMPLE 4.3

In the Walton–Braun model, plastic deformation of the contacting particles is represented by the residual overlap δ_0, of particles after unloading. Derive an expression for this residual overlap in terms of the loading and unloading stiffness constants k_l and k_u.

Answer

Assuming both a linear loading and linear unloading, both the loading and unloading curves can be represented as straight lines, as shown below.

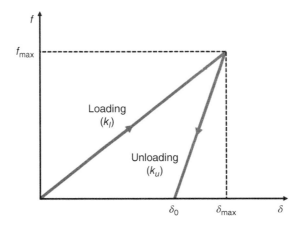

Under the assumption of a linear relationship, expressions for the loading and unloading force may be written as:

$$f_1 = k_1\delta + c_1$$

and

$$f_u = k_u\delta + c_2$$

From the above plot, we know that:

$$f_1(0) = 0 \rightarrow c_1 = 0 \rightarrow f_l = k_l\delta$$

$$f_u(\delta_0) = 0 \rightarrow k_u\delta_0 + c_2 = 0 \rightarrow c_2 = -k_u\delta_0 \rightarrow f_u = k_u(\delta - \delta_0)$$

$$f_1(\delta_{max}) = f_1(\delta_{max}) = f_{max}$$

$$\rightarrow k_l\delta_{max} = k_u(\delta_{max} - \delta_0) = k_u\delta_{max} - k_u\delta_0$$

$$\delta_0 = \delta_{max}\left(1 - \frac{k_l}{k_u}\right)$$

WORKED EXAMPLE 4.4

Consider the following systems:

(a) a salt shaker;

(b) a laboratory-scale pharmaceutical blender;

(c) a commercial-scale coffee roaster;

(d) a 500-m long, 10-cm wide, dense-phase pneumatic conveying system transporting fly ash

For each of the above, consider:

 (i) How feasible would it be to simulate this system on a PC?

 (ii) How feasible would it be to simulate this system on a high-performance computer?

(iii) If the system is too difficult to simulate directly, what would be the best way to reduce the computational expense?

For each case, make estimates regarding, for example, the size of the particles involved, the size of the system involved, the number of particles involved etc.

Example Answer:

(a) Salt Shaker

In order to assess the feasibility of simulation, the most important factor is the number of particles in the system.

Let us assume a grain of salt can be represented as a cube of side ~0.3 mm, giving a volume of ~0.03 mm^3. Similarly, let us assume a typical salt shaker to be a cylinder of height 70 mm and diameter 30 mm, yielding an approximate volume of 50 000 mm^3. Assuming a packing fraction of 0.6, the typical number of salt particles in a (full) salt shaker can thus be estimated as ~1 000 000.

As such, for this question, a salt shaker can feasibly be simulated on a desktop or a high-performance computer (HPC) with no need to implement any methods for speed up.

(b) Laboratory-scale blender

The parameters of this problem are deliberately wide to allow for a range of answers, so we will here provide two possible, perfectly reasonable answers which yield very different results.

Possibility 1

Let us take the size of a typical pharmaceutical powder as 10 μm = 1×10^{-5} m. Assuming the particles to be spherical, this would equate to a volume per particle of approximately 5×10^{-16} m^3. (Note: in reality, we would of course normally be mixing two or more distinct powder components in a pharmaceutical blender, but for the sake of this thought experiment let us make the improbable assumption that all components have both the same size and a perfectly uniform size distribution)

Let us assume that the capacity of the laboratory-scale blender is 10 litre, and that the blender is operated at 80% capacity, giving a volume of 8 litre = $0.008 \, \text{m}^3$.

Assuming a packing density of approximately 0.6, we can expect approximately 9 trillion(!) particles. Such a number would be well beyond the capabilities of any (current) PC, and the vast majority of HPCs, though it may be achievable with a top-end supercomputer.[11]

Possibility 2

In this instance let us consider a pharmaceutical powder of characteristic size 100 μm, corresponding to a volume of approximately $5 \times 10^{-13} \, \text{m}^3$, and a laboratory-scale blender with a maximum capacity of 1 litre, operated with a 20% fill ($V_{\text{blender}} = 0.0002 \, \text{m}^3$). Assuming the same packing density as before, we can expect in this case approximately 400 million particles (or ~230 million if we assume that 20% fill includes ~40% void fraction). While this is still likely to be a little too much for the vast majority of current home computers, such a number could be reasonably comfortably achieved (given a decent algorithm and enough time) on a standard HPC.

Suitable speed-up methods

The manner(s) in which one may reduce the computational expense of the above-described simulations will depend strongly on the system being modelled.

In the case of a cylindrical, horizontal, rotating-drum-type mixer, the computational expense may be reduced considerably by the use of periodic boundaries, as illustrated in the main text. However, the majority of real industrial mixers (v-blenders, conical blenders, bladed mixers, inclined continuous blenders, planetary mixers etc.) do not possess the required degree of axial uniformity for such an approach to be successful.

For systems which may be reasonably assumed to be axisymmetric (vertical stirred mixers, resonant acoustic mixers etc.) one may reduce computational expense by simulating only a segment of the system. However, the extent to which the simulated volume may be reduced in this manner is somewhat limited. As such, while it may be sufficient to reduce the 400 million particle simulation to a size which is just about manageable on a laptop, it is unlikely to render the 9 trillion particle simulation feasible.

In order to render the 9 trillion particle simulation viable, the most reliable course to take is almost certainly coarse-graining. A coarse-graining factor of 20 would be sufficient to reduce the number of simulated particles to the order of 1 billion, and thus make simulation on an HPC feasible. A coarse-graining factor of 200 (high, but used successfully in some prior works) would further reduce this to approximately 1 million, putting it into the realm of PC computation.

---

[11] For the students reading this worked example 20 years after publication, we very much hope that computing capabilities have evolved to the point that you are laughing derisively at this statement, as our current students regularly do regarding DEM papers written in the early twenty-first century.

(c) Commercial-scale coffee roaster

For an industrial coffee roaster, a batch size of 100 kg is a reasonable estimate. If a single coffee bean weighs 100 mg = 0.0001 kg, the simulation would require approximately 1 000 000 particles, which could be comfortably simulated on an HPC or even a decent laptop, without any need for coarse-graining or other speed-up measures.

In this case, one may also consider a few alternative factors which may affect simulation time, for example the non-spherical nature of the coffee beans (do we need to use more computationally expensive shape models such as superquadrics or multispheres?) or, as many coffee roasters use heated air, the need to include CFD. However, prior work on such systems has shown that modelling particles as simple spheres with increased rolling friction is sufficient to create an accurate model (Che *et al.*, 2023) and, if well implemented, CFD coupling does not necessitate a significant increase in computation time.

One may notice that the computational effort required to simulate this system is markedly lower than the previous example in (b), despite the former example being a small, laboratory-scale system and the latter being a full, commercial-scale system. The choice of these two examples is intended to emphasize the point that the feasibility of a DEM simulation is not simply determined by how big (in geometric terms) a system is.

(d) Pneumatic conveyor

To give our simulations the best chance, let us consider the upper end of the typical size distribution for fly ash (~100 µm) and the lower end of the range of solids loadings for dense-phase transport (30%).

The total available volume of the pipe used is approximately $4 \, \text{m}^3$, meaning the total volume of particles at 30% loading is approximately $1.2 \, \text{m}^3$. The volume of a single particle is approximately $5 \times 10^{-13} \, \text{m}^3$, meaning we must simulate approximately 2 trillion particles.

As we can see from the above, for this system, even the most conservative estimate results in a number of particles beyond the current capabilities of readily available computational facilities.

While the number of particles is comparable to (indeed lower than) the system considered in (b), in the present case coarse-graining cannot be so readily used due to the highly unequal dimensions of the system ($D \ll L$). More specifically, the narrow nature of the system means that we are limited in the extent to which particles can be coarse-grained before wall effects begin to influence the system's dynamics in an unrealistic manner. For example, if we implement the CG factor of 200 cited in (b), the modelled meso-particles would be 2 cm in diameter, or one-fifth of the system diameter, which is certainly likely to introduce some unrealistic dynamics!

Considering the axisymmetry of the cylindrical system, periodic boundaries may potentially be employed to extract some relevant information from the system, but the use of such a simplification to model the full system would be reliant on a number of questionable assumptions regarding the system, for example that it operates at steady state, and is free of any bends.

The take-home point from this problem is that DEM, though powerful, is not necessarily the best solution for *all* industrial problems. Alternative methods for modelling pneumatic transport systems are described in detail in Chapter 7.

EXERCISES

4.1 Consider the following systems:

   (a) A pharmaceutical tablet. What assumptions are you making in your calculation?

   (b) A laboratory-scale fluidized bed using Geldart group D particles.

   (c) The same bed as (b) but using Geldart group C particles.

   (d) A pilot-scale stirred-media mill ($D = 0.5$ m) processing 500 µm alumina particles using 10-mm grinding media. What additional assumptions or considerations might you make for this system?

For each of the above, consider:

   (i) How feasible would it be to simulate this system on a PC?

  (ii) How feasible would it be to simulate this system on a high-performance computer?

 (iii) If the system is too difficult to simulate directly, what would be the best way to reduce the computational expense?

Interactive Exercises

A series of interactive exercises providing deeper insight into the discrete element method can be found at www.wiley.com/go/rhodes/particle3e. These exercises will allow you to explore and thus better understand key elements of DEM modelling, such as the importance of setting the correct time step, the strengths and weaknesses of different integration methods, and the trade-off between simulation accuracy and simulation speed.

5

Colloids, Aerosols and Fine Particles

The behaviour of small particles is dominated by surface forces and both gravity and particle inertia become much less important. For these reasons, the science of small particles has become almost a distinct subject. This chapter describes the forces acting on particles due to the surrounding molecules and the consequences of these, including random motion due to molecular impacts and the various forms of particle–particle interactions which can stabilize or destabilize *colloidal* suspensions of particles in liquids. Molecular level interactions are also important in the study of *aerosol* particles in gases and determine how easy it is to remove them from a gas. The flow or *rheology* of suspensions depends on the same molecular interactions as determine their stability. Some of the many variants of flow behaviour are described, including flow-induced changes such as shear thinning and shear thickening, which provide important opportunities for the design of particulate products.

5.1 INTRODUCTION

The properties of particles depend very much on their size and there are examples of this throughout the other chapters in this book. Particle size is particularly important in products such as paints, ceramics, foods, pharmaceuticals and other consumer goods, some of which are described in Chapter 13. As the particle size decreases, the ratio of surface area to particle volume or mass increases. For a sphere:

$$\frac{\text{Surface area}}{\text{Volume}} = \frac{\pi x^2}{\frac{\pi}{6} x^3} = \frac{6}{x} \qquad (5.1)$$

Adapted from an original chapter for the 2nd Edition by George V. Franks of the University of Melbourne, Australia.

Introduction to Particle Technology, Third Edition. Martin Rhodes and Jonathan Seville.
© 2024 John Wiley & Sons Ltd. Published 2024 by John Wiley & Sons Ltd.
Website: www.wiley.com/go/rhodes/particle3e

The behaviour of fine particles therefore becomes dominated by surface forces rather than body forces such as gravity. *Colloids* are very fine particles, with one or more linear dimensions between about 1 nm and 10 μm suspended in a fluid. If the fluid is a gas, the particle suspension is generally known as an *aerosol*. The dominance of surface forces results in the cohesive nature of fine particles, the high viscosity of concentrated suspensions and the slow sedimentation of dispersed colloidal or aerosol suspensions. These properties can be advantageous or disadvantageous, depending on the situation. In many products, finer particles are preferred.

In Chapter 2, we showed how cohesive forces between particles can scale with the particle diameter x, or in some cases be independent of x, while the particle mass scales with x^3. The consequence is that for particles smaller than 10–100 μm (depending on conditions), surface forces exceed gravitational forces. In this chapter, we describe the chemical nature of those surface-dominated cohesive forces, which are the result of a number of physicochemical interactions such as van der Waals, electrical double layer (EDL), bridging and steric forces; more detail is available in the colloid and surface chemistry textbooks (Hiemenz and Rajagopolan, 1997; Hunter, 2001; Israelachvili, 2011; Shaw, 1992).

Surface forces may result in either attraction or repulsion between two particles depending on the material of which the particles are composed, the fluid type and the distance between the particles. Generally, if nothing is done to control the interaction between particles, they will be attracted to each other due to van der Waals forces which are always present. (The few rare cases where the van der Waals forces are repulsive are described in Section 5.3.1.) The dominance of attraction is the reason why fine powders in air are usually cohesive.

5.2 BROWNIAN MOTION

Molecules in a fluid are in constant motion and will collide with any particles which are present. For larger particles the effect is negligible, but smaller colloidal or aerosol particles are affected by these random collisions so that they themselves move randomly in what is called Brownian[1] motion, as shown in Figure 5.1.

A simple application of a kinetic model allows us to determine the influence of key parameters on the average velocity of particles in suspension. Consider that the thermal energy of the environment is transferred to the particles as kinetic energy. It is a consequence of the kinetic theory of gases that all suspended particles, regardless of their size, have the same translational kinetic energy (Shaw, 1992). The average thermal energy is $^3/_2 k_B T$ (where k_B is Boltzmann's constant = 1.381×10^{-23} J/K and T is the temperature in Kelvin). To a first approximation, therefore, the average velocity $\bar{v}$ of the particle can be estimated by equating the kinetic energy $\frac{1}{2} m v^2$ (where m is the mass of the particle) with the thermal energy as follows:

$$\bar{v} = \sqrt{\frac{3k_B T}{m}} \tag{5.2}$$

---

[1] After Robert Brown (1773–1858), Scottish botanist, who first observed the behaviour in the motion of pollen grains in water in 1827.

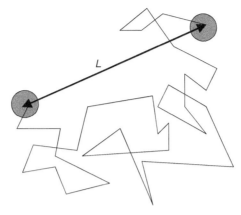

Figure 5.1 Illustration of the random walk of a Brownian particle. The distance the particle has moved over a period of time is L

This simple analysis cannot be used to determine the actual distance of the particle from its original position because it does not move in a straight line (see Figure 5.1 and below), but it does show that either increasing the temperature or decreasing the particle mass increases Brownian motion.

Thermodynamic principles dictate that the lowest free energy state (greatest entropy) of a suspension is a uniform distribution of particles throughout the volume of the fluid. Thus, the random walk of a particle due to Brownian motion provides a mechanism for the particles to arrange themselves uniformly throughout the volume of the fluid. The result is diffusion of particles from regions of high concentration to regions of lower concentration. Statistical analysis of the one-dimensional random walk enables us to determine the average (root mean square) distance that a Brownian particle moves as a function of time.

$$L = \sqrt{2\alpha t} \tag{5.3}$$

where α is the diffusion coefficient or *diffusivity*. This result is due to Einstein[2], who further developed a relationship for the diffusion coefficient that incorporates the hydrodynamic drag on a spherical particle (see Section 3.1.1 of Chapter 3):

$$\alpha f = k_B T \tag{5.4}$$

where the drag coefficient f is defined as F_D/U_{rel}, where U_{rel} is the relative velocity between the particle and fluid. Then for creeping laminar flow we find from Stokes' law [Equation (3.7) in Chapter 3] that $f = 3\pi x \mu$ so that:

$$\alpha = k_B T / 3\pi x \mu \tag{5.5}$$

---

[2] Albert Einstein (1879–1955), German-born physicist, better known for his work on relativity!

and

$$L = \sqrt{\frac{2k_B T t}{3\pi x \mu}} \tag{5.6}$$

Thus, the average distance that a particle will move over a period of time can be determined. Increasing temperature increases the distance travelled over a period of time while increasing particle size and fluid viscosity both reduce the distance travelled. Note that the distance scales with the square root of time.

Note that Equations (5.3)–(5.6) have been derived for the case of the one-dimensional random walk. This is because these equations will be used later in the analysis of sedimentation under gravity where motion in only one direction is of interest. In the case of a three-dimensional random walk, the analogous expression to Equation (5.3) would be: $L = \sqrt{6at}$.

5.3 SURFACE FORCES

Surface forces between particles ultimately arise from the interaction between the molecules which make up the material of the particles, modified by the nature of the fluid which surrounds them.

In general, the force between two particles (F) may be either attractive or repulsive. The force depends upon the surface–surface separation distance (D) between the particles and the potential energy (V) at that separation distance. The relationship between the force and potential energy is that the force is the negative of the gradient of the potential energy with respect to distance.

$$F = -\frac{dV}{dD} \tag{5.7}$$

Typical relationships for the variation of potential energy and force with separation distance for fine particles are shown schematically in Figure 5.2. Thermodynamics dictates that a pair of particles move to the separation distance that results in the lowest energy configuration. A force between the particles will result if the particles are at any other separation distance. There is always a strong repulsive force at zero separation distance that prevents the particles from occupying the same space. When there is attraction (at all other separation distances), the particles reside in a potential energy well (an energy minimum) at an equilibrium separation distance [Figure 5.2(a) and (b)]. In some cases, a repulsive potential energy barrier exists that prevents the particles from moving to the minimum energy separation because they do not have enough thermal or kinetic energy to surmount it. (In terms of force, there is not enough force applied to the particles to exceed the repulsive force field.) In this case the particles cannot touch each other and reside at a separation distance greater than the extent of the repulsive barrier, which is typically several nanometres [Figure 5.2(c) and (d)].

The relationships between force and distance as well as the underlying physical and chemical mechanisms responsible for those forces are described below for several of the forces with the most significance in practice.

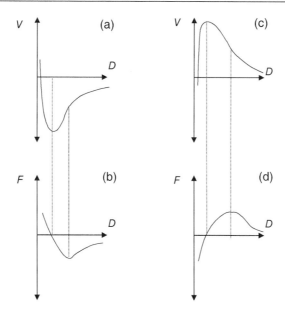

Figure 5.2 Schematic representations of interparticle potential energy (V) and force (F) versus particle surface–surface separation distance (D). (a) Energy versus separation distance curve for an attractive interaction. The particles will reside at the separation distance where the minimum in energy occurs. (b) Force versus separation distance for the attractive potential shown in (a). (The convention used in this book is that positive interparticle forces are repulsive.) The particles feel no force if they are at the equilibrium separation distance. An applied force greater than a maximum is required to pull the particles apart. (c) Energy versus separation distance curve for a repulsive interaction. When the potential energy barrier is greater than the available thermal and kinetic energy, the particles cannot come into contact and move away from each other to reduce their energy. (d) Force versus separation distance for the repulsive potential shown in (c). There is no force on the particles when they are very far apart. There is a maximum force that must be exceeded to push the particles into contact

5.3.1   van der Waals Forces

van der Waals forces is the term commonly used to refer to a group of electrodynamic inter-actions that occur between the atoms in two different particles, of which the dominant contribution is known as the *dispersion force*. The dispersion force is a result of electrical interactions between correlated fluctuating instantaneous dipole moments within the atoms that comprise the two particles. To understand this concept, imagine that each atom in a material contains a positively charged nucleus surrounded by orbiting negative electrons. The nucleus and the electrons are separated by a short distance, on the order of an Ångstrom (10^{-10} m). At any instant in time, a dipole moment exists between the nucleus and the centre of electron density. This dipole moment fluctuates very rapidly with time, revolving around the nucleus as the electrons move. The dipole moment of each atom creates an electric field that emanates from the atom and is felt by all other atoms in both particles. In order to lower the overall energy of the system, the dipole moments of all the atoms in both particles correlate their dipole moments (i.e. they align themselves like choreographed pairs of dancers who remain synchronized although they

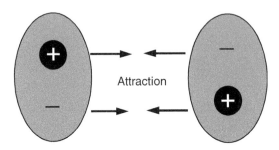

Figure 5.3 Schematic representation of the dipole–dipole attraction that exists between the instantaneous dipoles of two atoms in two particles. The + represents the nucleus of the atom and the − represents the centre of the electron density. Because the centre of electron density is typically not coincident with the nucleus, a dipole moment exists between the two separated opposite charges in each atom. The lowest free energy configuration is as shown in the figure. The resulting position of positive and negative charges leads to an attraction between the two atoms

are always moving). When the two particles are composed of the same material, the lowest energy configuration of the correlated dipoles is such that there is attraction between the dipoles as shown in Figure 5.3. The combined attraction between all the dipoles in the two particles results in an overall attraction between them. In general, the van der Waals interaction can be attractive or repulsive depending on the dielectric properties of the two particles and the medium between the particles.

Adding together the interactions between all atoms in both particles results in a surprisingly simple equation for the overall interaction between two spherical particles of the same size when the distance between the particles (D) is much less than the diameter of the particles (x).

$$V_{vdW} = -Ax/24D \tag{5.8a}$$

and

$$F_{vdW} = -Ax/24D^2 \tag{5.8b}$$

where V_{vdW} is the van der Waals interaction energy and F_{vdW} is the van der Waals force. More complicated relationships arise if the particles are not the same size or are small relative to the distance between them. The constant A is known as the Hamaker[3] constant and expresses the magnitude and direction of the interaction for a particular pair of particles in a given medium. When the Hamaker constant is greater than zero the interaction is attractive and when the Hamaker constant is less than zero the interaction is repulsive. Figure 5.4 shows the configuration of two particles (materials 1 and 3) interacting through an intervening medium (material 2). The Hamaker constant can be calculated from the dielectric properties of the three materials. Table 5.1 shows the Hamaker constants for several combinations of particles and intervening media. Note that oil droplets in emulsions and bubbles in foams may be considered in the same way as particles for

---

[3] After Dutch physicist Hugo Christiaan Hamaker (1905–1993).

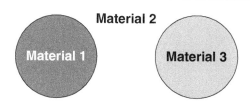

Figure 5.4 Notation used to indicate the type of material for each particle and the intervening medium

the purpose of understanding interaction forces and the stability of the emulsions and foams.

When materials 1 and 3 are the same, the van der Waals interaction is always attractive, as in the case of mineral oxides interacting across water or air. Note that the van der Waals interaction is reduced when the particles are in a denser medium (e.g. water compared with air). Thus, it is easier to separate (or disperse) fine particles in liquids than in air. When materials 1 and 3 are different, repulsion will result between the two particles if the dielectric properties of the intervening medium lie between those of the two particles, such as for silica particles and air bubbles interacting across water.

Table 5.1 Hamaker constants of some common material combinations

Material 1	Material 2	Material 3	Hamaker constant (approximate) (J)	Example
Alumina	Air	Alumina	15×10^{-20}	Oxide minerals in air are strongly attractive and cohesive
Silica	Air	Silica	6.5×10^{-20}	
Zirconia	Air	Zirconia	20×10^{-20}	
Titania	Air	Titania	15×10^{-20}	
Alumina	Water	Alumina	5.0×10^{-20}	Oxide minerals in water are attractive but less so than in air
Silica	Water	Silica	0.7×10^{-20}	
Zirconia	Water	Zirconia	8.0×10^{-20}	
Titania	Water	Titania	5.5×10^{-20}	
Metals	Water	Metals	40×10^{-20}	Conductivity of metals makes them strongly attractive
Air	Water	Air	3.7×10^{-20}	Foams
Octane	Water	Octane	0.4×10^{-20}	Oil in water emulsions
Water	Octane	Water	0.4×10^{-20}	Water in oil emulsions
Silica	Water	Air	-0.9×10^{-20}	Particle bubble attachment in mineral flotation, weak repulsion

5.3.2 Electrical Double-Layer Forces

When particles are immersed in a liquid they may develop a surface charge by one of a number of mechanisms. Consider the case of oxide particles immersed in aqueous solutions. The surface of a particle is comprised of atoms that have unsatisfied bonds. In a

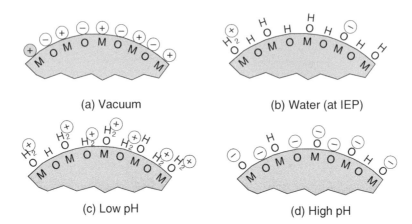

(a) Vacuum   (b) Water (at IEP)

(c) Low pH   (d) High pH

Figure 5.5 Schematic representation of the surface of metal oxides (a) in vacuum. Unsatisfied bonds lead to positive and negative sites associated with metal and oxygen atoms, respectively. (b) The surface sites react with water or water vapour in the environment to form surface hydroxyl groups (M-OH). At the isoelectric point (IEP) the neutral sites dominate, and the few positive and negative sites present exist in equal numbers. (c) At low pH, the surface hydroxyl groups react with H^+ in solution to create a positively charged surface composed mainly of the $\left(M-OH_2^+\right)$ species. (d) At high pH, the surface hydroxyl groups react with OH^- in solution to create a negatively charged surface composed mainly of (M-O$^-$) species

vacuum, these unfulfilled bonds result in an equal number of positively charged metal ions and negatively charged oxygen ions as shown in Figure 5.5(a). When exposed to ambient air (which usually has at least 15% relative humidity) or immersed in water, the surface reacts with water to produce surface hydroxyl groups (denoted M-OH) as shown in Figure 5.5(b).

The surface hydroxyl groups react with acid and base at low and high pH, respectively, via surface ionization reactions as follows:

$$M\text{-}OH + H^+ \overset{K_a}{\to} M\text{-}OH_2^+ \qquad\qquad (5.9a)$$

$$M\text{-}OH + OH^- \overset{K_b}{\to} M\text{-}O^- + H_2O \qquad\qquad (5.9b)$$

resulting in either a positively charged surface $M - OH_2^+$ as in Figure 5.5(c), or a negatively charged surface $(M - O^-)$ as in Figure 5.5(d). The values of the surface ionization reaction constants (K_a and K_b) depend upon the particular type of material (for example SiO_2, Al_2O_3 and TiO_2; the value of K expresses the equilibrium position for the ionization reaction). For each type of material, there is a pH known as the isoelectric point (IEP), where the majority of surface sites are neutral (M-OH) and the net charge on the surface is zero. At a pH below or above the IEP, the particle surfaces become positively $\left(M - OH_2^+\right)$ or negatively (M-O$^-$) charged due to the addition of either acid (H^+) or base (OH^-), respectively. Figure 5.6 shows how the concentration of surface sites changes with pH. Table 5.2 contains a listing of the IEPs of some common materials.

For each charged surface site, there is a counterion of opposite charge in solution, which maintains overall electrical neutrality. For example, the counterion for a positive

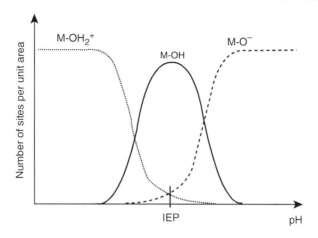

Figure 5.6 Number density per unit area of neutral (M-OH), positive $\left(M-OH_2^+\right)$ and negative (M-O⁻) surface sites as a function of pH

surface site is a Cl⁻ anion in the case when HCl is used to reduce the pH and the counterion for a negative surface site is a Na⁺ cation if NaOH is used to increase the pH. The entire system is electrically neutral. The separation of charge between the surface and the bulk solution results in a potential difference known as the surface potential (Ψ_0)

In order to maintain the electrical neutrality of the system, the counterions form a diffuse cloud that shrouds each particle. When two particles are forced together their counterion clouds begin to overlap, thus increasing the concentration of counterions in the gap between them, as shown in Figure 5.7.

If both particles have the same charge, this gives rise to a repulsive potential due to what is known as *electrical double layer (EDL)* repulsion. If the particles are of opposite charge an EDL attraction will result. It is important to realize that EDL interactions are not simply determined by the electrical interaction between the two charged spheres, but are due to the *osmotic pressure* (concentration-driven) effects of the counterions in the gap between the particles.

Table 5.2 Isoelectric points of some common materials

Material	pH at the isoelectric point (IEP)
Silica	2–3
Alumina	8.5–9.5
Titania	5–7
Zirconia	7–8
Hematite	7–9
Calcite	8
Oil	3–4
Air	3–4

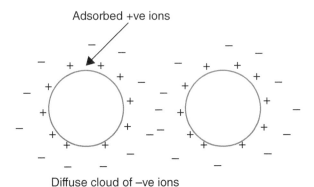

Adsorbed +ve ions

Diffuse cloud of –ve ions

Figure 5.7   Electrical double-layer repulsion

A measure of the thickness of the counterion cloud (and thus the range of the repulsion) is the Debye length[4] (κ^{-1}) where the Debye screening parameter (κ), for monovalent salts is (Israelachvili, 2011):

$$\kappa = 3.29\sqrt{[c]}\,(\mathrm{nm}^{-1}) \tag{5.10}$$

where [c] is the molar concentration of the monovalent electrolyte. When the Debye length is large (small counterion concentration), the particles are repulsive at large separation distances so that the van der Waals attraction is overwhelmed, as in Figure 5.2(c) and (d). The EDL is compressed (the Debye length is reduced) by adding a salt, which increases the concentration of the counterions around the particle. When sufficient salt is added, the range of the EDL repulsion is decreased sufficiently to allow the van der Waals attraction to dominate at large separation distances. At this point, an attractive potential energy well results, as shown in Figure 5.2(a) and (b). As described in Chapter 13, effects of this kind can have important consequences for the stability of products based on particle suspensions.

An approximate expression for the EDL potential energy (V_{EDL}) versus the surface–surface separation distance (D) between two spherical particles of diameter (x) with the same surface charge is (Israelachvili, 2011):

$$V_{EDL} = \pi\varepsilon\varepsilon_0 x\Psi_0^2 e^{-\kappa D} \tag{5.11}$$

where Ψ_0 is the surface potential (created by the surface charge), ε the relative permittivity of water (not void fraction as frequently used in other parts of the book), ε_0 the permittivity of free space ($8.854 \times 10^{-12}\,\mathrm{C}^2/\mathrm{J/m}$), and κ the inverse Debye length. This expression is valid when the surface potential is constant and below about 25 mV and the separation distance between the particles is small relative to their size (Israelachvili, 2011).

As we have seen, a layer of immobile ions and water molecules exists at the surface of particles so that it is not easy to measure their surface potential directly. Instead, a closely related quantity known as the *zeta potential* is usually measured. The zeta potential is the potential at the plane of shear between the immobilized surface layer and the bulk solution, as shown in Figure 5.8, and can be determined by measuring the particle velocity under the influence of an electric field. Commercial instruments are available in order

---
[4] After Peter Debye (1884–1966), Dutch-American physical chemist.

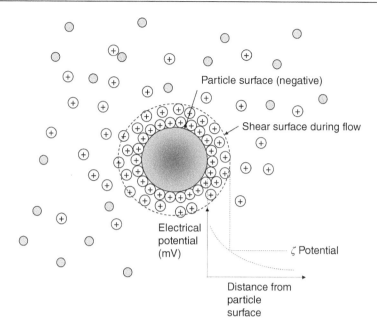

Figure 5.8 Ionic concentration and electrical potential as a function of distance from a particle surface, showing zeta potential (−ve ions shown as grey)

to make this measurement and relate it to zeta potential. The plane of shear is typically located only a few Angstroms from the surface so that there is little difference between the zeta potential and the surface potential. In practice, the zeta potential can be used in place of the surface potential in Equation (5.11) to predict the interparticle forces as a function of separation distance with little error. Addition of salt to a suspension reduces the magnitude of the zeta potential for the same reason as it compresses the range of the double layer as described above. Figure 5.9 shows an example of how pH and salt concentration influence the zeta potential of alumina particles.

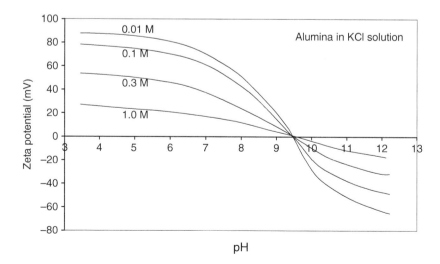

Figure 5.9 Zeta potential of alumina particles as a function of pH and salt concentration. Data from Johnson *et al.* (2000)

5.3.3   Adsorbing Polymers, Bridging and Steric Forces

Another way of controlling surface forces between particles in suspensions is through the addition of a soluble polymer to the solution. Consider the situation where the polymer has an affinity for the particle surface, on which it tends to adsorb. Either attraction by polymer *bridging* or repulsion due to *steric hindrance* can result, depending on the polymer molecular weight and the amount adsorbed, as shown in Figure 5.10. Steric interactions are molecular shape effects which arise at higher polymer concentrations.

Bridging flocculation is a method in which a polymer that adsorbs onto the particle surface is added in a quantity that is less than sufficient to fully cover it. The polymer chains adsorbed onto one particle surface can then extend and adsorb on another particle surface and hold them together. The optimum amount of polymer to add is usually just enough to cover half of the total particle surface area. The best bridging flocculation is usually found with polymers of high to very high molecular weight (typically 1×10^6 to 20×10^6 g/mol). Commercial polymeric *flocculants* are typically charged or nonionic copolymers of polyacrylamide. They are used extensively in water treatment, wastewater, paper and mineral processing industries to aid in solid/liquid separation. In that application, they operate by creating attraction between the fine particles, resulting in the formation of aggregates known as *flocs*. Since the mass of a floc is many times greater than that of an individual particle, separation by sedimentation under gravity becomes possible.

Steric repulsion occurs when the particle surfaces are completely covered with a thick layer of polymer. The polymer must adsorb to the surfaces of the particles and extend out into the solution. In a good solvent, as the separation distance becomes less than twice the extent of the adsorbed polymer, the polymer layers begin to overlap and a strong repulsion results as shown in Figure 5.2(c) and (d). The polymers that work best in creating steric repulsion are typically of low to moderate molecular weight (typically less than 1×10^6 g/mol). When the solvent quality is poor, the steric interaction can be attractive at moderate to long range. This can occur when a poorly soluble polymer is adsorbed to the particle surface. *Electrosteric stabilization* occurs when the polymer is a charged

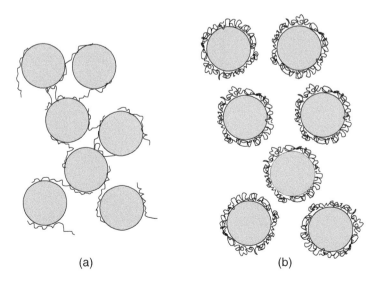

(a)   (b)

Figure 5.10   Schematic representation of (a) bridging flocculation and (b) steric repulsion

polyelectrolyte such that both EDL and steric repulsion are active. Steric and electrosteric stabilization are commonly used in the processing of ceramics to control suspension stability and viscosity. Other examples of stabilization of particulate suspension products are given in Chapter 13.

5.3.4   Net Interaction Force

For particles in suspension in a liquid, the most important surface interactions are often van der Waals and EDL. In the widely used *DVLO theory*[5], the total particle interaction is determined by simply summing the two contributions, and this approach has been widely verified experimentally. Furthermore, it has been found that many other forces may be combined in the same way to determine the overall interparticle interaction. Examples of some net interparticle interaction forces are shown in Figure 5.11.

5.4   EFFECT OF SURFACE FORCES ON BEHAVIOUR IN AIR AND WATER

From the equations for van der Waals forces [Equation (5.8a,b)] and EDL repulsion [Equation (5.11)] one can see that the magnitude of surface forces increases linearly with particle size. Body forces which depend on the mass of the particle, however, increase with the cube of the particle size. As discussed in Chapter 2, it is the *relative* values of body forces and interparticle surface forces that are important in determining how particulate systems will behave.

When particle surfaces interact through air, such as in dry fine powders, the dominant interaction is attraction, due either to van der Waals interactions or capillary bridges (see Chapter 2). In air, other gases and vacuum the only mechanism which can generate repulsion is electrostatic charging by *triboelectrification*, caused by repeated contacts between particles and with walls. These charges may be large and if not dissipated may lead to sparking and the resulting danger of explosion, as discussed in Chapter 14. Moderate humidification of air or other carrying gases may limit the capacity for triboelectrification and enhance electrical dissipation through increasing the surface conductivity of particles. Ionizing sources are sometimes used to neutralize charges.

Attraction between particles results in cohesive behaviour of the powder. (If the charge on the particles is of the same sign, repulsion will result.) Strong cohesion is the reason why fine particles are difficult to fluidize (Geldart's Group C powders described in Chapter 6). Strong cohesion is also the cause of the high unconfined yield stresses of powders described in Chapter 9. The high unconfined yield stress of these powders means that the powders are not free flowing and will require a larger hopper opening relative to free-flowing powders of the same bulk density. One solution is to increase the particle size, by granulation, for example, as discussed in Chapter 12. This decreases the ratio between the cohesive force and the gravity force, resulting in a more free-flowing powder.

---

[5] After Boris Derjaguin (1902–1994; Soviet-Russian chemist), Lev Landau (1908–1968; Soviet-Azerbaijani physicist), Evert Verwey (1905–1981; Dutch chemist) and Theodoor Overbeek (1911–2007; Dutch physicist).

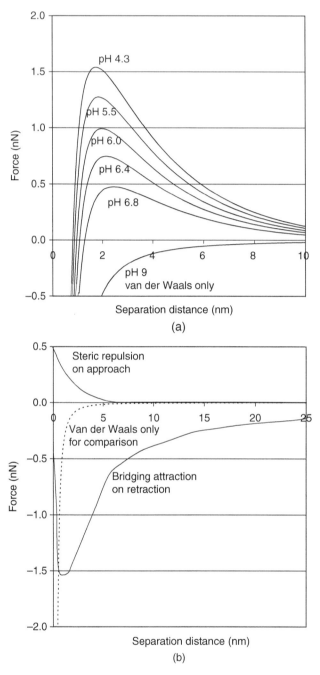

Figure 5.11   (a) Force versus distance curves for alumina at different pH values calculated from Equations (5.8b) and (5.11) with parameters as detailed in Franks *et al*. (2000). At pH 9 the van der Waals attraction dominates. As pH is decreased the range and magnitude of the EDL repulsion increases as zeta potential increases (see Figure 5.8). At very small separation distances the van der Waals attraction always dominates over the EDL repulsion. (b) Force versus distance curves for silica particles interacting with an adsorbed polymer (Zhou *et al*., 2008). Upon approach, the adsorbed polymer provides a weak steric repulsion. Upon separation (retraction) the polymer creates a strong long-range attraction because chains are adsorbed on both surfaces. The van der Waals only interaction is shown for comparison

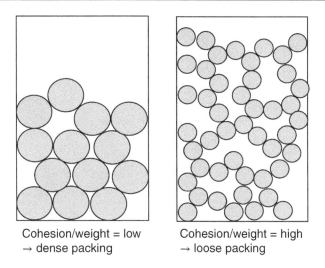

Cohesion/weight = low   Cohesion/weight = high
→ dense packing   → loose packing

Figure 5.12 Schematic of the effect of cohesion on packing density

The influence of attractive forces between fine dry powders is observed as the effect of particle size on bulk density, as shown in Figure 5.12. As the particle size decreases, both loose packed and tapped bulk densities also tend to decrease. This is because as the particle size becomes smaller, the influence of the attractive surface forces becomes stronger than that of the gravity force and structures of larger void fraction can thus be stabilized. Consolidation is aided by body forces (such as gravity) which allow the particles to rearrange into denser packing structures.

An obvious difference between the cases of particles in liquids and in gases is that the density and viscosity of liquids are much higher than those for gases. This means that some of the simplifications which can be applied for particles in gases cannot be applied to particles in liquids. For example, as we see in Chapter 9, in the flow of larger particles from storage hoppers the effect of the gas present can be ignored. For particles in liquid, the presence of the liquid phase can never be ignored!

When fine particles are suspended or dispersed in liquids, such as water, the interaction forces can be controlled by prudent choice of the solution chemistry. This control of interaction forces is of significant technological importance because we can thus control the suspension behaviour such as stability, sedimentation rate, viscosity, and sediment density. Additives such as acids, bases, polymers and surfactants can easily be used in formulations to develop the range and magnitude of either repulsion or attraction, as demonstrated in Figure 5.11. When fine particles are suspended or dispersed in liquids, such as water, there are several mechanisms that can produce repulsive forces between particles that can overwhelm the attractive van der Waals interaction in order to keep the particles dispersed. Examples of the beneficial use of these effects are given in Chapter 13.

Figure 5.13 summarizes how the suspension behaviour depends upon the interparticle forces which in turn depend upon the solution conditions. The suspension properties of interest such as stability, sedimentation, sediment density, particle packing and rheological (flow) behaviour are discussed in the following sections.

Low zeta potential (near IEP)   High zeta potential (away from IEP)
High salt (coagulation)   Low salt
Bridging polymers   Polymer cushions (steric repulsion)

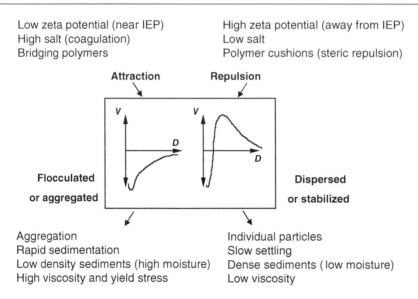

Aggregation   Individual particles
Rapid sedimentation   Slow settling
Low density sediments (high moisture) Dense sediments (low moisture)
High viscosity and yield stress   Low viscosity

Figure 5.13   The top section of the figure gives examples of how the solution conditions influence the forces between particles. The bottom section shows how attractive and repulsive forces influence some behaviour of suspensions

5.5   INFLUENCES OF PARTICLE SIZE AND SURFACE FORCES ON SOLID/ LIQUID SEPARATION BY SEDIMENTATION

The two primary factors that influence the efficiency of solid/liquid separation by gravity are the rate of sedimentation and the liquid fraction (1 – solids fraction) of the sediment. The rate of sedimentation should be maximized while the liquid fraction of the sediment should be minimized.

5.5.1   Sedimentation Rate

In a suspension, gravity causes particles to settle while Brownian motion tends to randomize their position. It is possible to estimate the time for which the suspension remains stable by comparing the average distance which a particle will travel due to Brownian motion with the distance that it settles over the same time period. The time t for which these two distances (L) are equal is given by:

$$L = \sqrt{\frac{2k_B T t}{3\pi x \mu}} = \frac{\left(\rho_p - \rho_f\right)x^2 g}{18\mu} t \tag{5.12}$$

where the distance travelled under Brownian motion is obtained from Equation (5.6) and the distance settled comes from Stokes' law [Equation (3.7)].
   Solving for (non-zero) time:

$$t = \frac{216 k_B T \mu}{\pi g^2 \left(\rho_p - \rho_f\right)^2 x^5} \tag{5.13}$$

Because the Brownian distance depends upon the square root of time and the distance settled depends linearly on time, given enough time all suspensions will eventually settle out if they are not disturbed. The time frame of stability is important when the engineering objective is solid/liquid separation because it is typically only economically viable to conduct solid/liquid separation by sedimentation in a unit operation such as a thickener when the residence times are on the order of hours rather than on the order of weeks or months.

In order to increase the sedimentation rate of colloidal suspensions which would otherwise remain stable for days or weeks, a polymeric flocculant which produces bridging attraction (see Section 5.3.3) is typically added to the suspension. As described earlier, the suspension will then flocculate and the resulting flocs will settle more rapidly than individual particles, making economical solid/liquid separation possible by gravity sedimentation. This is further described in Chapter 3.

5.5.2   Sediment Concentration and Consolidation

Sedimentation of particles in a liquid can be a problem; it reduces the shelf life in certain products, for example. However, it is also a means by which particles can be separated from liquids, either as products for further processing or as contaminants which need removal. Alternative methods for separating particles from liquids include filtration through a porous medium of some kind and inertial separation in a hydrocyclone.

The moisture content of the sediment and the extent to which the sediment consolidates in response to an applied consolidation pressure (due for instance to the weight of the sediment above it or the pressure difference applied to it) depend upon the interparticle forces. Figure 5.11 shows schematically the different effects which can occur during batch sedimentation. Although the sedimentation rate is slow when repulsion and Brownian motion dominate, the sediment bed that eventually forms is quite concentrated and approaches a value near random dense packing of monodisperse spheres (solids volume fraction = 0.64). This is because the repulsive particles joining the sediment bed are able to rearrange into a lower energy (lower height) position as illustrated in Figure 5.14 (a). As discussed earlier in association with the packing of cohesive particles, attractive particles (and aggregates) can form sediments that are quite open and contain high levels of residual liquid, as shown in Figure 5.14(b). This kind of structure may be compressed by further processing.

Pressure may be applied to a particle network in a number of ways including direct application as in a filter press or centrifuge and by the weight of the particles sitting above a particular level in a sediment. The response of the particle network to the applied pressure depends upon the interparticle force between individual particles. The difference between repulsive particles and attractive particles is demonstrated in Figure 5.15. The repulsive particles (dispersed suspensions) easily pack to near the maximum random close packing limit at all consolidation pressures. Strongly attractive particles have the lowest packing densities at a particular applied pressure and weakly attractive particles have intermediate behaviour.

It is apparent that the addition of a flocculant to a colloidal suspension produces both desirable and undesirable effects so far as solid–liquid separation is concerned. Flocculation accelerates sedimentation but the resulting sediment is of lower solids content. Improving solid/liquid separation processes focuses on controlling the interparticle interaction for each step of the separation process: that is, attraction when rapid sedimentation is required and repulsion when consolidation is desired.

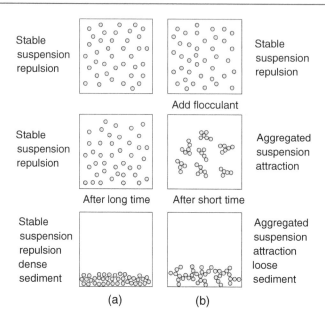

Figure 5.14   (a) Repulsive colloidal particles result in stable dispersions that only form sediments after extended periods. The sediment is quite concentrated. (b) When a flocculant is added to a stable dispersion, the resulting attraction causes aggregation of the particles and rapid sedimentation of the flocs. The sediment in this case is quite open

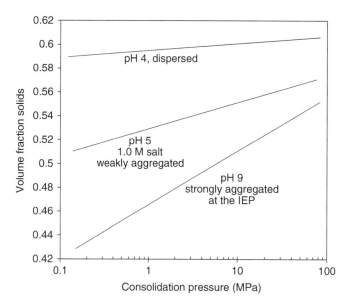

Figure 5.15   Equilibrium volume fraction as a function of consolidation pressure in a filter press (data from Franks and Lange, 1996) for 200 nm diameter alumina. At pH 4 the strong repulsion between particles results in consolidation to high densities over a wide range of pressures. At pH 9, which is the IEP of the powder, the strong attraction results in a filter cake which is initially of low solids fraction and therefore in pressure-dependent filtration behaviour. The weak attraction at pH 5, with added salt results in intermediate behaviour

5.6 GAS–SOLID SEPARATION

Separation of solid particles from a gas differs from solid–liquid separation in several respects. Most obviously, gases are normally several orders of magnitude less dense than liquids or solids, and less viscous, which are both factors which promote settling. Typical densities of solid dispersions in a gas are much lower than in liquids. A loading of $1 \, \text{g/m}^3$ of aerosol particles would be considered high in many processes, and legal limits in the general environment are expressed as $\mu\text{g/m}^3$, while concentrations of order $100 \, \text{kg/m}^3$ would not be unusual in liquid/solid separation.

As we have seen in Chapter 3, the movement of particles relative to a fluid is characterized by a particle Reynolds number $Re_p = Ux\rho_f/\mu$, where U is the relative velocity between the particle and the fluid, of density ρ_f and viscosity μ. In what follows, the fluid is taken to be a gas. For small values of Re_p, the drag force F_D can be estimated from Stokes' law [Equation (3.7)] with a correction to allow for 'slip effects' when the particle size is comparable with the mean free path of the gas λ (i.e. the average distance through which a gas molecule moves between changes of direction and/or energy):

$$F_D = \frac{3\pi x}{C}$$

where

$$C = \frac{\text{Drag on particle in continuum flow at same } Re_p}{\text{Drag on particle in presence of slip}} \tag{5.14}$$

C, known as the *Cunningham slip correction factor*, therefore accounts for the fact that for very small particles the gas does not appear continuous as it does for larger ones, so drag is reduced. Expressions for C are available as a function of (λ/x); under most conditions C is only significantly different from 1 for values of x below $1 \, \mu\text{m}$, as shown in Figure 5.16.

At low Re_p, therefore, the terminal velocity U_T is given by:

$$U_T = Cgx^2\left(\rho_p - \rho_f\right)/18\mu \tag{5.15}$$

where ρ_p is the particle density. Normally, ρ_p is much greater than ρ_f so that this simplifies to:

$$U_T = Cg\rho_p x^2/18\mu \tag{5.16}$$

Note the strong dependence on particle size. Typical values are shown in Figure 5.16 for the case of $\rho_p = 2500 \, \text{kg/m}^3$ in air at ambient conditions. Note that the single particle settling velocity goes from approximately $0.8 \, \text{m/s}$ for particles of $100 \, \mu\text{m}$ to less than $1 \, \text{cm/s}$ at $10 \, \mu\text{m}$ and less than $0.1 \, \text{mm/s}$ at $1 \, \mu\text{m}$. Separation by settling is therefore unlikely to be successful for particles much below $100 \, \mu\text{m}$, especially since thermally generated air currents will easily exceed $1 \, \text{cm/s}$.

As seen earlier, the Brownian diffusivity α of particles in a fluid is given by:

$$\alpha = Ck_B T/3\pi\mu x \tag{5.17}$$

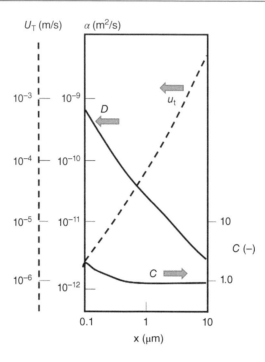

Figure 5.16   Variation of terminal velocity U_T, diffusion coefficient α and Cunningham slip correction factor C with particle diameter x. Clift *et al.* (1981) / with permission of Elsevier

where k_B is Boltzmann's constant, 1.380×10^{-23} J/K. [This is the same as Equation (5.5) except for the addition of the factor C.] If the effect of C is neglected, which as noted earlier is approximately valid for particle sizes above 1 μm, α is therefore inversely proportional to particle size. Typical values are also given in Figure 5.16.

The overall result, as seen in Figure 5.16, is that the effects of particle size on settling velocity and diffusion work in different directions: as the particle size decreases, the settling velocity reduces strongly but the diffusivity increases. This suggests two different approaches to particle separation from a gas. If the particle size is sub-micron then they can be separated by *diffusional collection*, in a depth filter, for example. Note that naturally occurring cohesion due to van der Waals forces will usually be sufficient to hold such particles onto a surface once they have made initial contact. If the size is much above 1 μm and certainly above 10 μm, the usual approach is to use *inertial collection* in a cyclone or some other device in which the gas is forced to change direction, causing the particles to separate out through their inertia. The explanation above also explains why filters tend to show a 'most penetrating size', often around 1 μm, for which collection is poor by either mechanism.

This subject is considered in more detail in Chapter 8.

5.7   SUSPENSION RHEOLOGY

Rheology is the study of flow; here we consider the flow of particle suspensions in a liquid. Consider a simple shear geometry as shown in Figure 5.17; the top plate is moving to the right at a velocity u and h is the height of the sheared liquid layer. For a *Newtonian* liquid, the shear rate is proportional to the shear stress, or

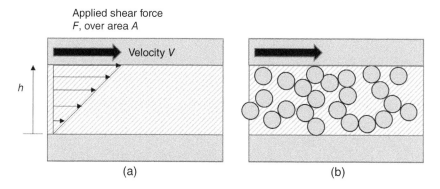

Figure 5.17 Simple shear of (a) a continuous liquid and (b) a suspension

$$\tau = \mu\dot{\gamma} \qquad\qquad (5.18)$$

where τ (=F/A) is the shear stress, μ is the fluid viscosity, and $\dot{\gamma}$ (=u/h) is the shear rate. This relationship is shown in Figure 5.18(a), together with three other common types of characteristic stress *versus* shear rate behaviour which can be observed in different types of suspension. These are (i) *yield stress* behaviour, in which a minimum shear stress must be imposed before flow can occur; (ii) *shear thinning*, in which the effective viscosity

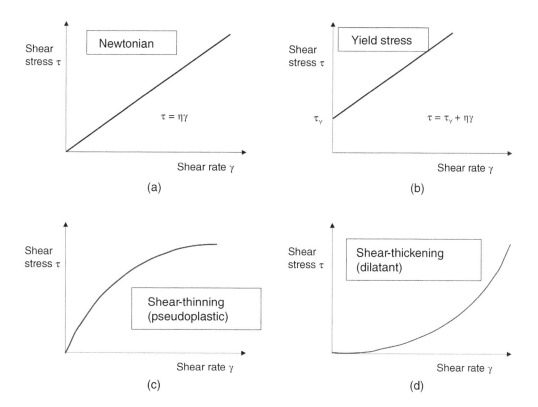

Figure 5.18 Typical shear stress versus shear rate curves: (a) Newtonian; (b) yield stress; (c) shear thinning and (d) shear thickening. Seville and Wu (2016) / with permission of Elsevier

decreases with an increase in shear rate; and (iii) *shear thickening*, in which the effective viscosity increases with an increase in shear rate.

Suspensions of particles can exhibit this entire range of behaviour from Newtonian liquids with viscosities near that of water to high yield stress and high viscosity pastes such as mortar or toothpaste. The main parameters that influence the rheological behaviour of suspensions are the volume fraction of solids, the original viscosity of the carrier fluid, surface forces between particles, particle size and particle shape.

First, consider the case when only hydrodynamic forces and Brownian motion are important and there are no interparticle forces, which is known as the *non-interacting hard-sphere model*. The influence of surface forces is considered in Section 5.8.

Consider a molecular liquid with Newtonian behaviour, such as water, benzene or alcohol. The addition of a spherical particle to the liquid will increase its viscosity due to the additional energy dissipation related to the hydrodynamic interaction between the liquid and the sphere. Further addition of spherical particles increases the viscosity of the suspension linearly. The relationship between the viscosity of a dilute suspension and the volume fraction of solid spherical particles was first obtained by Einstein:

$$\mu_s = \mu_1(1 + 2.5\phi) \tag{5.19}$$

where μ_s is the suspension viscosity μ_1 is the liquid viscosity and ϕ is the volume fraction of solids ($\phi = 1 - \varepsilon$ where ε is the void fraction). Note that the viscosity of the suspension remains Newtonian and that this relationship applies only for dilute suspensions; in practice it can be used when the volume fraction of solids is less than about 7%. This relationship has been verified extensively. Figure 5.19 shows an example of the relationship as measured for silica spheres.

Einstein's analysis was based on the assumption that the particles are far enough apart that they do not influence each other. Once the volume fraction of solids reaches about 10%, the average separation distance between particles is about equal to their diameter.

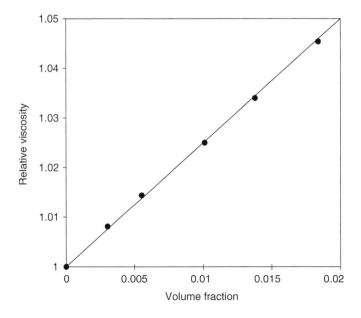

Figure 5.19 Relative viscosity (μ_s/μ_1) of hard-sphere silica particle suspensions (black circles) and Einstein's relationship, Equation (5.14) (line). Adapted from Jones *et al.* (1991)

At this separation, the hydrodynamic disturbance of the liquid by one sphere begins to influence the behaviour of other spheres. In the semi-dilute concentration regime around 10 vol% (about 7–15 vol% solids), the hydrodynamic interactions between spheres result in positive deviation for Einstein's relationship. This equation has been extended to include higher order terms in volume fraction: at higher concentrations the suspension viscosity is still Newtonian but increases with volume fraction according to:

$$\mu_s = \mu_1\left(1 + 2.5\phi + 6.2\phi^2\right) \tag{5.20}$$

This relationship is due to Batchelor; details of this and other such relationships are given by Barnes *et al.* (1989). At higher concentrations of particles, the particle–particle hydrodynamic interactions become more significant and the suspension viscosity becomes even more sensitive to concentration; the suspension rheology becomes *shear thinning* (see Figure 5.18) rather than Newtonian.

Brownian motion dominates the behaviour of concentrated suspensions at rest and at a low shear rate such that a random particle structure results. As shown in Figure 5.20 there is typically a range of low shear rates over which the viscosity is independent of shear rate, which is a characteristic of Newtonian behaviour. At high shear rates, hydrodynamic interactions are more significant than Brownian motion and preferred flow structures develop, such as sheets and strings of particles, as shown schematically in Figure 5.20. The viscosity of suspensions with such preferred flow structures is much lower than the viscosity of a suspension with the same volume fraction but with a randomized structure, because the preferred flow structure minimizes the particle–particle hydrodynamic interaction. These structures develop naturally as the shear rate is increased. Typically, there is a limit to the extent of shear thinning so that at even higher shear rates the viscosity reaches a plateau again. The shear thinning behaviour observed in concentrated hard-sphere suspensions can therefore be considered as the transition

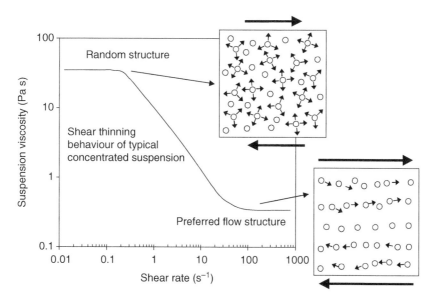

Figure 5.20 The transition from Brownian-dominated random structures to preferred flow structures as the shear rate is increased is the mechanism for the shear thinning behaviour of concentrated suspensions of hard-sphere colloids

from the randomized structure of the low shear rate Newtonian plateau to the fully developed flow structure of the high shear rate Newtonian plateau as illustrated in Figure 5.20.

As the volume fraction of particles continues to increase, the effective viscosity of a suspension becomes difficult to predict. Ultimately, a limiting concentration is reached at which particles become jammed and no flow is possible. Although there is currently no first principles model that is able to predict the rheological behaviour of highly concentrated suspensions, there are a number of semi-empirical expressions which are widely used. The Kreiger–Dougherty model takes the form:

$$\mu_s^* = \mu_1 \left(1 - \frac{\phi}{\phi_{max}}\right)^{-[\eta]\phi_{max}} \tag{5.21}$$

where μ_s^* can represent the viscosity at either the low shear rate Newtonian plateau or the high shear rate Newtonian plateau. The parameter ϕ_{max} is a fitting parameter that is considered an estimate of the maximum packing faction of the powder. $[\eta]$ is known as the intrinsic viscosity and represents the dissipation due to a single particle. Its value is 2.5 for spherical particles, which is consistent with the Einstein relationship, and increases for particles of non-spherical geometry. For real powders, the exact values of ϕ_{max} and $[\eta]$ are not easy to determine and since the index $(\phi_{max} \times [\eta] = {\sim}2.5 \times {\sim}0.64)$ is approaching 2, a simpler version of the Kreiger–Dougherty model known as the Quemada equation is often used.

$$\mu_s^* = \mu_1 \left(1 - \frac{\phi}{\phi_{max}}\right)^{-2} \tag{5.22}$$

Figure 5.21 shows the good correlation of the Quemada equation with $\phi_{max} = 0.631$ to the experimental results for low shear rate viscosity for silica hard-sphere suspensions.

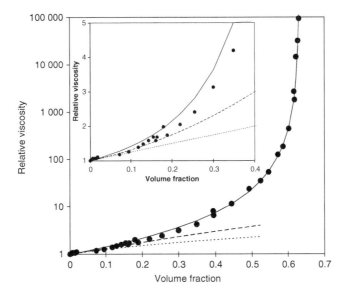

Figure 5.21   Relative viscosity (μ_s/μ_1) at low shear rate of hard-sphere silica suspensions (circles). Quemada's model (solid line) with $\phi_{max} = 0.631$; Batchelor's model (dashed line) and Einstein's model (dotted line). Adapted from Jones *et al.* (1991)

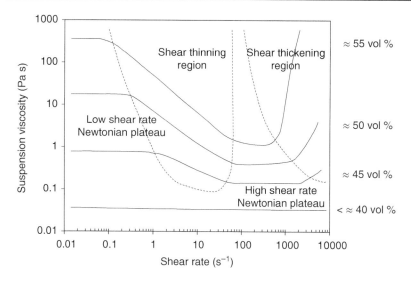

Figure 5.22 Map of typical rheological behaviour of hard-sphere suspensions as a function of shear rate for suspensions with volume fractions between about 40 and 55 vol% solid particles. The dashed lines indicate the approximate location of the boundaries between Newtonian and non-Newtonian behaviour

When the volume fraction of solid particles is very near the maximum packing fraction and the shear rate is high, the preferred flow structures that have developed become unstable. The resulting large-magnitude hydrodynamic interactions push particles together into clusters that begin to jam the entire flowing suspension. Depending on the conditions, the resulting *shear thickening* (increase in viscosity with shear rate) may be gradual or abrupt. Figure 5.22 shows the behaviour of typical hard-sphere suspensions over a wide range of particle concentrations and shear rates.

Note that for hard spheres, perhaps surprisingly, there is no influence of the particle size on the viscosity. The only concern about particle size in hard-sphere suspensions is that if the particles are too big they will settle out. If the particles are near to being neutrally buoyant (of similar density to the fluid) sedimentation issues are not significant.

5.8   INFLUENCE OF SURFACE FORCES ON SUSPENSION FLOW

As we have seen in earlier sections, interparticle forces can have a strong effect on fine particle and colloidal suspensions and they can also be expected to be important in rheology. The sense (attractive or repulsive), range and magnitude of the surface forces all influence the suspension rheological behaviour.

5.8.1   Repulsive Forces

Particles that interact through long-range repulsive forces behave much like hard spheres when the distance between the particles is larger than the range of the repulsive force. This is usually the case when the volume fraction is low (the average distance between the particles is large) and/or when the particles are relatively large (so that the range of the repulsion is small compared with the particle size). If the volume fraction is high, the repulsive force fields of the particles overlap and the viscosity of the suspension is increased, as discussed in Section 5.7. If the particles are very small (typically 100 nm or less) the average distance between the particles (even at a moderate volume fraction) is of the order of the range of the repulsion so the repulsive force fields again overlap and viscosity is increased.

Even dilute suspensions of repulsive particles will have greater viscosity than suspensions of particles showing no interaction forces, because of the additional viscous dissipation related to the flow of fluid through the repulsive regions around each particle. For particles with EDL repulsion this is known as the *primary electro-viscous effect* (Hunter, 2001). Another way of looking at this is to consider the total drag on the particle and the double layer to be greater than the drag on a hard sphere with no double layer. The increase in viscosity due to the primary electro-viscous effect is typically minimal.

Concentrated suspensions of repulsive particles can have significantly elevated viscosities (relative to hard spheres at the same volume fraction) due to the interaction between overlapping EDLs. For particles to push past each other, the double layer must be distorted. This effect is known as the *secondary electro-viscous effect* (Hunter, 2001). Similar effects occur when the repulsion is by a steric mechanism.

The effect of repulsive forces on suspension viscosity can be estimated by considering the effective volume fraction of the particles to include the volume fraction of the particles plus the fraction of volume occupied by the repulsive region around each particle.

$$\phi_{\text{eff}} = \frac{\text{volume of solid} + \text{excluded volume}}{\text{total volume}} \tag{5.23}$$

The effective volume fraction accounts for the volume fraction of fluid that cannot be occupied by particles because they are excluded from that region by the repulsive force as illustrated in Figure 5.23. The suspension rheology can then be estimated reasonably well using the Kreiger–Dougherty or Quemada model with ϕ_{eff} in place of ϕ.

5.8.2   Attractive Forces

Attractive forces may be imagined as generating temporary bonds between particles which thus form a network which is first distorted and then broken in order for flow to occur. The attractive bonding results in material behaviour that is characterized by *viscoelasticity*, a *yield stress* (the minimum stress required for flow) and *shear thinning* (see Figure 5.18).

The shear thinning of an attractive particle network is more pronounced than for hard-sphere suspensions of the same particles at the same volume fraction and is caused by a different mechanism, which is illustrated in Figure 5.24. At rest, the particle network spans the entire volume of the container and resists flow. At low shear rates, the particle network is broken up into large clusters that flow as units. Much of the liquid is trapped within the particle clusters and the viscosity is high. As the shear rate is increased and hydrodynamic forces overcome interparticle attraction, the particle clusters are broken down into smaller and smaller flow units, releasing more and more liquid and reducing the viscosity. At very high shear rates the particle network is completely broken down and the particles can again flow as individuals, almost as if they were non-interacting.

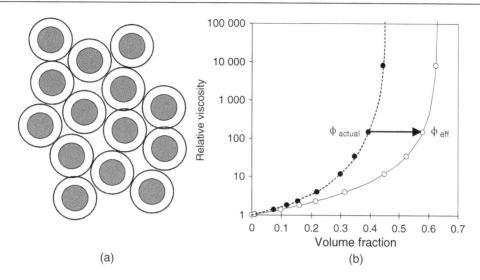

Figure 5.23 (a) Illustration of suspension of particles with volume fraction 0.4 (grey circles) with repulsive interaction extending to the dotted line, resulting in an effective volume fraction of 0.57. (b) Relative viscosity of suspensions of repulsive particles (black dots and dotted line) as a function of actual volume fraction. When the rheological results are plotted as a function of effective volume fraction (open dots) the data map onto the Quemada model (solid line)

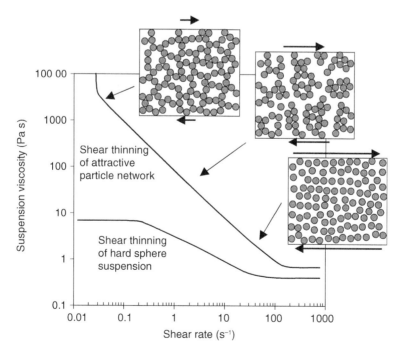

Figure 5.24 Comparison of typical shear thinning behaviour of an attractive particle network with less pronounced shear thinning of hard-sphere suspensions. The attractive particle network is broken down into smaller flow units as the shear rate is increased

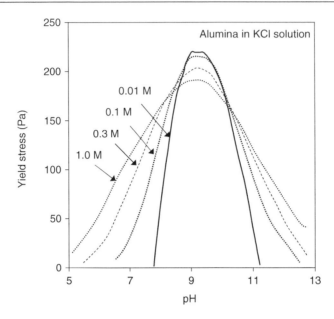

Figure 5.25   Yield stress of 25 vol% alumina suspensions (0.3 μm diameter) as a function of pH and salt concentration. Data from Johnson *et al.* (1999)

A greater magnitude of interparticle attraction results in increased viscosities at all shear rates. Attractive particle networks also exhibit a yield stress (minimum stress required for flow) because there is an attractive force [as shown in Figure 5.2(a) and (b)] which must be exceeded in order to pull two particles apart. The yield stress of the suspension also depends upon the magnitude of the attraction, with stronger attraction resulting in higher yield stresses. Figure 5.25 shows an example of the yield stress of alumina suspensions as a function of pH. The maximum in yield stress corresponds with the IEP of the powder. At low salt concentrations, as the pH is adjusted away from the IEP, the EDL repulsion increases as the zeta potential increases (see Figure 5.9) thus reducing the overall attraction and decreasing the yield stress. As the salt concentration is increased at pH values away from the IEP, the magnitude of the zeta potential decreases (see Figure 5.9) and the EDL repulsion decreases. As such, the resulting overall interaction is attractive and the attraction increases as salt content is increased. The result presented in Figure 5.25 indicates that the yield stress increases as salt is increased at pH values away from the IEP, which is consistent with the force predictions.

It was pointed out in Section 5.7 that the rheological behaviour of hard-sphere suspensions (non-interacting particles) is not influenced by the size of the particles. This is not true of attractive particle networks. When the particles in a suspension are attractive, smaller particle size results in increased values of rheological properties such as yield stress, viscosity, and elastic modulus. The influence of particle size on rheology can be considered to depend upon a combination of the strength of the bond between particles and the number of bonds per unit volume. For example, consider the shear yield stress:

$$\tau_Y \propto \frac{\text{Number of bonds}}{\text{Unit volume}} \times \text{Strength of bond} \tag{5.24}$$

To a first approximation, the strength of the bond of an attractive particle network increases linearly with particle size as indicated by Equations (5.8a,b) and (5.11).

$$\text{Strength of bond} \propto x \tag{5.25}$$

The number of bonds per unit volume depends upon the structure of the particle network and the size of the particles. In the first instance we assume that the structure of the particle network does not vary with particle size. (Details of the aggregate and particle network structure are beyond the scope of the present text.) Then the number of bonds per unit volume simply varies with the inverse cube of the particle size:

$$\frac{\text{Number of bonds}}{\text{Unit volume}} \propto \frac{1}{x^3} \tag{5.26}$$

When the contributions of the strength of the bond and the number of bonds are considered, one finds that the rheological properties such as yield stress, viscosity and elastic modulus vary inversely with the square of the particle size:

$$\tau_Y \propto \left(\frac{1}{x^3} \times x\right) \propto \frac{1}{x^2} \tag{5.27}$$

Many experimental measurements confirm this result although there is considerable variation from the inverse square dependence. One example of a well-controlled experiment that confirms the inverse square dependence of the yield stress on the particle size is shown in Figure 5.26.

When a stress less than the yield stress is applied to an attractive particle network, the network shows an elastic response. The attractive bonds between the particles are

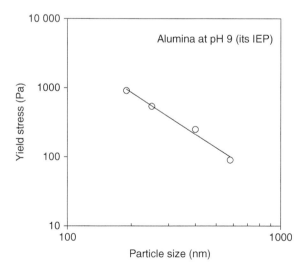

Figure 5.26 Yield stress of alumina suspensions at their IEP as a function of particle size. The best-fit line has a slope of −2.01, correlating quite well with the predicted inverse square particle size dependence as per Equation (5.27). Data from Zhou *et al.* (2001)

stretched rather than broken and when the stress is removed the particles are pulled back together by the attractive bonds and the suspension returns to near its original shape. Because the stretching and breaking of bonds is a statistical phenomenon, pure elasticity is not usually achieved; the attractive particle network is better considered to be a visco-elastic material, displaying behavioural characteristics of both solids and fluids. Some of the work done to deform the material is stored elastically and some of the energy is dissipated by a viscous mechanism.

5.9   NANOPARTICLES

Nanoparticles are defined as particles of any shape with dimensions in the 1–100 nm range (0.001–0.1 μm), so that the upper range of nanoparticles might be regarded as the lower range of the fine or colloidal particles considered in the earlier parts of this chapter, and the lower range of nanoparticles overlaps with larger molecules.

As mentioned earlier, the special properties of colloidal particles often result from their high surface–volume ratio and this is even more true of nanoparticles, for which a large proportion of the atoms in a particle "lies" within only a few atomic diameters of the surface. In effect, there is no 'bulk'; these particles are all surface. This means that nanoparticles can often have markedly different optical, electronic, and other properties than larger particles of the same substance. Reactivity and catalytic effects can be extremely strong, for example, and quantum effects can occur. Some nanoparticles can pass through cell membranes, which can lead to the development of beneficial therapies but also raises concerns about effects on health (see Chapter 14).

Because of the unusual properties of nanoparticles, there are numerous emerging applications and processes where nanoparticles are or may be used. Examples of such applications include:

anti-reflective coatings;

fluorescent labels for biotechnology;

drug delivery systems;

clear inorganic sunscreens;

high-performance solar cells;

catalysts;

high-density magnetic storage media;

high energy density batteries;

self-cleaning glass;

improved light-emitting diodes (LEDs);

high-performance fuel cells;

nanostructured materials.

The success of these potential applications for nanoparticles relies heavily on our ability to produce, transport, separate and safely handle nanoparticles. The concepts presented in this chapter on fine particles and colloids provide a starting point for dealing with these issues.

5.10 WORKED EXAMPLES

WORKED EXAMPLE 5.1

Brownian Motion and Settling
Estimate the amount of time that each of the following suspensions will remain stabilized against sedimentation due to Brownian motion at room temperature (300 K).

(a) 200 nm diameter alumina ($\rho = 3980$ kg/m^3) in water (typical ceramic processing suspension);

(b) 200 nm diameter latex particles ($\rho = 1060$ kg/m^3) in water (typical paint formulation);

(c) 150 nm diameter fat globules ($\rho = 780$ kg/m^3) in water (homogenized milk);

(d) 1000 nm diameter fat globules ($\rho = 780$ kg/m^3) in water (non-homogenized milk).

Solution

The time that the suspension remains stable against gravity can be approximated by equating the average distance moved by a particle due to Brownian motion to the distance settled due to gravity. This time is presented in Equation (5.13) as follows:

$$t = \frac{216kT\mu}{\pi g^2 \left(\rho_p - \rho_f\right)^2 x^5}$$

where $k = 1.381 \times 10^{-23}$ J/K, $\mu_{water} = 0.001$ Pas, $g = 9.8$ m/s^2, $\rho_{water} = 1000$ kg/m^3.

$$\text{then} \quad t = \frac{216\left(1.381 \times 10^{-23} \text{ J/K}\right)300 \text{ K}(0.001 \text{ Pa s})}{\pi(9.8 \text{ m/s}^2)^2 \left(\rho_p - 1000 \text{ kg/m}^3\right)^2 x^5}$$

$$t = \frac{2.96 \times 10^{-24} \text{ kg}^2 \text{ s/m}}{\left(\rho_p \text{ kg/m}^3 - 1000 \text{ kg/m}^3\right)^2 x^5 \text{ m}^5}$$

(a) For the alumina suspension

$$t = \frac{2.96 \times 10^{-24} \text{ kg}^2 \text{ s/m}}{\left(3980 \text{ kg/m}^3 - 1000 \text{ kg/m}^3\right)^2 \left(200 \times 10^{-9}\right)^5 \text{ m}^5} = 1042 \text{ s} = 17.4 \text{ min}$$

(b) For the latex particles in paint

$$t = \frac{2.96 \times 10^{-24} \text{ kg}^2 \text{ s/m}}{\left(1060 \text{ kg/m}^3 - 1000 \text{ kg/m}^3\right)^2 \left(200 \times 10^{-9}\right)^5 \text{ m}^5} = 2.57 \times 10^6 \text{ s} = 30 \text{ days}$$

(c) For the homogenized milk

$$t = \frac{2.96 \times 10^{-24} \text{ kg}^2 \text{ s/m}}{\left(780 \text{ kg/m}^3 - 1000 \text{ kg/m}^3\right)^2 \left(150 \times 10^{-9}\right)^5 \text{ m}^5} = 8.06 \times 10^5 \text{ s} = 9.3 \text{ days}$$

(d) For the non-homogenized milk

$$t = \frac{2.96 \times 10^{-24} \text{ kg}^2 \text{ s/m}}{\left(780 \text{ kg/m}^3 - 1000 \text{ kg/m}^3\right)^2 \left(1000 \times 10^{-9}\right)^5 \text{ m}^5} = 61 \text{ s}$$

These characteristic times correspond best with the time for the first particle to settle out. The time for all the particles to settle out depends upon the height of the container. Nonetheless, one can understand the reasons why the alumina suspension needs to be mixed to keep all of the particles suspended for extended periods of time, why latex paints must be stirred if kept for a month and why milk is homogenized to prevent cream from forming while the milk is in your fridge for a week.

WORKED EXAMPLE 5.2

van der Waals and EDL Forces

Use the DLVO equation, $F_T = \pi \varepsilon \varepsilon_0 x \Psi_o^2 \kappa e^{-\kappa D}\left(Ax/24D^2\right)$, to plot the total interparticle force (F_T) versus interparticle separation distance (D) for two alumina particles and for two oil droplets under the following conditions. The particles are spherical, 1 μm in diameter and suspended in water that contains 0.01 M NaCl. Plot three conditions for each material: (a) at the IEP; (b) with $\zeta = 30$ mV; and (c) with $\zeta = 60$ mV. Comment on the differences in the behaviour of the two different materials. Which particles are easier to disperse and why?

Solution

Assume the surface potential equals the zeta potential ($\Psi_o = \zeta$).

Calculate the inverse Debye length (κ) with Equation (5.10).

$$\kappa = 3.29\sqrt{[c]}\left(\text{nm}^{-1}\right) = \kappa = 3.29\sqrt{0.01}\left(\text{nm}^{-1}\right) = 0.329\left(\text{nm}^{-1}\right) = 3.29 \times 10^8 \text{ m}^{-1}$$

The relative permittivity of water (ε) is 80 and the permittivity of free space (ε_0) is 8.854×10^{-12} $C^2/\text{J/m}$.

The diameter of the particles is 1×10^{-6} m.

Then

$$F_T = \pi 80\left(8.854 \times 10^{-12} C^2/\text{J/m}\right)\left(1 \times 10^{-6} \text{ m}\right)\Psi_o^2\left(3.29 \times 10^8 \text{ m}^{-1}\right)e^{-\left(3.29 \times 10^8 \text{ m}^{-1}\right)D} - \frac{A\left(1 \times 10^{-6} \text{ m}\right)}{24D^2}$$

$$F_T = \left(7.32 \times 10^{-7} C^2/\text{J/m}\right)\Psi_o^2 e^{-\left(3.2 \times 10^8 \text{ m}^{-1}\right)D} - \frac{\left(1 \times 10^{-6} \text{ m}\right)A}{24D^2}$$

where Ψ_o is in volts and D in metres.

From Table 5.1, the Hamaker constants (A) are:

For alumina $A = 5.0 \times 10^{-20}$ J

For oil $A = 0.4 \times 10^{-20}$ J

Then for alumina

$$F_T = \left(7.32 \times 10^{-7} C^2/J/m\right)\Psi_o^2 e^{-\left(3.29 \times 10^8 \text{ m}^{-1}\right)D} - \frac{\left(1 \times 10^{-6} \text{ m}\right)\left(5 \times 10^{-20} \text{ J}\right)}{24D^2}$$

$$F_T = \left(7.32 \times 10^{-7} C^2/J/m\right)\Psi_o^2 e^{-\left(3.29 \times 10^8 \text{ m}^{-1}\right)D} - \frac{\left(5 \times 10^{-26} \text{J m}\right)}{24D^2}$$

and for oil

$$F_T = \left(7.32 \times 10^{-7} C^2/J/m\right)\Psi_o^2 e^{-\left(3.29 \times 10^8 \text{ m}^{-1}\right)D} - \frac{\left(1 \times 10^{-6} \text{ m}\right)\left(0.4 \times 10^{-20} \text{ J}\right)}{24D^2}$$

$$F_T = \left(7.32 \times 10^{-7} C^2/J/m\right)\Psi_o^2 e^{-\left(3.29 \times 10^8 \text{ m}^{-1}\right)D} - \frac{\left(0.4 \times 10^{-26} \text{J m}\right)}{24D^2}$$

These equations can be plotted by a standard data plotting software such as Excel, KG, or Sigmaplot, but first the units and typical values must be checked by hand to ensure no errors are made while writing the equation to the spreadsheet.

Unit analysis

$$F_T = C^2/J/m V^2 e^{m^{-1} \times m} - \frac{J \text{ m}}{m^2} \quad \text{where } V = J/C$$

$$F_T = C^2/J/m \ J^2/C^2 e^{m^{-1} \times m} - \frac{J \text{ m}}{m^2} \quad \text{so} \quad F_T = m^{-1} J - \frac{J}{m} = \frac{J}{m} \text{ and } J = N \text{ m so}$$

F_T is in Newtons so the units are OK.

Figures 5.W2.1 and 5.W2.2 are the plotted results.

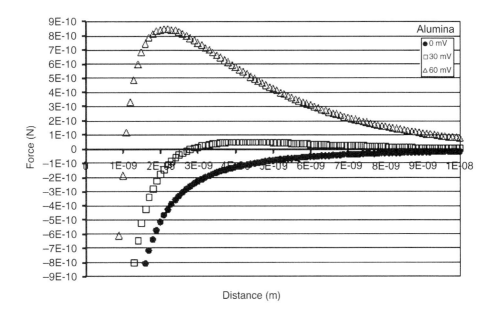

Figure 5.W2.1 Force versus separation distance curves for alumina particles

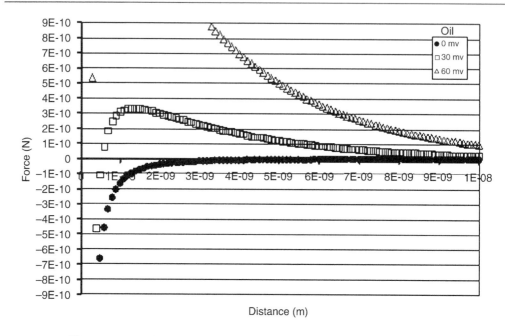

Figure 5.W2.2   Force versus separation distance curves for oil droplets

The difference between the two materials is that the Hamaker constant for the alumina is much greater than for the oil and thus the attraction between alumina particles is much stronger between alumina than between oil droplets. Thus, it is possible to just stabilize the oil droplets with 30 mV zeta potential, whereas when the alumina has 30 mV zeta potential, there is still attraction between the particles. Hence 60 mV is needed to stabilize the alumina to the same extent that 30 mV was able to stabilize oil droplets.

TEST YOURSELF

5.1   What is the typical size range of colloidal particles?

5.2   What two influences are more important for colloidal particles than body forces?

5.3   What is the influence of Brownian motion on a suspension of colloidal particles?

5.4   What is the relationship between surface forces and the potential energy between a pair of particles?

5.5   Under what conditions can you expect van der Waals interactions to be attractive and under what conditions can you expect van der Waals interactions to be repulsive? Which set of conditions is more commonly encountered?

5.6   What are surface hydroxyl groups? What are surface ionization reactions? What is the isoelectric point?

5.7   What is the physical basis for the electrical double-layer repulsion between similarly charged particles?

5.8 What is bridging flocculation? What type of polymers are most suitable to induce bridging attraction? What relative surface coverage of the polymer on the particles surface is typically optimum for flocculation?

5.9 What is steric repulsion? What type of polymers are most suitable to induce steric repulsion? What relative surface coverage of the polymer on the particle's surface is typically optimum for steric stabilization?

5.10 What is meant by the DLVO theory?

5.11 Why are fine particles in typical atmospheric conditions cohesive? What would happen if all humidity were removed from the air?

5.12 How does separation of solids from a gas differ from separation of solids from a liquid? In each case, how does the approach differ according to the size of the particles?

5.13 Explain why a suspension of repulsive colloidal particles cannot be economically separated from a liquid by sedimentation. What is the important parameter that is changed when the particles are flocculated that allows the particles to be economically separated from the liquid by sedimentation?

5.14 How do interparticle interaction forces influence suspension consolidation?

5.15 Why does Einstein's prediction of suspension rheology break down as solids concentration increases above about 7 vol%?

5.16 What is the mechanism for the shear thinning of hard-sphere suspensions?

5.17 What happens to suspension viscosity as the volume fraction of solids is increased?

5.18 How do repulsive forces influence suspension rheology?

5.19 How do attractive forces influence suspension rheology?

5.20 What three types of rheological behaviour are typical of attractive particle networks?

5.21 What is the mechanism for shear thinning in attractive particle networks?

5.22 Describe the influence of particle size on rheological properties of attractive particle networks.

5.23 How do you think the concepts presented in this chapter will be important for producing products from nanoparticles?

EXERCISES

5.1 Colloidal particles may be either 'dispersed' or 'aggregated'.

(a) What causes the difference between these two cases? Answer in terms of interparticle interactions.

(b) Name and describe at least two methods to create each type of colloidal dispersion.

(c) Describe the differences in the behaviour of the two types of dispersions (including but not limited to rheological behaviour, settling rate, sediment bed properties.)

5.2 (a) What forces are important for colloidal particles? What forces are important for non-colloidal particles?

   (b) What is the relationship between interparticle potential energy and interparticle force?

   (c) Which three types of rheological behaviour are characteristic of suspensions of attractive particles?

5.3 (a) Describe the mechanism responsible for shear thinning behaviour observed for concentrated suspensions of micrometre-sized hard-sphere suspensions.

   (b) Consider the same suspension as in (a) except instead of hard-sphere interactions, the particles are interacting with a strong attraction such as when they are at their isoelectric point. In this case, describe the mechanism for the shear thinning behaviour observed.

   (c) Draw a schematic plot (log–log) of the relative viscosity as a function of shear rate comparing the behaviour of the two suspensions described in (a) and (b). Be sure to indicate the relative magnitude of the low shear rate viscosities.

   (d) Consider two suspensions of particles. All factors are the same except for the particle shape. One suspension has spherical particles and the other elongated particles like grains of rice.

   (i) Which suspension will have a higher viscosity?

   (ii) What two physical parameters does the shape of the particles influence that affect the suspension viscosity?

5.4 (a) Explain why the permeability of the sediment from a flocculated mineral suspension (less than 5 μm) is greater than the permeability of the sediment of the same mineral suspension that settles while dispersed.

   (b) Fine clay particles (approximately 0.15 μm diameter) wash from a farmer's soil into a river due to rain.

   (i) Explain why the particles will remain suspended and be carried downstream in the fast-flowing freshwater.

   (ii) Explain what happens to the clay when the river empties into the ocean.

5.5 Calculate the effective volume fraction for a suspension of 150 nm silica particles at 40 vol% solids in a solution of 0.005 M NaCl.

(Answer: 0.473.)

5.6 You are a sales engineer working for a polymer supply company selling poly acrylic acid (PAA). PAA is a water-soluble anionic (negatively charged polymer) that comes in different molecular weights: 10 000, 100 000, 1 million and 10 million. You have two customers. The first customer is using 0.8 μm alumina to produce ceramics. This customer would like to reduce the viscosity of the suspension of 40 vol% solids suspensions. The second customer is trying to remove 0.8 μm alumina from wastewater. There is about 2 vol% alumina in the water and she wants to remove it by settling. What would you recommend to each customer? Consider if PAA is the right material to use, what molecular weight should be used and how much should be used. Figure 5.E6.1 shows the adsorption isotherms for PAA with different molecular weights.

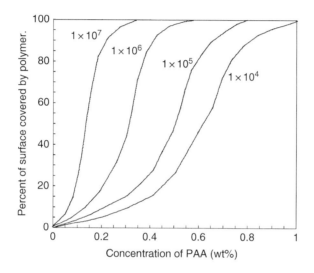

Figure 5.E6.1 Adsorption isotherms for PAA of various molecular weights

5.7 (a) Draw the typical $\log \mu$ versus $\log \dot{\gamma}$ plot for suspensions of hard spheres of approximately micrometre-sized particles at 40, 45, 50 and 55 vol% solids.

(b) Draw the relative viscosity (μ_s/μ_1) versus volume fraction curve for the low shear viscosities of a typical hard-sphere suspension.

6

Packed Beds and Fluidized Beds

There are many applications of packed beds and fluidized beds in industry and so they warrant special attention here. Examples of packed beds include filtration of solids from liquids and gases, removal of chemical components from gas and pressure swing adsorption in gas separation processes. Applications of fluidized beds are found in the manufacture and recycling of plastics, oil refining, mineral processing, energy conversion and pharmaceutical manufacture. In this chapter we begin with the fundamentals of the flow of fluids through a packed bed of particles and of fluidization before studying heat transfer and chemical reaction in fluidized beds. Finally, some attention is given to the many applications of fluidized beds.

6.1 FLUID FLOW THROUGH A PACKED BED OF PARTICLES

6.1.1 Pressure Drop–Flow Relationship

Laminar flow

In the nineteenth century, Darcy[1] (1856) observed that the flow of water through a packed bed of sand was governed by the relationship:

$$\left(\begin{array}{c} \text{pressure} \\ \text{gradient} \end{array} \right) \propto \left(\begin{array}{c} \text{liquid} \\ \text{velocity} \end{array} \right) \text{ or } \frac{(-\Delta p)}{H} \propto U \tag{6.1}$$

where U is the superficial fluid velocity through the bed and $(-\Delta p)$ is the pressure drop across a bed of depth H. (Superficial velocity = fluid volumetric flow rate/cross-sectional area of bed, Q/A.)

---

[1] Henry Darcy (1803–1858), French engineer.

Introduction to Particle Technology, Third Edition. Martin Rhodes and Jonathan Seville.
© 2024 John Wiley & Sons Ltd. Published 2024 by John Wiley & Sons Ltd.
Website: www.wiley.com/go/rhodes/particle3e

The flow of a fluid through a packed bed of solid particles may be analysed by analogy with the fluid flow through tubes. The starting point is the Hagen–Poiseuille equation for laminar flow through a tube:

$$\frac{(-\Delta p)}{H} = \frac{32\mu U}{D^2} \tag{6.2}$$

where D and L are the tube diameter and length, respectively, and μ is the fluid viscosity.

Consider the packed bed to be equivalent to many tubes of equivalent diameter D_e following tortuous paths of equivalent length H_e and carrying fluid with a velocity U_i. Then, from Equation (6.2):

$$\frac{(-\Delta p)}{H_e} = K_1 \frac{\mu U_i}{D_e^2} \tag{6.3}$$

U_i is the actual velocity of fluid flowing through the void space between the particles and is related to the superficial fluid velocity by:

$$U_i = U/\varepsilon \tag{6.4}$$

where ε is the *void fraction* (sometimes referred to as the *voidage*) of the packed bed.

Although the paths of the tubes are tortuous, we can assume that their actual length is proportional to the bed depth, that is:

$$H_e = K_2 H \tag{6.5}$$

The tube equivalent diameter is defined as:

$$\frac{4 \times \text{flow area}}{\text{wetted perimeter}}$$

where flow area $= \varepsilon A$, where A is the cross-sectional area of the vessel holding the bed; wetted perimeter $= S_B A$, where S_B is the particle surface area per unit volume of the bed.

That this is so may be demonstrated by comparison with pipe flow:
Total particle surface area in the bed $= S_B A H$. For a pipe,

$$\text{wetted perimeter} = \frac{\text{wetted surface}}{\text{length}} = \frac{\pi D L}{L}$$

and so, for the packed bed, wetted perimeter $= \dfrac{S_B A H}{H} = S_B A$.

Now if S_v is the surface area per unit volume of particles, then:

$$S_V (1 - \varepsilon) = S_B \tag{6.6}$$

since:

$$\left(\frac{\text{surface of particles}}{\text{volume of particles}} \right) \times \left(\frac{\text{volume of particles}}{\text{volume of bed}} \right) = \left(\frac{\text{surface of particles}}{\text{volume of bed}} \right)$$

and so:

$$\text{equivalent diameter, } D_e = \frac{4\varepsilon A}{S_B} = \frac{4\varepsilon}{S_v(1-\varepsilon)} \tag{6.7}$$

Substituting Equations (6.4), (6.5) and (6.7) in Equation (6.3):

$$\frac{(-\Delta p)}{H} = K_3 \frac{(1-\varepsilon)^2}{\varepsilon^3} \mu U S_v^2 \tag{6.8}$$

where $K_3 = K_1 K_2$. Equation (6.8) is known as the Carman–Kozeny[2] equation describing laminar flow through randomly packed particles. The constant K_3 depends on particle shape and surface properties and has been found by experiment to have a value of about 5. Taking $K_3 = 5$, for laminar flow through a randomly packed bed of monosized spheres of diameter x (for which $S = 6/x$) the Carman–Kozeny equation becomes:

$$\frac{(-\Delta p)}{H} = 180 \frac{\mu U}{x^2} \frac{(1-\varepsilon)^2}{\varepsilon^3} \tag{6.9}$$

This is the most common form in which the Carman–Kozeny equation is quoted.

Turbulent flow

For turbulent flow through a randomly packed bed of monosized spheres of diameter x the equivalent equation is:

$$\frac{(-\Delta p)}{H} = 1.75 \frac{\rho_f U^2}{x} \frac{(1-\varepsilon)}{\varepsilon^3} \tag{6.10}$$

General equation for turbulent and laminar flow

Based on extensive experimental data covering a wide range of size and shape of particles, Ergun[3] (1952) suggested the following general equation for any flow conditions:

$$\frac{(-\Delta p)}{H} = \underbrace{150 \frac{\mu U}{x^2} \frac{(1-\varepsilon)^2}{\varepsilon^3}}_{\substack{\text{laminar} \\ \text{component}}} + \underbrace{1.75 \frac{\rho_f U^2}{x} \frac{(1-\varepsilon)}{\varepsilon^3}}_{\substack{\text{turbulent} \\ \text{component}}} \tag{6.11}$$

This is known as the Ergun equation and applies to flow through a randomly packed bed of spherical particles of diameter x. Ergun's equation additively combines the laminar

---

[2] After the work of Philip Carman and Josef Kozeny (1889–1967), Austrian hydraulic engineer and physicist.
[3] After the work of Sabri Ergun (1918–2006), Turkish chemical engineer.

and turbulent components of the pressure gradient. Under laminar conditions, the first term dominates and the equation reduces to the Carman–Kozeny equation [Equation (6.9)], but with the constant 150 rather than 180. (The difference in the values of the constants is probably due to differences in shape and packing of the particles.) In laminar flow the pressure gradient increases linearly with superficial fluid velocity and is independent of fluid density. Under turbulent flow conditions, the second term dominates; the pressure gradient increases as the square of superficial fluid velocity and is independent of fluid viscosity. In terms of the Reynolds number defined in Equation (6.12), fully laminar conditions exist for Re^* less than about 10 and fully turbulent flow exists at Reynolds numbers greater than around 2000.

$$Re^* = \frac{xU\rho_f}{\mu(1-\varepsilon)} \tag{6.12}$$

In practice, the Ergun equation is often used to predict the packed bed pressure gradient over the entire range of flow conditions. For simplicity, this practice is followed in the Worked Examples and Exercises in this chapter.

Ergun also expressed the flow through a packed bed in terms of a friction factor defined in Equation (6.13):

$$\text{Friction factor, } f^* = \frac{(-\Delta p)}{H} \frac{x}{\rho_f U^2} \frac{\varepsilon^3}{(1-\varepsilon)} \tag{6.13}$$

(Compare the form of this friction factor with the familiar Fanning friction factor for flow through pipes.)

Equation (6.11) then becomes:

$$f^* = \frac{150}{Re^*} + 1.75 \tag{6.14}$$

with

$$f^* = \frac{150}{Re^*} \text{ for } Re^* < 10 \quad \text{and} \quad f^* = 1.75 \text{ for } Re^* > 2000$$

(see Figure 6.1).

Non-spherical particles

The Ergun and Carman–Kozeny equations also accommodate non-spherical particles if x is replaced by x_{sv} the diameter of a sphere having the same surface–volume ratio as the non-spherical particles in question. Use of x_{sv} gives the correct value of specific surface S (surface area of particles per unit volume of particles). The relevance of this will be apparent if Equation (6.8) is recalled.

Thus, in general, the Ergun equation for flow through a randomly packed bed of particles of surface–volume diameter x_{sv} becomes:

$$\frac{(-\Delta p)}{H} = 150 \frac{\mu U}{x_{sv}^2} \frac{(1-\varepsilon)^2}{\varepsilon^3} + 1.75 \frac{\rho_f U^2}{x_{sv}} \frac{(1-\varepsilon)}{\varepsilon^3} \tag{6.15}$$

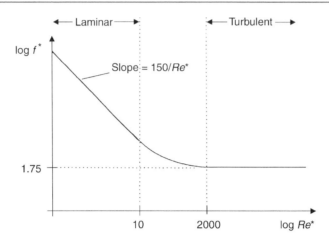

Figure 6.1 Friction factor versus Reynolds number plot for fluid flows through a packed bed of spheres

and the Carman–Kozeny equation for laminar flow through a randomly packed bed of particles of surface–volume diameter x_{sv} becomes:

$$\frac{(-\Delta p)}{H} = 180 \frac{\mu U}{x_{SV}^2} \frac{(1-\varepsilon)^2}{\varepsilon^3} \tag{6.16}$$

It was shown in Chapter 1 that if the particles in the bed are not monosized, then the correct mean size to use in these equations is the surface–volume mean $\bar{x}_{sv}$.

6.1.2 Further Reading

For further information on fluid flow through packed beds and its applications the reader is referred to *Perry's Chemical Engineering Handbook* (Green and Southard 2018).

6.1.3 Worked Examples

WORKED EXAMPLE 6.1

Water flows through 3.6 kg of glass particles of density 2590 kg/m³ forming a packed bed of depth 0.475 m and diameter 0.0757 m. The variation in frictional pressure drop across the bed with water flow rate in the range 200–1200 cm³/min is shown in columns one and two in Table 6.W1.1.

(a) Demonstrate that the flow is laminar.

(b) Estimate the mean surface–volume diameter of the particles.

(c) Calculate the relevant Reynolds number.

Table 6.W1.1   Pressure drop versus flow rate data for the flow of water through a packed bed

Water flow rate (cm³/min)	Pressure drop (mmHg)	U(m/s × 10⁴)	Pressure drop (Pa)
200	5.5	7.41	734
400	12.0	14.81	1600
500	14.5	18.52	1935
700	20.5	25.92	2735
1000	29.5	37.00	3936
1200	36.5	44.40	4870

Solution

(a) First, convert the volumetric water flow rate values into superficial velocities and the pressure drop in millimetres of mercury into Pascal. These values are shown in columns 3 and 4 of Table 6.W1.1.

If the flow is laminar then the pressure gradient across the packed bed should increase linearly with superficial fluid velocity, assuming constant bed void fraction and fluid viscosity. Under laminar conditions, the Ergun equation [Equation (6.15)] reduces to:

$$\frac{(-\Delta p)}{H} = 150 \frac{\mu U}{x_{sv}^2} \frac{(1-\varepsilon)^2}{\varepsilon^3}$$

Hence, since the bed depth H, the water viscosity μ and the packed bed void fraction ε may be assumed constant, then $(-\Delta p)$ plotted against U should give a straight line of gradient:

$$150 \frac{\mu H}{x_{sv}^2} \frac{(1-\varepsilon)^2}{\varepsilon^3}$$

This plot is shown in Figure 6.W1.1. The data points fall reasonably on a straight line confirming laminar flow. The gradient of the straight line is 1.12×10^6 Pa s/m and so:

$$150 \frac{\mu H}{x_{sv}^2} \frac{(1-\varepsilon)^2}{\varepsilon^3} = 1.12 \times 10^6 \text{ Pa·s/m}$$

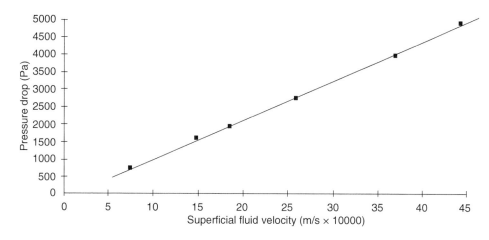

Figure 6.W1.1   Plot of packed bed pressure drop versus superficial fluid velocity

(b) Knowing the mass of particles in the bed, the density of the particles and the volume of the bed, the void fraction may be calculated:

$$\text{mass of bed} = AH(1-\varepsilon)\rho_{\mathrm{p}}$$
$$\text{giving } \varepsilon = 0.3497$$

Substituting $\varepsilon = 0.3497$, $H = 0.475$ m and $\mu = 0.001$ Pa.s in the expression for the gradient of the straight line, we have:

$$x_{\mathrm{sv}} = 792\mu\mathrm{m}$$

(c) The relevant Reynolds number is $Re^* = \dfrac{xU\rho_{\mathrm{f}}}{\mu(1-\varepsilon)}$ [Equation (6.12)] giving $Re^* = 5.4$ (for the maximum velocity used). This is less than the limiting value for laminar flow (10), a further confirmation of laminar flow.

TEST YOURSELF – FLUID FLOW THROUGH PACKED BEDS

6.1 For low Reynolds number (<10) flow of a fluid through a packed bed of particles how does the frictional pressure drop across the bed depend on (a) superficial fluid velocity, (b) particle size, (c) fluid density, (d) fluid viscosity and (e) void fraction?

6.2 For high Reynolds number (>500) flow of a fluid through a packed bed of particles how does the frictional pressure drop across the bed depend on (a) superficial fluid velocity, (b) particle size, (c) fluid density, (d) fluid viscosity and (e) void fraction?

6.3 What is the correct mean particle diameter to be used in the Ergun equation? How can this diameter be derived from a volume distribution?

6.2 FLUIDIZATION

6.2.1 Fundamentals

When a fluid is passed upwards through a bed of particles the pressure loss in the fluid increases with increasing fluid flow. A point is reached when the upward drag force exerted by the fluid on the particles is equal to the apparent weight of particles in the bed (weight of the particles less the buoyancy force). At this point the particles are mobilized by the fluid, the separation of the particles increases, and the bed becomes fluidized. The force balance across the fluidized bed dictates that the fluid pressure loss across the bed of particles is equal to the apparent weight of the particles per unit area of the bed. Thus:

$$Pressure\ drop = \frac{weight\ of\ particles - buoyancy\ force\ on\ particles}{bed\ cross\text{-}sectional\ area}$$

For a bed of particles of density ρ_p, fluidized by a fluid of density ρ_f, to form a bed of depth H and void fraction ε in a vessel of cross-sectional area A:

$$\Delta p = \frac{HA(1-\varepsilon)\left(\rho_p - \rho_f\right)g}{A} \tag{6.17}$$

or

$$\Delta p = H(1-\varepsilon)(\rho_P - \rho_f)g \tag{6.18}$$

A plot of fluid pressure loss across the bed versus superficial fluid velocity through the bed would have the appearance of Figure 6.2. In this figure, the straight-line region OA is the packed bed region, in which the particles do not move relative to one another, and their separation is constant. The pressure loss versus fluid velocity relationship in this region is described by the Carman–Kozeny equation [Equation (6.9)] in the laminar flow regime and the Ergun equation in general [Equation (6.11)]. (See Section 6.1 for a detailed analysis of packed bed flow.)

The region BC is the fluidized bed region where Equation (6.17) applies. At point A, it will be noticed that the pressure loss rises above the value predicted by Equation (6.17). This rise is more marked in small vessels and in powders which have been compacted to some extent before the test and is associated with the extra force required to overcome wall friction and adhesive forces between the bed and the distributor. On decreasing the fluid velocity from point C, the bed pressure drop might typically follow a curve such as CBO as the particles rearrange, eventually forming a packed bed of greater void fraction than the original. This would especially be the case for fine particles, where interparticle forces are strong compared to particle weight, or for a powder that had been slightly compacted when charged to the vessel. When measuring the minimum fluidization velocity of a powder it is therefore recommended that the bed be taken through the cycle of increasing and decreasing fluid velocity in order to achieve a representative packed bed void fraction. The minimum fluidization velocity, U_{mf}, is taken as the fluid velocity corresponding to the intersection of the packed bed line and the fluidized bed line in Figure 6.2.

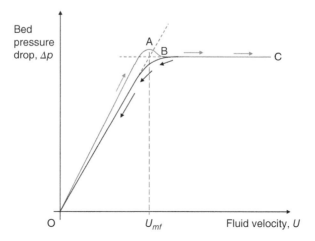

Figure 6.2 Pressure drop versus fluid velocity for packed and fluidized beds

The superficial fluid velocity at which the packed bed becomes a fluidized bed is known as the *minimum fluidization velocity*, U_{mf}. This is also sometimes referred to as the velocity at *incipient* fluidization (incipient meaning 'beginning'). U_{mf} increases with particle size and particle density and is affected by the fluid properties. It is possible to derive an expression for U_{mf} by equating the expression for pressure loss in a fluidized bed [Equation (6.18)] with the expression for pressure loss across a packed bed. Thus, recalling the Ergun equation [Equation (6.11)]:

$$\frac{(-\Delta p)}{H} = 150\frac{(1-\varepsilon)^2}{\varepsilon^3}\frac{\mu U}{x_{sv}^2} + 1.75\frac{(1-\varepsilon)}{\varepsilon^3}\frac{\rho_f U^2}{x_{sv}} \qquad (6.19)$$

Substituting the expression for $(-\Delta p)$ from Equation (6.18):

$$(1-\varepsilon)\left(\rho_p - \rho_f\right)g = 150\frac{(1-\varepsilon)^2}{\varepsilon^3}\frac{\mu U_{mf}}{x_{sv}^2} + 1.75\frac{(1-\varepsilon)}{\varepsilon^3}\frac{\rho_f U_{mf}^2}{x_{sv}} \qquad (6.20)$$

Rearranging,

$$(1-\varepsilon)\left(\rho_p - \rho_f\right)g = 150\frac{(1-\varepsilon)^2}{\varepsilon^3}\left(\frac{\mu^2}{\rho_f x_{sv}^3}\right)\left(\frac{U_{mf}x_{sv}\rho_f}{\mu}\right) + 1.75\frac{(1-\varepsilon)}{\varepsilon^3}\left(\frac{\mu^2}{\rho_f x_{sv}^3}\right)\left(\frac{U_{mf}^2 x_{sv}^2 \rho_f^2}{\mu^2}\right)$$
$$(6.21)$$

and so

$$(1-\varepsilon)\left(\rho_p - \rho_f\right)g\left(\frac{\rho_f x_{sv}^3}{\mu^2}\right) = 150\frac{(1-\varepsilon)^2}{\varepsilon^3}Re_{mf} + 1.75\frac{(1-\varepsilon)}{\varepsilon^3}Re_{mf}^2 \qquad (6.22)$$

or

$$Ar = 150\frac{(1-\varepsilon)}{\varepsilon^3}Re_{mf} + 1.75\frac{1}{\varepsilon^3}Re_{mf}^2 \qquad (6.23)$$

where Ar is the dimensionless number known as the *Archimedes number*,

$$Ar = \frac{\rho_f\left(\rho_p - \rho_f\right)gx_{sv}^3}{\mu^2}$$

and Re_{mf} is the Reynolds number at minimum fluidization,

$$Re_{mf} = \left(\frac{U_{mf}x_{sv}\rho_f}{\mu}\right)$$

In order to obtain a value of U_{mf} from Equation (6.23) we need to know the void fraction of the bed at minimum fluidization, $\varepsilon = \varepsilon_{mf}$. Taking ε_{mf} as the void fraction of the packed bed, we can obtain a crude value for U_{mf}. However, in practice void fraction at the onset of fluidization may be considerably greater than the packed bed void fraction. A typical and often used value of ε_{mf} is 0.4. Using this value, Equation (6.23) becomes:

$$Ar = 1406\, Re_{mf} + 27.3\, Re_{mf}^2 \tag{6.24}$$

Wen and Yu (1966) produced an empirical correlation for U_{mf} with a form similar to Equation (6.24):

$$Ar = 1652\, Re_{mf} + 24.51\, Re_{mf}^2 \tag{6.25}$$

The Wen and Yu correlation is often expressed in the form:

$$Re_{mf} = 33.7\left[\left(1 + 3.59 \times 10^{-5} Ar\right)^{0.5} - 1\right] \tag{6.26}$$

and is valid for spheres (and valid approximately for non-spheres of non-extreme shape) in the range $0.01 < Re_{mf} < 1000$.

For gas fluidization the Wen and Yu correlation is often taken as being most suitable for particles larger than 100 µm, whereas the correlation of Baeyens and Geldart (1974), shown in Equation (6.27), is a better fit to experimental data for particles less than 100 µm.

$$U_{mf} = \frac{\left(\rho_p - \rho_f\right)^{0.934} g^{0.934} x_p^{1.8}}{1110 \mu^{0.87} \rho_f^{0.066}} \tag{6.27}$$

6.2.2   Relevant Powder and Particle Properties

The correct density for use in fluidization equations is the *particle density*, defined as the mass of a particle divided by its hydrodynamic volume. This is the volume 'seen' by the fluid in its fluid dynamic interaction with the particle and includes the volume of all the open and closed pores (see Figure 6.3):

$$\text{particle density} = \frac{\text{mass of particle}}{\text{hydrodynamic volume of particle}}$$

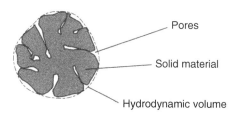

Figure 6.3   Hydrodynamic volume of a particle

For non-porous solids, this is easily measured by a gas pycnometer or specific gravity bottle, but these devices should not be used for porous solids since they give the true or *absolute density* ρ_{abs} of the material of which the particle is made and this is not appropriate where interaction with fluid flow is concerned:

$$\text{absolute density} = \frac{\text{mass of particle}}{\text{volume of solids material making up the particle}}$$

Another name for absolute density is *skeletal density*; the two are equivalent.

For porous particles, the particle density ρ_p (also called *apparent* or *envelope density*) is not easy to measure directly although several methods are given in Geldart (1990).

Bed density is another term used in connection with fluidized beds; it is defined as:

$$\text{bed density} = \frac{\text{mass of particles in a bed}}{\text{volume occupied by particles and voids between them}}$$

For example, 600 kg of powder is fluidized in a vessel of cross-sectional area $1\,\text{m}^2$ and achieves a bed height of 0.5 m. What is the bed density?

Mass of particles in the bed = 600 kg.

Volume occupied by particles and voids = $1 \times 0.5 = 0.5\,\text{m}^3$.

Hence, bed density = $600/0.5 = 1200\,\text{kg/m}^3$.

If the particle density of these solids is $2700\,\text{kg/m}^3$, what is the bed void fraction?

Bed density ρ_B is related to particle density ρ_p and bed void fraction ε by Equation (6.28):

$$\rho_B = (1 - \varepsilon)\rho_p \tag{6.28}$$

Hence

$$\text{void fraction} = 1 - \frac{1200}{2700} = 0.555$$

Another density often used when dealing with powders is the *bulk density*. It is defined in a similar way to fluid bed density:

$$\text{bulk density} = \frac{\text{mass of particles}}{\text{volume occupied by particles and voids between them}}$$

For some powders the bulk density depends strongly on the degree of compaction. Two extreme measures of bulk density are the loose-poured bulk density (or settling density) and the tapped (or tap) bulk density. Loose-poured bulk density may be measured by pouring the powder slowly through a mesh into a small cylindrical container of known volume. Tapped bulk density is obtained by 'tapping' the container in a reproducible manner until a constant density is achieved. (Commercial machines are available in order to make this measurement in a reproducible way.) The Hausner ratio is the ratio of tapped bulk density to loose-poured bulk density. It is sometimes used as a measure of flowability, but the connection is not always reliable. A Hausner ratio near to 1.0 would suggest a free-flowing powder. A ratio greater than 1.3 would indicate the likelihood of poor flowability. In fluidization research the Hausner ratio has been used as a measure of the cohesiveness of powders – the greater the ratio, the more cohesive the powder, because cohesion initially stabilizes structures of higher void fraction (see Figure 5.12), which then collapse on repeated tapping.

The most appropriate particle size to use in equations relating to fluid-particle interactions is a *hydrodynamic diameter*, i.e. an equivalent sphere diameter derived from a measurement technique involving hydrodynamic interaction between the particle and fluid. In practice, however, in most industrial applications sizing is done using sieving or image analysis and many of the older correlations use either sieve diameter, x_p or equivalent volume sphere diameter, x_v but do not specify which mean is to be used to represent the powder as a whole.

For use in fluidization applications, starting from a sieve analysis the mean size of the powder is often calculated from:

$$\text{mean } x_p = \frac{1}{\sum m_i/x_i} \tag{6.29}$$

where x_i is the arithmetic mean of adjacent sieves between which a mass fraction m_i is collected. This is the harmonic mean of the mass distribution, which was shown in Chapter 1 to be equivalent to the arithmetic mean of a surface distribution. Many size measurement methods permit calculation of means that are relevant to fluidized bed applications. However, since this involves conversion between distributions, the results should be treated with caution (see Chapter 1).

6.2.3 Bubbling and Non-Bubbling Fluidization

Beyond the minimum fluidization velocity, bubbles or particle-free voids may appear in the fluidized bed. Figure 6.4 shows bubbles in a gas fluidized bed. It is important to note that although these look very much like bubbles of gas in a liquid, they do not have a

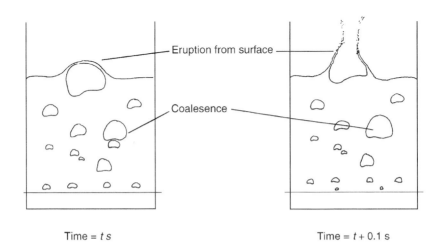

Time = t s   Time = t + 0.1 s

Figure 6.4 Sequence showing bubbles in a 'two-dimensional' fluidized bed of Group B powder. Images taken from video

surface tension as bubbles of gas in a liquid do. Their surfaces are completely permeable to gas, which flows invisibly through them as well as around them.

The equipment used in Figure 6.4 is a so-called 'two-dimensional fluidized bed'. A favourite tool of researchers looking at bubble behaviour, this is actually a vessel of a rectangular cross section, whose shortest dimension (into the page) may be 2–5 cm depending on the mean size of the solids being studied. Semicylindrical beds with a transparent faceplate are also used for visualization of bed behaviour.

As Figure 6.4 shows, bubbles in fluidized beds have a characteristic shape, which in 3D is known as a spherical cap [Figure 6.5(a)]. Bubbles carry particles with them as they rise, in the filled part of the sphere known as the bubble *wake*. The wake fraction is defined as the ratio of the volumes of the wake and the bubble and varies from as high as 0.5 for powders of 60 μm in size to about 0.3 for particles of 200 μm and less for larger particles. Particles also follow the rising bubble in a trail known as the *drift* region. The capacity of bubbles to move particles in this way is an important feature of fluidization: it gives rise to the rapid mixing and high rates of heat transfer which are advantageous in many applications.

As they rise, bubbles combine or coalesce, so that the average bubble size rises with distance above the gas distributor, as shown in Figure 6.5(b and c). Bubbles can move sideways into the track of a leading bubble; near the wall of the bed this can only occur by moving inwards, towards the axis. The result is that the region of most vigorous bubbling concentrates towards the axis of a cylindrical bed, as shown in Figure 6.6(b). As discussed above, the rising bubbles carry solids upward in their wake and drift regions, which must return to the base of the bed; this occurs preferentially close to the walls. The resulting overall circulation pattern is as shown. In the upper part of the bed, solids motion is strongly upward at the centre and downward at the walls. Close to the distributor, a more complex pattern may develop like the one shown.

At superficial velocities above the minimum fluidization velocity, fluidization may in general be either bubbling or non-bubbling. Some combinations of fluid and particles give rise to *only bubbling* fluidization and some combinations give *only non-bubbling*

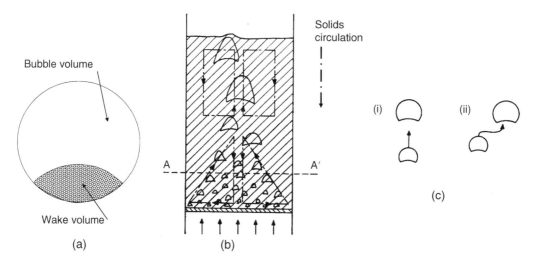

Figure 6.5   (a) Bubble and wake volumes, (b) bubble and solids flow patterns in a fluidized bed and (c) bubble coalescence. (Clift, 1986)

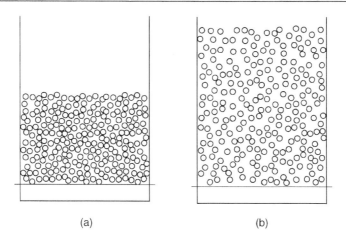

Figure 6.6 Expansion of a liquid fluidized bed: (a) just above U_{mf} and (b) liquid velocity several times U_{mf}. Note the uniform increase in void fraction. Images taken from video

fluidization. Most liquid–fluidized systems, except those involving very dense particles, do not give rise to bubbling. Figure 6.6 shows a bed of glass spheres fluidized by water exhibiting non-bubbling fluidized bed behaviour, where the increase in void fraction with liquid flow rate is uniform. Gas fluidized systems, however, give either only bubbling fluidization or non-bubbling fluidization beginning at U_{mf}, followed by bubbling fluidization as fluidizing velocity increases. Non-bubbling fluidization is also known as *particulate or homogeneous fluidization* and bubbling fluidization is often referred to as *aggregative or heterogeneous fluidization*.

6.2.4 Classification of Powders

Geldart (1973) classified powders into four groups according to their fluidization properties at ambient conditions. The Geldart classification of powders is now used widely in all fields of powder technology. Powders which when fluidized by air at ambient conditions show a region of non-bubbling fluidization beginning at U_{mf}, followed by bubbling fluidization as fluidizing velocity increases, are classified as Group A. Powders which under these conditions give only bubbling fluidization are classified as Group B. Geldart identified two further groups: Group C powders – very fine, cohesive powders which are incapable of fluidization in the strict sense, and Group D powders – large particles distinguished by their ability to produce deep *spouting* beds (see Figure 6.7). (Note that spouted beds do not fully satisfy the definition of fluidization because the pressure drop across a spouted bed does not fully support the particle weight.) Figure 6.8 shows how the group classifications are related to the particle and gas properties. This figure, often referred to as the Geldart diagram, was developed for systems involving air at ambient conditions. The particle size, x, is the *Sauter mean* – the diameter of a sphere with the same surface–volume ratio as the particle of interest.

   The fluidization properties of a powder in air may be predicted by establishing in which group it lies. Note, however, that changes in temperature and pressure have some effect on fluidization behaviour, which effectively move the boundaries shown in

Figure 6.7   A spouted fluidized bed of rice

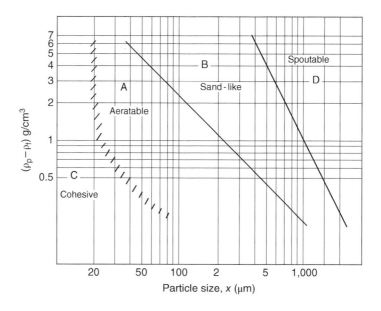

Figure 6.8   Simplified diagram showing Geldart's classification of powders according to their fluidization behaviour in air under ambient conditions. Adapted from Geldart (1973)

Table 6.1 Geldart's classification of powders

	Group C	Group A	Group B	Group D
Most obvious characteristic	Cohesive, difficult to fluidize	Ideal for fluidization. Exhibits a range of non-bubbling fluidization	Starts bubbling at U_{mf}	Coarse solids
Typical solids	Flour, cement	Cracking Catalyst	Building sand	Gravel, coffee beans
Property				
Bed expansion	Low because of channelling	High	Moderate	Low
De-aeration rate	Initially fast, then exponential	Slow, linear	Fast	Fast
Bubble properties	No bubbles-only channels	Bubbles split and coalesce. Maximum bubble size.	No limit to size	No limit to size
Solids mixing	Very low	High	Moderate	Low
Gas back mixing	Very low	High	Moderate	Low
Spouting	No	No	Only in shallow beds	Yes, even in deep beds

Figure 6.8. In particular, higher temperatures can cause increased cohesion, causing a powder that is freely bubbling at ambient temperature to behave more like a Group C powder at higher temperature.

Table 6.1 presents a summary of the typical properties of the different powder classes.

Since the range of gas velocities over which non-bubbling fluidization occurs in Group A powders is small, bubbling fluidization is the type most commonly encountered in gas-fluidized systems in commercial use. The superficial gas velocity at which bubbles first appear is known as the *minimum bubbling velocity* U_{mb}. Premature bubbling can be caused by poor distributor design or protuberances inside the bed, either of which can cause mal-distribution of the flow. Abrahamsen and Geldart (1980) correlated the maximum values of U_{mb} with gas and particle properties using the following correlation:

$$U_{mb} = 2.07\, exp(0.716F) \left(\frac{x_p \rho_f^{0.06}}{\mu^{0.347}} \right) \tag{6.30}$$

where F is the fraction of powder less than 45 µm.

The most important feature of Group A powders is that they show a region of bubble-free expansion at gas velocities between U_{mf} and U_{mb}. As the gas velocity is increased above U_{mb}, bubbling is observed and bubbles grow in size, accompanied by frequent splitting and coalescing, until a *maximum stable bubble size* is achieved, which is typically 1–10 cm. Many small bubbles make for good quality, smooth fluidization. Figure 6.9 shows bubbles in a Group A powder in a two-dimensional fluidized bed.

In Groups B and D powders, bubbling is observed at the minimum fluidization velocity, i.e. $U_{mb} = U_{mf}$. Bubbles continue to grow, limited only by the size of the apparatus (see

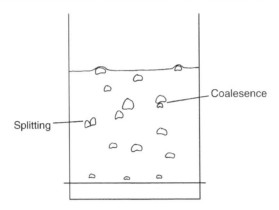

Figure 6.9 Bubbles in a 'two-dimensional' fluidized bed of Group A powder. Image taken from video

Figure 6.4). Fewer, larger bubbles make for rather poor-quality fluidization associated with large pressure fluctuations and less effective mixing.

In Group C powders the interparticle forces are large compared with the inertial forces on the particles. As a result, the particles are unable to achieve the separation they require to be totally supported by drag and buoyancy forces and true fluidization does not occur. Bubbles as such do not appear; instead, the gas flow forms channels through the powder (see Figure 6.10). Since the particles are not fully supported by the gas, the pressure loss

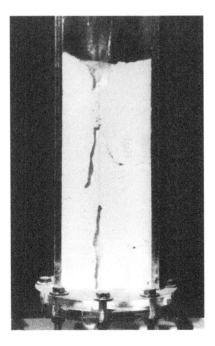
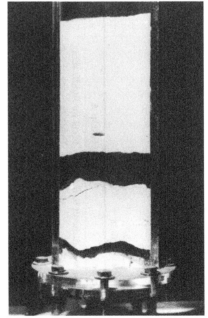

Figure 6.10 Attempts to fluidize Group C powder-producing cracks and channels or discrete solid plugs

across the bed is always less than the apparent weight of the bed per unit cross-sectional area. Consequently, measurement of bed pressure drop is one means of detecting this Group C behaviour if a visual observation is inconclusive. Fluidization, of sorts, can be achieved with the assistance of a mechanical stirrer or vibration. In some situations, a special bubble cap distributor giving high-velocity horizontal gas jets may be used to break up the agglomerates and achieve satisfactory fluidization.

When the size of the bubbles is greater than about one-third of the diameter of the equipment their rise velocity is controlled by the equipment dimensions and they become *slugs* of gas. *Slugging* is attended by large pressure fluctuations and so it is generally avoided in large units since it can cause vibration to the plant. Slugging is unlikely to occur at any velocity if the bed is sufficiently shallow. According to Yagi and Muchi (1952), slugging will not occur provided the following criterion is satisfied:

$$\left(\frac{H_{mf}}{D}\right) \leq \frac{1.9}{\left(\rho_p x_P\right)^{0.3}} \tag{6.31}$$

This criterion works well for most powders. If the bed is deeper than this critical height then slugging will occur when the gas velocity exceeds U_{ms} as given by (Baeyens and Geldart, 1974):

$$U_{ms} = U_{mf} + 0.16\left(1.34D^{0.175} - H_{mf}\right)^2 + 0.07(gD)^{0.5} \tag{6.32}$$

6.2.5 Expansion of a Fluidized Bed

Non-bubbling fluidization

In a non-bubbling fluidized bed at gas velocity $U > U_{mf}$ the particle separation increases with increasing fluid superficial velocity whilst the pressure loss across the bed remains constant. This increase in bed void fraction with fluidizing velocity is referred to as bed expansion (see Figure 6.6). The relationship between superficial fluid velocity U and bed void fraction ε was determined by Richardson and Zaki (1954) and is shown in Equation (6.33).

$$U = U_T \varepsilon^n \tag{6.33}$$

$$\text{For } Re_p \leq 0.3, n = 4.65 \tag{6.34}$$

$$\text{For } Re_p \geq 500, n = 2.4 \tag{6.35}$$

where the single-particle Reynolds number, Re_p is calculated at terminal velocity, U_T.

Khan and Richardson (1989) suggested the correlation given in Equation (3.59) which permits the determination of the exponent n at intermediate values of Reynolds number (although it is expressed in terms of the Archimedes number Ar there is a direct relationship between Re_p and Ar). This correlation also incorporates the effect of the vessel diameter on the exponent. Thus Equations (6.33)–(6.35) in conjunction with Equation (3.59) permit calculation of the variation in bed void fraction with fluid velocity beyond U_{mf}.

Knowledge of the bed void fraction allows calculation of the fluidized bed height as illustrated below:

$$\text{mass of particles in the bed} = M_B = (1 - \varepsilon)\rho_p A H \tag{6.36}$$

If packed bed depth (H_1) and void fraction (ε_1) are known, then if the mass remains constant the bed depth at any void fraction can be determined:

$$(1 - \varepsilon_2)\rho_p A H_2 = (1 - \varepsilon_1)\rho_p A H_1 \tag{6.37}$$

Hence:

$$H_2 = \frac{(1 - \varepsilon_1)}{(1 - \varepsilon_2)} H_1 \tag{6.38}$$

Bubbling fluidization

The simplest description of the expansion of a bubbling fluidized bed is derived from the two-phase theory of fluidization of Toomey and Johnstone (1952). This theory considers the bubbling fluidized bed to be composed of two phases: the *bubble phase* (the gas bubbles) and the *particulate phase* (the fluidized solids around the bubbles). The particulate phase is also sometimes referred to as the *emulsion* phase (although this usage is misleading because a fluidized bed does not form an emulsion in the sense described in Chapter 5; we therefore avoid using this term for fluidized beds). The theory states that any gas in excess of that required for minimum fluidization will pass through the bed in the form of bubbles. Figure 6.11 shows the effect of fluidizing gas velocity on bed

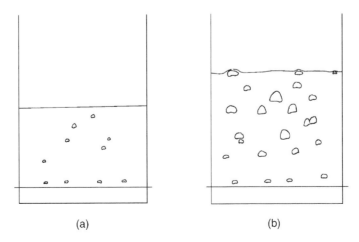

(a)   (b)

Figure 6.11 Bed expansion in a 'two-dimensional' fluidized bed of Group A powder: (a) just above U_{mb}; (b) fluidized at several times U_{mb}. Images taken from video

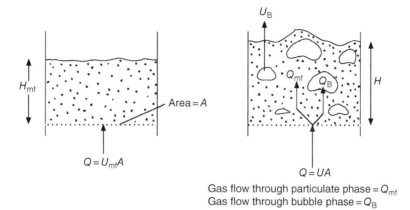

Gas flow through particulate phase = Q_{mf}
Gas flow through bubble phase = Q_B

Figure 6.12 Gas flows in a fluidized bed according to the two-phase theory

expansion of a Group A powder fluidized by air. Referring to Figure 6.12, Q is the actual gas flow rate to the fluid bed and Q_{mf} is the gas flow rate at minimum fluidization, then:

$$\text{gas passing through the bed as bubbles} = Q - Q_{mf} = (U - U_{mf})A \qquad (6.39)$$

$$\text{gas passing through the particulate phase } Q_{mf} = U_{mf}A \qquad (6.40)$$

Expressing the bed expansion in terms of the fraction of the bed occupied by bubbles, ε_B:

$$\varepsilon_B = \frac{H - H_{mf}}{H} = \frac{Q - Q_{mf}}{AU_B} = \frac{(U - U_{mf})}{U_B} \qquad (6.41)$$

where H is the bed height at U, H_{mf} is the bed height at U_{mf} and U_B is the mean rise velocity of a bubble in the bed (obtained from correlations; see below). The void fraction of the particulate phase is taken to be that at minimum fluidization ε_{mf}. The mean bed void fraction is then given by:

$$(1 - \varepsilon) = (1 - \varepsilon_B)(1 - \varepsilon_{mf}) \qquad (6.42)$$

In practice, the elegant two-phase theory overestimates the volume of gas passing through the bed as bubbles (the visible bubble flow rate) and better estimates of bed expansion may be obtained by replacing $(Q - Q_{mf})$ in Equation (6.41) with:

$$\text{visible bubble flow rate, } Q_B = YA(U - U_{mf}) \qquad (6.43)$$

where,

$$0.8 < Y < 1.0 \text{ for Geldart Group A powders}$$

$$0.6 < Y < 0.8 \text{ for Geldart Group B powders}$$

$$0.25 < Y < 0.6 \text{ for Geldart Group D powders}$$

Strictly the equations should be written in terms of U_{mb} rather than U_{mf} and Q_{mb} rather than Q_{mf}, so that they are valid for both Group A and Group B powders. Here they have been written in their original form. In practice, however, it makes little difference, since in practical applications, both U_{mb} and U_{mf} are usually much smaller than the superficial fluidizing velocity, U [so $(U - U_{mf}) = (U - U_{mb})$]. In rare cases where the operating velocity is not much greater than U_{mb}, then U_{mb} should be used in place of U_{mf} in the equations.

The above analysis requires a knowledge of the bubble rise velocity U_B, which depends on the bubble size d_{Bv} and bed diameter D. The bubble diameter at a given height above the distributor depends on the orifice density in the distributor N, the distance above the distributor L and the excess gas velocity $(U - U_{mf})$.

For group B powders

Darton et al. (1977) suggest the following expression for bubble size:

$$d_{Bv} = \frac{0.54}{g^{0.2}}(U - U_{mf})^{0.4}\left(L + 4[A/N_{or}]^{0.5}\right)^{0.8} \tag{6.44}$$

And for the bubble rise velocity, Harrison and Davidson (1963) suggest:

$$U_B = U - U_{mf} + 0.71(gd_{Bv})^{0.5} \tag{6.45}$$

For group A powders

Bubbles reach a maximum stable size which may be estimated from (Geldart, 1992):

$$d_{Bvmax} = \frac{2U_{T2.7}^2}{g} \tag{6.46}$$

where $U_{T2.7}$ is the terminal free fall velocity for particles of diameter 2.7 times the actual mean particle diameter.

Bubble velocity for Group A powders is given by:

$$U_B = \Phi_A\left(gd_{Bv}\right)^{0.5}(\text{Werther}, 1983) \tag{6.47}$$

where

$$\left\{\begin{array}{lll}\Phi_A = 1 & \text{for} & D \leq 0.1\,\text{m} \\ \Phi_A = 2.5D^{0.4} & \text{for} & 0.1 < D \leq 1\,\text{m} \\ \Phi_A = 2.5 & \text{for} & D > 1\,\text{m}\end{array}\right\} \tag{6.48}$$

6.2.6   Entrainment

The term *entrainment* will be used here to describe the ejection of particles from the surface of a bubbling bed and their removal from the vessel in the fluidizing gas. In the literature on the subject other terms such as *carry-over* and *elutriation* are often used to describe the same process. In this section we will study the factors affecting the rate of entrainment of solids from a fluidized bed and develop a simple approach to the estimation of the entrainment rate and the size distribution of entrained solids.

Consider a single particle falling under gravity in a static gas in an infinite space, i.e. not near any solid boundaries. We know that this particle will reach a terminal velocity when the forces of gravity, buoyancy and drag are balanced (see Chapter 3). If the gas is now considered to be moving upwards at a velocity equal to the terminal velocity of the particle, the particle will be stationary with respect to the observer. If the gas is moving upwards within a pipe, however, at a superficial velocity equal to the particle's terminal velocity, then:

(a) In laminar flow: the particle may move up or down depending on its radial position because of the parabolic velocity profile of the gas in the pipe.

(b) In turbulent flow: the particle may move up or down depending on its radial position. In addition, the random velocity fluctuations superimposed on the time-averaged velocity profile make the actual particle motion less predictable.

If we now introduce into the moving gas stream a number of particles with a range of particle sizes, some particles may fall and some may rise depending on their size and their radial position. Thus, the entrainment of particles in an upward-flowing gas stream is a complex process. We can see that the rate of entrainment and the size distribution of entrained particles will in general depend on particle size and density, gas properties, gas velocity, gas flow regime (including radial velocity profile and fluctuations) and vessel diameter. In addition (i) the mechanisms by which the particles are ejected into the gas stream from the fluidized bed are dependent on the characteristics of the bed – in particular bubble size and velocity at the surface, and (ii) the gas velocity profile immediately above the bed surface is distorted by the bursting bubbles. It is not surprising then that prediction of entrainment from first principles is not possible and in practice an empirical approach must be adopted.

This empirical approach defines *coarse* particles as particles whose terminal velocities are greater than the superficial gas velocity ($U_T > U$) and *fine* particles as those for which $U_T < U$, and it considers the region above the fluidized bed surface to be composed of several zones, as shown in Figure 6.13:

- *Freeboard*: Region between the bed surface and the gas outlet.

- *Splash zone*: Region just above the bed surface in which coarse particles fall back down.

- *Disengagement zone*: Region above the splash zone in which the upward flux and suspension concentration of fine particles decrease with increasing height.

- *Dilute-phase transport zone*: Region above the disengagement zone in which all particles are carried upwards; particle flux and suspension concentration are constant with height.

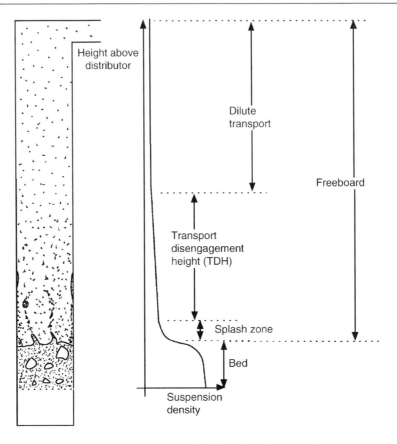

Figure 6.13   Zones in fluidized bed freeboard

Note that, although in general fine particles will be entrained and leave the system and coarse particles will remain, in practice some fine particles may stay in the system and some coarse particles may be entrained. Clustering may cause fine particles to stay in the system – especially for the cohesive Group C particles. Also, recalling that the terms *coarse* and *fine* are defined relative to the *superficial* gas velocity, there is actually a range of gas velocities in the freeboard – very low velocities near the walls allowing fine particles to shelter there and higher velocities in the middle of the freeboard which may be responsible for the carry-over of coarse particles.

The height from the bed surface to the top of the disengagement zone is known as the *transport disengagement height (TDH)*. Above TDH the entrainment flux and concentration of particles are constant. Thus, from the design point of view, in order to gain maximum benefit from the effect of gravity in the freeboard, the gas exit should be placed above the TDH. Many empirical correlations for TDH are available in the literature; those of Horio et al. (1980) presented in Equation (6.49) and Zenz (1983) presented graphically in Figure 6.14 are included here.

$$\text{TDH} = 4.47 d_{\text{bvs}}^{0.5} \qquad\qquad (6.49)$$

where d_{bvs} is the equivalent volume diameter of a bubble at the surface.

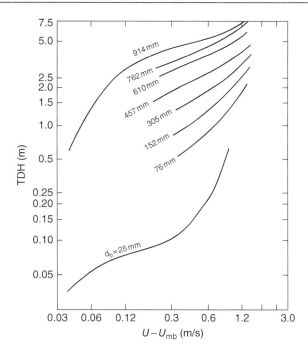

Figure 6.14 Graph for determination of transport disengagement height after the method of Zenz (1983). Adapted from Zenz (1983)

Any of the available correlations for TDH should be used with caution since their predictions can vary by as much as five orders of magnitude for the same conditions (Cahyadi et al., 2012).

The empirical estimation of entrainment rates from fluidized beds is based on the following rather intuitive equation:

$$\left(\begin{array}{c} \text{instantaneous rate of loss} \\ \text{of solids of size } x_i \end{array} \right) \propto \text{bed area} \times \left(\begin{array}{c} \text{fraction of bed with} \\ \text{size } x_i \text{ at time t} \end{array} \right)$$

$$\text{i.e. } R_i = -\frac{d}{dt}(M_B m_{Bi}) = K^*_{i\,h} A m_{Bi} \tag{6.50}$$

where $K^*_{i\,\mathrm{h}}$ is the elutriation rate constant (the entrainment flux at height h above the bed surface for the solids of size x_i, when $m_{Bi} = 1.0$), M_B is the total mass of solids in the bed, A is the area of bed surface and m_{Bi} is the fraction of the bed mass with size x_i at time t.

For continuous operation, m_{Bi} and M_B are constant and so:

$$R_i = K^*_{\mathrm{ih}} A m_{Bi} \tag{6.51}$$

and

$$\text{total rate of entrainment, } R_{\mathrm{T}} = \sum R_i = \sum K^*_{\mathrm{ih}} A m_{Bi} \tag{6.52}$$

The solids loading of size x_i in the off-gases is $\rho_i = R_i/UA$ and the total solids loading of the gas leaving the freeboard is $\rho_T = \sum \rho_i$.

For batch operation, the rates of entrainment of each size range, the total entrainment rate and the particle size distribution of the bed change with time. The problem can best be solved by writing Equation (6.50) in a finite increment form:

$$-\Delta(m_{Bi}M_B) = K^*_{ih}Am_{Bi}\Delta t \tag{6.53}$$

where $\Delta(m_{Bi}M_B)$ is the mass of solids in size range i entrained in time increment Δt.

$$\text{Total mass entrained in time } \Delta t = \sum_{i=1}^{k}[\Delta(m_{Bi}M_B)] \tag{6.54}$$

and mass of solids remaining in the bed at time:

$$t + \Delta t = (M_B)_t - \sum_{i=1}^{k}\left[\Delta(m_{Bi}M_B)_t\right] \tag{6.55}$$

where subscript t refers to the value at time t.

$$\text{Bed composition at time } t + \Delta t = (m_{Bi})_{t+\Delta t} = \frac{(m_{Bi}M_B)_t - [\Delta(m_{Bi}M_B)_t]}{(M_B)_t - \sum_{i=1}^{k}\{\Delta(m_{Bi}M_B)_t\}} \tag{6.56}$$

Solution of a batch entrainment problem proceeds by sequential application of Equations (6.53)–(6.56) for the required time period.

The elutriation rate constant K^*_{ih} cannot be predicted from first principles and so it is necessary to rely on the available correlations which differ significantly in their predictions. Correlations are usually in terms of the carry-over rate above TDH, $K^*_{i\infty}$. Two of the more reliable correlations are given below.

Geldart et al. (1979) (for particles >100 µm and $U > 1.2$ m/s):

$$\frac{K^*_{i\infty}}{\rho_g U} = 23.7\exp\left(-5.4\frac{U_{Ti}}{U}\right) \tag{6.57}$$

Zenz and Weil (1958) (for particles <100 µm and $U < 1.2$ m/s):

$$\frac{K^*_{i\infty}}{\rho_g U} = \begin{cases} 1.26 \times 10^7 \left(\dfrac{U^2}{gx_i\rho_p^2}\right)^{1.88} & \text{when } \left(\dfrac{U^2}{gx_i\rho_p^2}\right) < 3 \times 10^{-4} \\ 4.31 \times 10^4 \left(\dfrac{U^2}{gx_i\rho_p^2}\right)^{1.18} & \text{when } \left(\dfrac{U^2}{gx_i\rho_p^2}\right) > 3 \times 10^{-4} \end{cases} \tag{6.58}$$

6.2.7 Heat Transfer in Fluidized Beds

The transfer of heat between fluidized solids, gas and internal surfaces of equipment is very good. This makes for uniform temperatures and ease of control of bed temperature, which are important advantages of fluidized beds as chemical reactors, for example.

Gas–particle heat transfer

Gas–particle heat transfer coefficients are typically small, of the order of 5–20 W/m^2K. However, because of the very large heat transfer surface area provided by a mass of small particles (1 m^3 of 100 μm particles has a surface area of 60 000 m^2), the heat transfer between gas and particles is rarely limiting in fluid bed heat transfer. One of the most commonly used correlations for gas–particle heat transfer coefficient is that of Kunii and Levenspiel (1991):

$$Nu = 0.03\, Re_p^{1.3}\ (Rep < 50) \tag{6.59}$$

where Nu is the Nusselt number [$h_{gp}x/k_g$] and the single particle Reynolds number is based on the relative velocity between fluid and particle as in Chapter 3.

Gas–particle heat transfer is relevant where a hot fluidized bed is fluidized by cold gas. The fact that particle–gas heat transfer presents little resistance in bubbling fluidized beds can be demonstrated by the following example:

Consider a fluidized bed of solids held at a constant temperature T_s. Hot fluidizing gas at temperature T_{g0} enters the bed. At what distance above the distributor is the difference between the inlet gas temperature and the bed solids temperature reduced to half its original value?

Consider an element of the bed of height δL at a distance L above the distributor (Figure 6.15). Let the temperature of the gas entering this element be T_g and the change in gas temperature across the element be δT_g. The particle temperature in the element is T_s:

Neglecting heat losses through the wall, the energy balance across the element gives:

rate of heat loss from the gas = rate of heat transfer to the solids

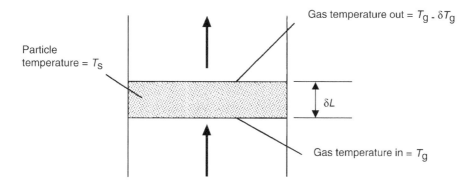

Figure 6.15 Analysis of gas–particle heat transfer in an element of a fluidized bed

Or

$$-\left(C_g U \rho_g\right)\mathrm{d}T_g = h_{gp}a\left(T_g - T_s\right)\mathrm{d}L \tag{6.60}$$

where a is the surface area of solids per unit volume of bed, C_g is the specific heat capacity of the gas, ρ_p is particle density, h_{gp} is the particle–gas heat transfer coefficient and U is superficial gas velocity.

Integrating with the boundary condition $T_g = T_{g0}$ at $L = 0$:

$$\ln\left(\frac{T_g - T_s}{T_{g0} - T_s}\right) = -\left(\frac{h_{gp}a}{U_{rel}\rho_g C_g}\right)L \tag{6.61}$$

The distance over which the temperature difference is reduced to half its initial value, $L_{0.5}$, is then:

$$L_{05} = -\ln(0.5)\frac{C_g U_{rel}\rho_g}{h_{gp}a} = 0.693\frac{C_g U_{rel}\rho_g}{h_{gp}a} \tag{6.62}$$

For a bed of spherical particles of diameter x, the surface area per unit volume of bed, $a = 6(1-\varepsilon)/x$, where ε is the bed void fraction.

Using the correlation for h_{gp} in Equation (6.59), then:

$$L_{0.5} = 3.85\frac{\mu^{1.3}x^{0.7}C_g}{U_{rel}^{0.3}\rho_g^{0.3}(1-\varepsilon)k_g} \tag{6.63}$$

As an example, we will take a bed of particles of mean size 100 μm, particle density 2500 kg/m³, fluidized by air of density 1.2 kg/m³, viscosity 1.84×10^{-5} Pa s, conductivity 0.0262 W/m/K and specific heat capacity 1005 J/kg/K.

Using the Baeyens and Geldart equation for U_{mf} [Equation (6.27)], $U_{mf} = 9.3 \times 10^{-3}$ m/s. The relative velocity between particles and gas under fluidized conditions can be approximated as U_{mf}/ε under these conditions.

Hence, assuming a fluidized bed void fraction of 0.47, $U_{rel} = 0.02$ m/s.

Substituting these values in Equation (6.63), we find $L_{0.5} = 0.95$ mm. So, within 1 mm of entering the bed the difference in temperature between the gas and the bed will be reduced by half. Typically for particles less than 1 mm in diameter the temperature difference between the hot bed and cold fluidizing gas would be reduced by half within the first 5 mm of the bed depth.

Bed-surface heat transfer

In a bubbling fluidized bed, the coefficient of heat transfer between the bed and any immersed surfaces (vertical bed walls or tubes) can be considered to be made up of three components which are approximately additive (Botterill, 1975).

$$\text{bed-surface heat transfer coefficient}, h = h_{pc} + h_{gc} + h_r$$

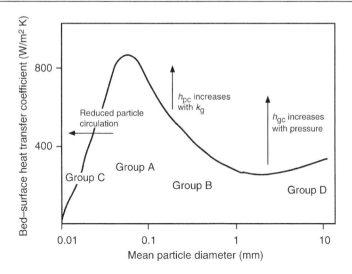

Figure 6.16 Range of bed-surface heat transfer coefficients. Botterill (1986) / John Wiley & Sons

where h_{pc} is the particle convective heat transfer coefficient and describes the heat transfer due to the motion of packets of solids carrying heat to and from the surface, h_{gc} is the gas convective heat transfer coefficient describing the transfer of heat by motion of the gas between the particles and h_r is the radiant heat transfer coefficient. Figure 6.16 gives an indication of the range of bed-surface heat transfer coefficients and the effect of particle size on the dominant heat transfer mechanism.

The largest contributor to heat transfer between the bed and immersed surfaces is usually *particle convective heat transfer*. On a volumetric basis the solids in the fluidized bed have about 1000 times the heat capacity of the gas and so, since the solids are continuously circulating within the bed, they transport the heat around the bed and to or from surfaces. Bubble motion ensures that solids adjacent to surfaces are frequently replaced. For heat transfer between the bed and a surface the limiting factor is the gas conductivity, since all the heat must ultimately be transferred through a gas film between the particles and the surface (Figure 6.17). The particle–surface contact area

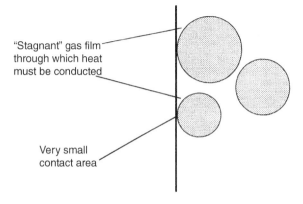

Figure 6.17 Heat transfer from bed particles to an immersed surface

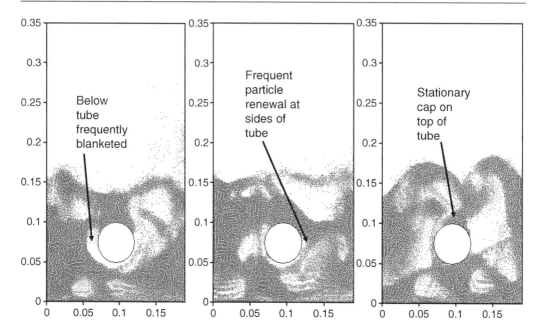

Figure 6.18 Particle motion around a horizontal heat transfer tube: discrete element simulation showing blanketing and capping of the tube (Y.S.Wong, PhD thesis, University of Birmingham, 2004)

is too small to allow significant heat transfer by direct conduction through the solids. Factors affecting the gas film thickness or the gas conductivity will therefore influence the heat transfer under particle convective conditions. Decreasing particle size, for example, decreases the mean gas film thickness and so improves h_{pc}. However, reducing particle size into the Group C range will reduce particle mobility and so reduce particle convective heat transfer. Increasing gas temperature increases gas conductivity and so improves h_{pc}.

Particle convective heat transfer is dominant in Groups A and B powders. Increasing gas velocity beyond minimum fluidization improves particle circulation and so increases particle convective heat transfer. The heat transfer coefficient increases with fluidizing velocity up to a broad maximum h_{max} and then declines as the heat transfer surface becomes blanketed by bubbles (see Figure 6.18 for images from a DEM simulation). This is shown in Figure 6.19 for powders in Groups A, B and D. The maximum in h_{pc} occurs relatively closer to U_{mf} for Groups B and D powders since these powders give rise to bubbles at U_{mf} and the size of these bubbles increases with increasing gas velocity. As noted earlier, Group A powders exhibit non-bubbling fluidization between U_{mf} and U_{mb} and achieve a maximum stable bubble size.

Botterill (1986) recommends the Zabrodsky (1966) correlation for h_{max} for Group B powders:

$$h_{max} = 35.8 \frac{k_g^{0.6} \rho_p^{0.2}}{x^{0.36}} \quad \text{W/m}^2/\text{K} \tag{6.64}$$

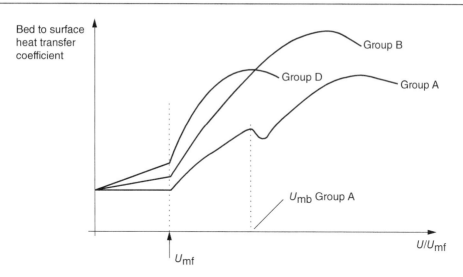

Figure 6.19 Effect of fluidizing gas velocity on bed-surface heat transfer coefficient in a fluidized bed (Botterill, 1986 / John Wiley & Sons)

and the correlation of Khan et al. (1978) for Group A powders:

$$Nu_{max} = 0.157 Ar^{0.475} \tag{6.65}$$

Gas convective heat transfer is not important in Groups A and B powders where the flow of interstitial gas is laminar but becomes significant in Group D powders, which fluidize at higher velocities and give rise to transitional or turbulent flow of interstitial gas.

Based on theoretical modelling and extensive experimental evidence, Molerus (2000) concluded that the relative importance of particle convective heat transfer and gas convective heat transfer between submerged surfaces and a fluidized bed was dependent on Archimedes number, Ar. Molerus suggested that for $Ar < 10^2$, the heat transfer mechanism was purely particle convective and gave the expression:

$$\frac{h_{pc}L_l}{k_f} = \frac{0.19(1-\varepsilon_{mf})}{1+0.5Pr^{-1}} \times \frac{1}{1 + 25\left[\left(\frac{U-U_{mf}}{U_{mf}}\right)^{1/3}\left(\frac{\rho_p C_p}{k_f g}\right)^{1/3}(U-U_{mf})\right]^{-1}} \tag{6.66}$$

For $Ar > 10^5$, Molerus showed that gas convective heat transfer was the dominant mechanism and recommended the expression:

$$\frac{h_{gc}L_t}{k_f} = 0.165 Pr^{1/3}\left(\frac{\rho_f}{\rho_p-\rho_f}\right)^{1/3}\left\{1 + 0.05\frac{U_{mf}}{U-U_{mf}}\right\}^{-1} \tag{6.67}$$

For the intermediate regime ($10^2 < Ar < 10^5$), Molerus presented a strategy for interpolation between the two extremes of purely particle convective at low Ar values and purely gas convective at high Ar values. Short residence times at the heat transfer surface result in gas convective behaviour and long residence times result in particle convective behaviour. Residence times at the surfaces decrease with increasing excess gas velocity due to increased bubble frequency. Increasing particle size, and hence U_{mf}, also shifts the behaviour from particle convective to gas convective. In combining the expressions for purely gas convective and purely particle convective heat transfer, Molerus introduced a damping function to take into account the observation that particle convective heat transfer must decrease with increasing U_{mf} and also with increasing excess gas velocity (U–U_{mf}). Molerus suggested the following expression for heat transfer in this intermediate regime, which although complex is justified by its predictive capability over a wide range of particle and gas properties.

$$\frac{hL_i}{k_f} = \text{Damped Particle Convective term} + \text{Gas Convective term} \qquad (6.68)$$

Damped Particle Convective term:

$$\frac{0.125\left(1-\varepsilon_{mf}\right)\left\{1+33.3\left(\left[\frac{U-U_{mf}}{U_{mf}}\right]^{1/3}\left[\frac{\rho_p C_p}{k_f g}\right]^{1/3}\left(U-U_{mf}\right)\right)^{-1}\right\}^{-1}}{1+0.5Pr^{-1}\left\{1+0.28\left(1-\varepsilon_{mf}\right)^2\left(\frac{\rho_f}{\rho_p-\rho_f}\right)^{0.5}\left(\left[\frac{\rho_p C_p}{k_f g}\right]^{1/3}\left(U-U_{mf}\right)\right)^2\left(\frac{U_{mf}}{U-U_{mf}}\right)\right\}} \qquad (6.69)$$

Gas Convective term:

$$0.165Pr^{1/3}\left(\frac{\rho_f}{\rho_p-\rho_f}\right)^{1/3}\left\{1+0.05\frac{U_{mf}}{U-U_{mf}}\right\}^{-1} \qquad (6.70)$$

where

h = combined heat transfer coefficient (excluding radiation)		W/m^{2}K
L_l = laminar flow length scale, $\left[\dfrac{\mu}{\left(\rho_p-\rho_f\right)\sqrt{g}}\right]^{2/3}$		m
L_t = turbulent flow length scale, $\left[\dfrac{\mu^2}{\rho_f\left(\rho_p-\rho_f\right)g}\right]^{1/3}$		m
C_p = specific heat capacity of particle material		J/kg.K
k_f = thermal conductivity of the gas		W/m K.

For temperatures beyond 600°C radiative heat transfer plays an increasing role and must be accounted for in calculations. The reader is referred to Botterill (1986) or Kunii and Levenspiel (1991) for the treatment of radiative heat transfer or for a more detailed look at heat transfer in fluidized beds.

6.2.8 A Simple Model for the Bubbling Fluidized Bed Reactor

As mentioned earlier, fluidized beds are attractive for carrying out certain kinds of chemical reaction because they are generally well mixed, heat transfer between the solids and in-bed surfaces is high and it is relatively easy to add and withdraw solids and gas continuously. In general, models for the fluidized bed reactor consider:

- The division of gas between the bubble phase and the particulate phase.

- The degree of mixing in the particulate phase.

- The transfer of gas between the phases.

It is outside the scope of this chapter to review in detail the models available for the fluidized bed as a reactor. However, in order to demonstrate the key components of such models, we will use the model of Orcutt et al. (1962). Although simple, this model allows the key features of a fluidized bed reactor for a gas-phase catalytic reaction to be explored.
The approach assumes the following:

- The two-phase theory applies.

- Perfect mixing takes place in the particulate phase.

- There is no reaction in the bubble phase, which is justified by the absence of catalyst particles within the bubbles.

The model is one-dimensional and assumes a steady state. The structure of the model is shown diagrammatically in Figure 6.20 The following notation is used: C_0 is the concentration of reactant at the distributor; C_p is the concentration of reactant in the particulate phase; C_B is the concentration of the reactant in the bubble phase at height h above the distributor; C_{BH} is the concentration of reactant leaving the bubble phase; and C_H is the concentration of reactant leaving the reactor.
In steady state, the concentration of reactant is constant throughout the particulate phase because of the assumption of perfect mixing within it. Throughout the bed, the gaseous reactant is assumed to pass between the particulate phase and the bubble phase, since, as noted in Section 6.2.3, the bubbles have permeable interfaces with the particulate phase.
The overall mass balance on the reactant is:

$$
\begin{pmatrix} \text{molar flow of} \\ \text{reactant into} \\ \text{reactor} \\ (1) \end{pmatrix} = \begin{pmatrix} \text{molar flow out} \\ \text{in the bubble phase} \\ (2) \end{pmatrix} + \begin{pmatrix} \text{molar flow out in} \\ \text{the particulate phase} \\ (3) \end{pmatrix} + \begin{pmatrix} \text{rate of} \\ \text{conversion} \\ (4) \end{pmatrix}
$$

$$(6.71)$$

Term (1) = UAC_0

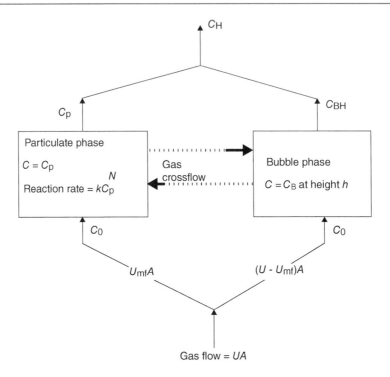

Figure 6.20   Schematic diagram of the Orcutt fluidized bed reactor model

Term (2): molar reactant flow in the bubble phase changes with height L above the distributor as gas is exchanged with the particulate phase. Consider an element of bed of thickness ΔL at a height L above the distributor. In this element:

$$\left(\begin{array}{c} \text{rate of increase of} \\ \text{reactant in bubble phase} \end{array} \right) = \left(\begin{array}{c} \text{rate of transfer of} \\ \text{reactant from particulate phase} \end{array} \right) \qquad (6.72)$$

$$\text{i.e.} (U - U_{\mathrm{mf}}) A \Delta C_{\mathrm{B}} = - K_{\mathrm{C}} (\varepsilon_{\mathrm{B}} A \Delta L)(C_{\mathrm{B}} - C_{\mathrm{p}})$$

in the limit as $\Delta L \rightarrow 0$:

$$\frac{\mathrm{d} C_{\mathrm{B}}}{\mathrm{d} L} = - \frac{K_{\mathrm{C}} \varepsilon_{\mathrm{B}} (C_{\mathrm{B}} - C_{\mathrm{p}})}{(U - U_{\mathrm{mf}})} \qquad (6.73)$$

where K_{C} is the mass transfer coefficient per unit bubble volume and ε_{B} is the bubble fraction. Integrating with the boundary condition that $C_{\mathrm{B}} = C_0$ at $L_0 = 0$:

$$C_{B} = C_{P} + (C_0 - C_P) \exp\left(- \frac{K_{C} L}{U_{B}} \right) \qquad (6.74)$$

since $\varepsilon_{\mathrm{B}} = (U - U_{\mathrm{mf}})/U_{\mathrm{B}}$ [Equation (6.41)].

At the surface of the bed, $L = H$ and so the reactant concentration in the bubble phase at the bed surface is given by:

$$C_{BH} = C_p + (C_0 - C_p) \exp\left(-\frac{K_C H}{U_B}\right)$$ (6.75)

Term (2) $= C_{BH}(U - U_{mf})A$
Term (3) $= U_{mf}AC_p$
Term (4): For a reaction which is jth order in the reactant under consideration,

$$\left(\begin{array}{c} \text{molar rate of conversion} \\ \text{per unit volume of solids} \end{array}\right) = kC_p^j$$

where k is the reaction rate constant per unit volume of solids.
Therefore,

$$\left(\begin{array}{c} \text{molar rate of} \\ \text{conversion in bed} \end{array}\right) = \left(\begin{array}{c} \text{molar rate of} \\ \text{conversion per unit} \\ \text{volume of solids} \end{array}\right) \times \left(\begin{array}{c} \text{volume of solids} \\ \text{per unit volume of} \\ \text{particulate phase} \end{array}\right)$$

$$\times \left(\begin{array}{c} \text{volume of particulate} \\ \text{phase per unit} \\ \text{volume of bed} \end{array}\right) \times \left(\begin{array}{c} \text{volume} \\ \text{of bed} \end{array}\right)$$

Hence, term (4),

$$\left(\begin{array}{c} \text{molar rate of} \\ \text{conversion in bed} \end{array}\right) = kC_p^j (1 - \varepsilon_p)(1 - \varepsilon_B)AH$$ (6.76)

where ε_p is the particulate phase void fraction.
Substituting these expressions for terms (1)–(4), the mass balance becomes:

$$UAC_0 = \left[C_p + (C_0 - C_p) \exp\left(-\frac{K_C H}{U_B}\right)\right](U - U_{mf})A + U_{mf}AC_p + kC_p^j (1 - \varepsilon_p)(1 - \varepsilon_B)AH$$ (6.77)

From this mass balance C_p may be found. The reactant concentration leaving the reactor C_H is then calculated from the reactant concentrations and gas flows through the bubble and particulate phases:

$$C_H = \frac{U_{mf}C_p + (U - U_{mf})C_{BH}}{U}$$ (6.78)

In the case of a first-order reaction ($j = 1$), solving the mass balance for C_p gives:

$$C_p = \frac{C_0[U - (U - U_{mf})e^{-\chi}]}{kH_{mf}(1 - \varepsilon_p) + [U - (U - U_{mf})e^{-\chi}]}$$ (6.79)

where $\chi = K_C H / U_B$, equivalent to a number of mass transfer units for gas exchange between the phases. It may be imagined as indicating the number of times the gas volume within a bubble is replaced by throughflow from the surrounding particulate phase. χ is related to bubble size and correlations are available. Generally χ decreases as bubble size increases and so small bubbles are preferred.

Thus, from Equations (6.78) and (6.79), we obtain an expression for the conversion in the reactor:

$$1 - \frac{C_H}{C_0} = (1 - \beta e^{-\chi}) - \frac{(1 - \beta e^{-\chi})^2}{\dfrac{kH_{mf}(1 - \varepsilon_p)}{U} + (1 - \beta e^{-x})} \tag{6.80}$$

where $\beta = (U - U_{mf})/U$, the fraction of gas passing through the bed as bubbles. It is interesting to note that although the two-phase theory does not always hold exactly, Equation (6.80) often holds with β still taken as the fraction of gas passing through the bed as bubbles, even if not necessarily equal to $(U - U_{mf})/U$.

Readers interested in reactions of order different from unity, solids reactions and more complex reactor models for the fluidized bed, are referred to Kunii and Levenspiel (1991).

Although the Orcutt model is simple, it does allow us to explore the effects of operating conditions, reaction rate and degree of interphase mass transfer on the performance of a fluidized bed as a gas-phase catalytic reactor. Figure 6.21 shows the variation of conversion with reaction rate (expressed as $kH_{mf}[1 - \varepsilon_p/U]$) with excess gas velocity (expressed as β) calculated using Equation (6.80) for a first-order reaction.

Noting that the value of χ is dictated mainly by the bed hydrodynamics, we see that:

- *For slow reactions,* overall conversion is insensitive to bed hydrodynamics and so reaction rate k is the rate-controlling factor.

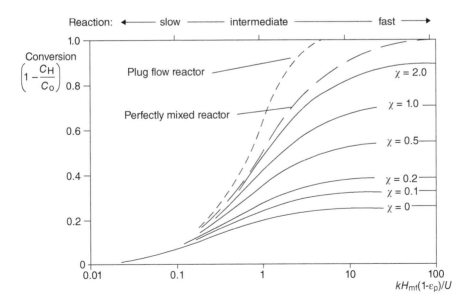

Figure 6.21 Conversion as a function of reaction rate and interphase mass transfer for $\beta = 0.75$ for a first-order gas-phase catalytic reaction [based on Equation (6.80)]

- *For intermediate reactions,* both reaction rate and bed hydrodynamics affect the conversion.

- *For fast reactions,* the conversion is determined by the bed hydrodynamics.

These results are typical for a gas-phase catalytic reaction in a fluidized bed and emphasize the importance of understanding *both* the chemical kinetics and the bed hydrodynamics; good conversion demands that both should be optimized.

6.2.9 High Velocity Fluidization

Increasing fluidizing velocity

What happens if we increase the flow rate of gas to a bubbling fluidized bed? Starting with a bubbling bed with some minimal entrainment of solids, there is a distinct interface between the dense region of the bubbling bed and the dilute region of the freeboard (Figure 6.22). An increase in the gas velocity causes the entrainment rate of solids to increase. To maintain the bed in the vessel, which we shall here call a *riser*, the entrained solids must be captured and continuously recycled. A further increase in gas velocity gives rise to a further increase in the rate of solids entrainment and the requirement for increased recycle or feed rate of solids to the base of the riser. At higher gas velocities a denser region may still exist at the base, but it will not have the characteristics of a bubbling bed – i.e. no discrete gas bubbles – and the interface between the dense and dilute regions becomes less distinct. This is often referred to as a *turbulent fluidization* regime (Figure 6.23) and is primarily exhibited by Geldart Group A powders. With a further increase in gas velocity and circulation flux of solids we see a more and more gradual change in suspension density with height – sometimes referred to as the S-shaped profile. Under these conditions, a stable coexistence of dense and dilute regions is possible. The minimum gas velocity for this to occur is termed the *transport velocity* for those solids. A *core-annulus flow structure* exists in both dilute and dense regions, with the fraction of cross section occupied by the core increasing with height. In the dilute region there

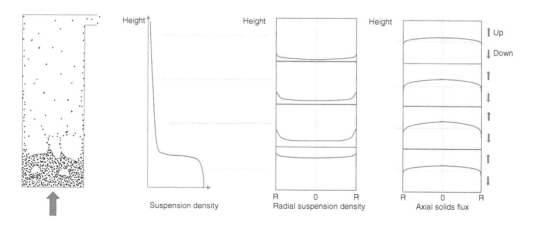

Figure 6.22 Bubbling fluidization

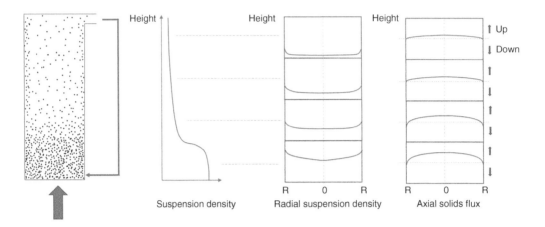

Figure 6.23   Turbulent fluidization

is solids downflow in the annulus and upflow in the core. In the dense flow region solids flow upwards in both core and annulus. In the interface between the two regions, solids from the annulus flow inwards to the core (Rhodes, Wang and Sollaart, 1998). This regime, characterized by core-annulus flow, stable co-existence of dense and dilute regions and the S-shaped axial suspension density profile is known as the *fast fluidization* regime, after Yerushalmi et al. (1976) – Figure 6.24. In the fast fluidization regime, an increase in solids circulation flux or a decrease in gas velocity results in an increase in the proportion of the riser height occupied by the dense flow region (Figure 6.25). Starting in the fast fluidization regime and increasing solids flux at constant gas velocity, the dense region may eventually occupy the entire height of the riser. This is known as the *dense-phase upflow* regime (Bi and Grace, 1999). In this regime, radial profiles of solids flux and suspension density are still quite strong, but all flows are upwards – Figure 6.26). Starting in the *fast fluidization* regime and increasing gas velocity at constant solids flux, the proportion of riser height occupied by the dense region diminishes. At a sufficiently

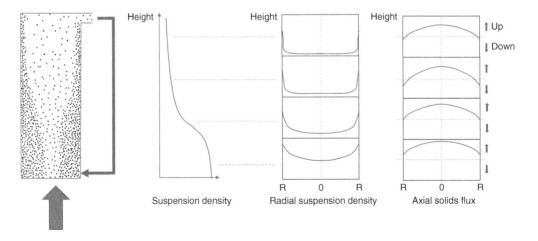

Figure 6.24   Fast fluidization

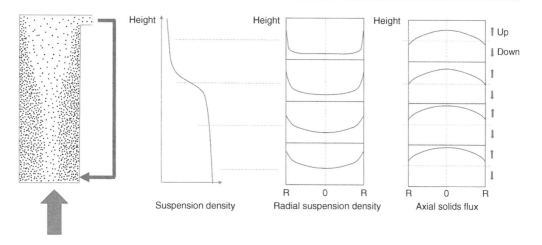

Figure 6.25   Fast fluidization at higher solids flux

high velocity the dense region disappears, and the entire height of the riser is occupied by a dilute region. This is the *dilute phase pneumatic transport* regime, where there is very little change in suspension density with height, little or no downward flux of solids near the walls, and a much flatter radial suspension density profile. The structure may be core-annulus flow with downward flux of solids near to the wall (core-annular dilute-phase flow – Figure 6.27) or, at high velocities, uniform upflow with no downward flux at the wall (homogeneous dilute-phase flow – Figure 6.28). The velocity at the boundary between the *fast fluidization* regime and the *dilute phase pneumatic transport* regime is often referred to as the *choking velocity* (or type A choking velocity, Bi and Grace, 1995). Figure 6.29 summarizes the fluidization regime changes with changes in gas velocity and solids flux.

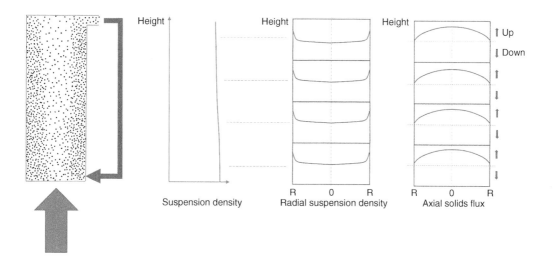

Figure 6.26   Dense-phase upflow

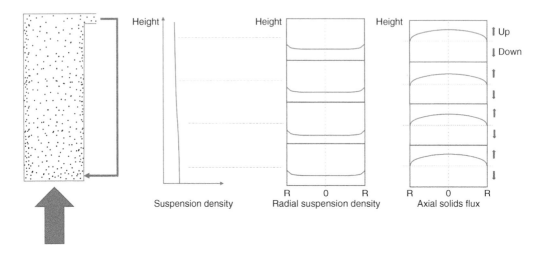

Figure 6.27   Dilute-phase core-annular flow

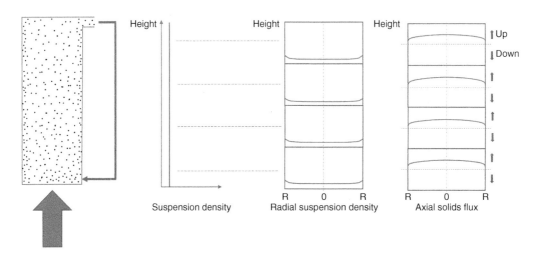

Figure 6.28   Dilute-phase homogenous flow

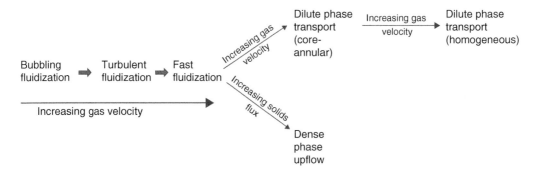

Figure 6.29   Fluidization regime changes with gas velocity and solids flux

Fluidization regime diagrams

The purpose of the regime diagram is to predict, for particular particle properties, the regime likely to be achieved for a given combination of gas velocity and imposed solids flux. Over the years there have been many attempts to show how the regimes of fluidization and transport are linked (for example, Grace, 1986; Leung, 1980; Squires et al., 1985; Rhodes, 1989; Hirama et al., 1992; Zenz, 1949). There are several forms of regime diagrams relevant to fluidization. Here we will focus on three. One type of diagram shows the types of expected behaviour on a plot of dimensionless gas velocity $(Re_p/Ar^{1/3})$ versus dimensionless particle size $(Ar^{1/3})$. The original diagram (Grace, 1986), reproduced in Figure 6.30, identified the regions where conventional fluidized beds, circulating fluidized beds, transport reactors and dilute transport operate. This form of diagram has been adapted (e.g. Bi and Grace, 1995; Grace et al., 2020) to show turbulent fluidization and fast fluidization. The diagram is helpful

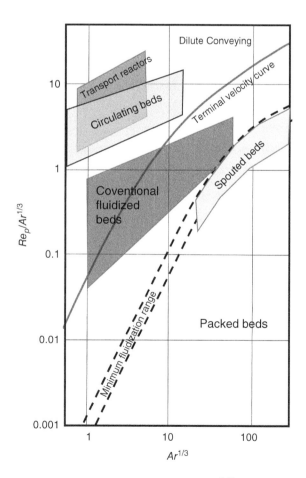

Figure 6.30 Flow regime diagram after Grace (1986). $Re_p/Ar^{1/3}$ is the dimensionless gas velocity $(U[(\rho_f^2)/(\rho_p - \rho_f)g\mu]^{1/3})$. $Ar^{1/3}$ is the dimensionless particle size $(x[(\rho_f (\rho_p - \rho_f)g)/\mu^2]^{1/3})$

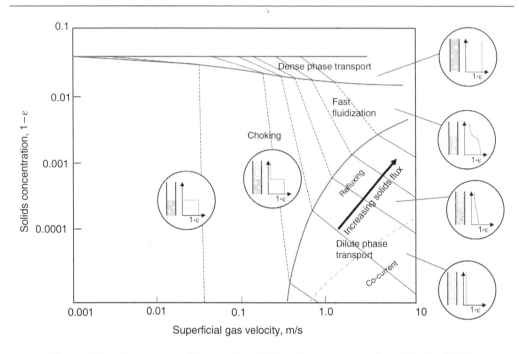

Figure 6.31   Flow regime diagram for Geldart Group A particles - Rhodes (1989).

in giving the big picture but has the limitation that it does not show the effect of imposed solids flux on the regimes and transitions between them.

A second type of diagram plots gas velocity versus suspension density with imposed solids flux as a parameter (for example, Squires et al., 1985, Rhodes, 1989). The diagram of Rhodes 1989 is reproduced in Figure 6.31 for Geldart Group A particles.

A third type is the plot of superficial solids flux versus superficial gas velocity (e.g. Bi and Grace, 1995) – designed to give an indication of the regime achieved under given conditions. Figure 6.32 is such a diagram modified to indicate where the dense-phase upflow regime might be located.

6.2.10   Some Practical Considerations

Gas distributor

The distributor is a device designed to ensure that the fluidizing gas is always uniformly distributed across the cross section of the bed. It is a critical part of the design of a fluidized bed system and many operating problems can be traced back to poor gas distribution. Good design is based on achieving a pressure drop which is a sufficient fraction of the bed pressure drop; readers are referred to Geldart (1986) for guidelines. It should be noted that while the pressure difference across a fluidized bed is limited to the weight per unit area of the contents (W/A; Section 6.2.1), the pressure difference across the

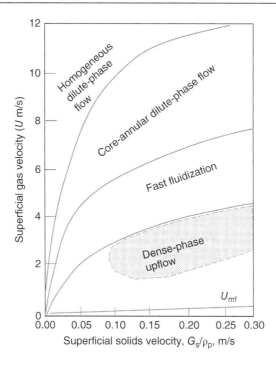

Figure 6.32 Flow regime diagram for Geldart Group A particles. Based on Bi and Grace (1995) and modified to include the approximate location of the dense-phase upflow regime. Adapted from Bi and Grace (1995)

distributor, which usually includes some sort of orifice, is generally proportional to the square of the superficial gas velocity (U^2). Therefore, a design which gives good distribution at higher gas velocities may not do so at lower ones. Some distributor designs in common use are shown in Figure 6.33.

Loss of fluidizing gas

Loss of fluidizing gas will lead to the collapse of the fluidized bed. If the process involves the emission of heat, then this heat will not be dissipated as well from the packed bed as it was from the fluidized bed. The consequences, particularly for an exothermic reaction, should be considered at the design stage.

Erosion

All parts of the fluidized bed unit are subject to erosion by the solid particles. Heat transfer tubes within the bed or the freeboard are particularly at risk and erosion here may lead to tube failure. Erosion of the distributor may lead to enlargement of the gas entry points and therefore poor fluidization, with areas of the bed becoming de-aerated.

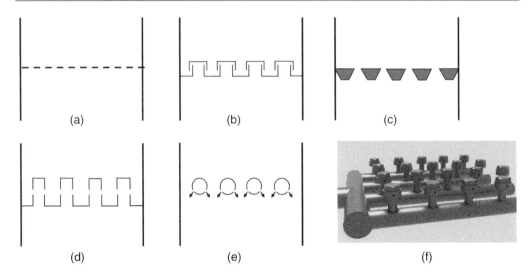

Figure 6.33 Some distributor designs in common industrial use: (a) drilled plate; (b) cap design; (c) continuous horizontal slots; (d) standpipe design; (e) sparge tube design with holes pointing downwards and allowing solids downflow and (f) sparge tube layout allowing solids downflow, with horizontal gas injection from nozzle elements; with acknowledgments to Dominik Werner

Attrition

Attrition of particles can occur at various locations in a fluidized bed, but particularly at gas entry points at or near the distributor and in ancillary equipment such as cyclones. Attrition is discussed in Chapter 11.

Loss of fine solids

Loss of fine solids from the bed reduces the quality of fluidization and reduces the area of contact between the solids and the gas in the process. In a catalytic process this means lower conversion.

Cyclones

Cyclone separators are often used in fluidized beds for separating entrained solids from the gas stream (see Chapter 8). Cyclones installed within the fluidized bed vessel would be fitted with a dip-leg and seal to prevent gas entering the solids exit. Fluidized systems may have two or more stages of cyclone in series to improve separation efficiency. Cyclones are also subject to erosion and must be designed to cope with this.

Solids feeders

Various devices are available for feeding solids into the fluidized bed. The choice of device depends largely on the nature of the solids feed and whether the bed is operated at elevated pressure and temperature. Screw conveyors, rotary valves, spray feeders, L-valves and pneumatic conveying are in common use.

6.2.11   Applications of Fluidized Beds

Fluidization technology has a surprisingly wide application in the process industries. In fact, fluidized bed reactors are at the heart of many of today's high-volume chemical production processes. Fluidized bed technology is also central to many unit operations, such as mixing, granulation, drying, heating and cooling, heat treatment and transport. The technology features in all the process industries: from oil and gas to pharmaceuticals and from mineral processing and power generation to the food industry.

The advantages of fluidized beds have been mentioned earlier in this chapter. The fluidized bed has excellent gas–solid contacting capabilities and large surface areas – good for gas–solid reactions, for heat transfer and for drying. The bubble motion in the bed creates good solids circulation, which gives rise to good mixing and good heat transfer. Good mixing and heat transfer in turn give rise to uniform temperature and ease of control. Since the solids are fluidized and therefore highly mobile they are able to be readily removed or transferred.

There are some disadvantages, however. The solids circulation patterns established in fluidized beds sometimes cause gas backing. Gas back mixing in gas-phase catalytic reactors gives rise to the possibility of secondary reactions resulting in undesired products. Because the gas passes through the bed as bubbles, under some conditions part of the gas can bypass the solids. Gas bypassing in gas-phase catalytic reactors reduces conversion, as demonstrated in the example on fluidized bed reactor design. Also, all fluidized beds, and particularly those using fine or friable materials, have the disadvantage that fine particles are entrained with the fluidizing gas as it leaves the bed. The entrained fines must be captured and returned to the bed or removed. This results in additional separation costs, and very small fines can evade capture devices and need subsequent removal from the product.

Fluid catalytic cracking

Fluid catalytic cracking (FCC) is probably the most celebrated application of fluidized bed technology. FCC was developed during the Second World War for cracking larger petroleum molecules into small molecules useful for gasoline and aviation fuel. More recent trends towards alternatives to fossil fuels, such as those derived from biological sources or recycling of waste plastics, still require cracking of large molecules into more useful smaller ones.

In the modern form of FCC a zeolite catalyst, typically a Geldart Group A powder, is circulated between a short contact time riser reactor (endothermic cracking reaction operating in the *fast fluidization* or *dense suspension upflow* regime) and the regenerator (exothermic reaction burning off carbon deposited in the cracking reaction operating in the *bubbling* bed or *turbulent fluidization* regime). See Figure 6.34 for an example of one type of FCC.

The FCC process typically operates at relatively low pressures (less than 3.5 atm), moderate temperatures (in the range 470–550°C) and cracker reactor residence times of less

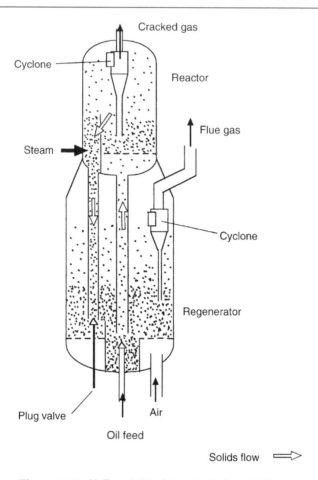

Figure 6.34   Kellogg's Model A Orthoflow FCC unit

than one second. The FCC is an enormous piece of process plant: over 30 m high and 8 m in diameter, with a catalyst circulation rate of 30 t/minute.

Acrylonitrile

Acrylonitrile is also made in a fluidized bed reactor. Acrylonitrile is the raw material used to make acrylic fibre (used to make clothing and carpeting), ABS plastic (which has many everyday applications) and nitrile rubber (which is used in the manufacture of hoses for pumping fuel).

The fluidized bed process for acrylonitrile production is based around the highly exo-thermic gas-phase catalytic reaction of propylene with oxygen and ammonia. Here again, the particulate fluidized solid is the catalyst. The fluidized bed, operating in the *bubbling* or *turbulent* regime, is favoured here because it ensures good gas-catalyst contact, uni-form temperature and ease of heat removal via immersed cooling coils.

Combustion

Fluidized bed technology has been applied to the combustion of solid fuels, biomass and waste materials. The earliest fluidized bed combustors used *bubbling* beds of inert sand (with Geldart Group B particles in the size range 200–1000 μm).

The solid fuel (usually in the form of particles of several millimetres to a few centimetres) is injected into this fluidized bed of sand. The fluidized bed provides good access to oxygen for the combustion process and a means of supplying and removing heat from the burning fuel particles.

In the 1980s and 1990s, circulating fluidized bed technology (CFB), which had already been applied to alumina calcination by the German engineering company Lurgi, was applied with great success to the combustion of solid fuels. CFBs replaced bubbling bed technology and allowed fluidized beds to be developed on much greater scales for power generation (up to 450 MWe). In the CFB combustor (Figure 6.35) the circulating solids are usually inert materials such as sand (typical mean size 175 μm) acting as a carrier for the solid fuel. The typical range of gas velocity is 4–6 m/s and the typical range of solids circulation flux would be 50–100 kg/m^2s and the regime of operation is *fast fluidization*. The heat of combustion is removed for steam generation via tubes which form part of the riser walls. Some designs also withdraw heat for steam generation through an external fluidized bed heat exchanger within the solids return leg. The CFB combustor is attractive because of its environmental advantages. The low operating temperature (typically 850°C) keeps NOx emissions low and the ability to add limestone directly to

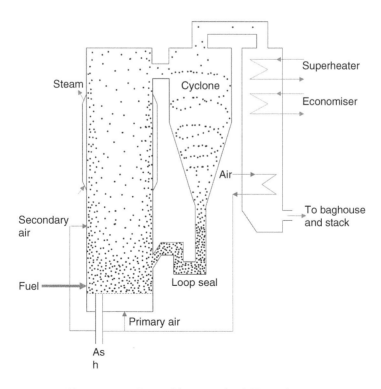

Figure 6.35 Typical layout of a CFB combustor

the combustor allows for sulphur to be removed as part of the process, thus maintaining low SOx emissions in the flue gases. The CFB combustor also has operating advantages over alternative technology – these include the ability to handle a range of fuels with little fuel preparation, good turndown capability, high bed–surface heat transfer coefficients, and small footprint for a given capacity.

Mineral processing

Bubbling fluidized beds are used in roasting of ores in the processes for the production of zinc, copper, gold and nickel. Circulating fluidized beds are used in the calcination of alumina, in gold ore roasting, in sulphur recovery from roaster off-gases and in the calcination of limestone and dolomite.

A typical simple fluidized bed roaster design operates in the *bubbling* regime at atmospheric pressure and at a gas velocity of around 0.5 m/s. The typical operating temperature is in the range 650–700 °C and these roasters, which may be up to 13 m in diameter and over 7 m high, may treat up to 700 t of ore per day.

Alumina (or aluminium oxide) is the starting point for aluminium production and the current global annual production is over 100 million tonnes per year. Much of the bauxite ore today is calcined in a circulating fluidized bed process operating in the *fast fluidization* regime – one of the very early applications of the CFB developed by Lurgi (see Reh, 1971).

Alumina leaving the calciner is often cooled in a bubbling fluidized bed cooler with water cooling (Figure 6.36).

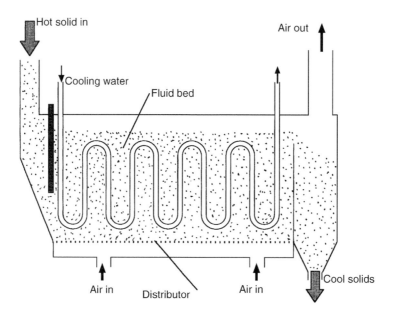

Figure 6.36   Schematic diagram of a fluidized bed solid cooler

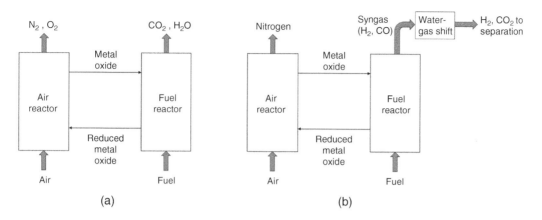

Figure 6.37   Chemical looping for (a) combustion with carbon capture, (b) hydrogen production

Chemical looping combustion

Chemical looping combustion (CLC) involves the use of a metal oxide to transfer oxygen from combustion air to the fuel, avoiding direct contact between air and fuel. A common form of CLC uses two interconnected fluidized beds – a fuel reactor and an air reactor (Figure 6.37). In the fuel reactor the metal oxide is reduced by reaction with the fuel, producing an outlet gas which is primarily carbon dioxide and water. The water is removed by condensation leaving carbon dioxide for storage. The metal oxide is continuously circulated to the air reactor where it is re-oxidized by contact with air. The exit gas from the air reactor is largely nitrogen with some unreacted oxygen. This technology was originally developed for the production of carbon dioxide, but is recently being applied to carbon capture in the combustion of solid, liquid and gaseous fuels. CLC has also been applied to the production of hydrogen from solid fuels with inherent carbon capture. The fuel reactor produces syngas (mostly hydrogen and carbon monoxide, with some methane and carbon dioxide) by partial oxidation, which, after further processing by a water-gas shift reaction, can be separated into hydrogen and carbon dioxide. Many other variants of chemical looping continue to be developed.

Gasification and pyrolysis

Gasification refers to processes in which a substance containing carbon reacts at high temperature (typically greater than 700 °C) in such a way that the quantity of oxygen is insufficient for combustion; the result is a gaseous mixture consisting mainly of carbon monoxide, carbon dioxide, hydrogen and nitrogen, which is sometimes known as *synthesis gas* or *syngas*. This can be used as a fuel or feedstock to subsequent chemical processes. *Pyrolysis* refers to the thermal degradation of a substance containing carbon in the absence of oxygen, which results in a mixture of hydrocarbons, usually including gases, liquids and solids. In each case, a wide variety of starting materials can be used, particularly including waste products. In the case of pyrolysis of plastic waste, the product could in principle be used as a feedstock to make more plastic, an example of a *circular economy* process. Both gasification and pyrolysis can be undertaken in a fluidized bed, which exploits the same advantages as are listed above: good mixing and heat transfer

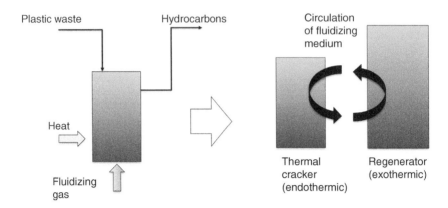

Figure 6.38   A two fluidized bed circulatory design for cracking plastic waste

and ease of solids handling into and out of the bed. In order to maintain good fluidization properties, it is usual for the bed to consist of an inert material such as sand, into which the fuel is added in a controlled way.

Pyrolysis and, to a lesser extent, gasification are endothermic processes, so energy must be added in order to maintain their operation. It is possible to do this using a twin fluidized bed arrangement, with bed material circulated between the main reaction bed and a separate bed which reheats the material before returning it, as shown schematically in Figure 6.38.

Polyethylene

Polyethylene, the largest volume production plastic, is made in a fluidized bed operating in *bubbling* or *turbulent* regimes. In Union Carbide's UNIPOL fluidized bed process (which requires moderate conditions of 20 atm and 80–100°C) the polythene grows on the catalyst surface to form granules (of the order of 250–1000 μm). The catalyst is chromium–titanium on a silica support. The reaction is highly exothermic, and since the conversion per pass is low, gases are recycled, condensed and reinjected into the bed as liquid.

Mixing, drying, coating and granulation

Fluidized beds are used in the food processing and pharmaceutical industries for the physical processes of drying, mixing, coating and granulation. Some of these processes are considered in more detail in Chapter 12. What differentiates fluidized beds used in the pharmaceutical industry from those in other process industries is the fact that they operate in batch mode, rather than continuously. This creates special difficulties, especially for drying and granulation, since the design needs to accommodate a wide range of bed properties, from surface-wet cohesive material to dry powders. Hence such beds can incorporate unusual features, such as conical-shaped vessels and specially designed distributors to create strong circulation patterns, examples of which are shown in Figure 12.22.

Often mixing, granulation and drying happen in the same fluidized bed. The cycle begins with mixing of the excipients and active drug, followed by the periodic addition of liquid binder by spraying and finishing with a drying period. The dried granulated product is fed to the tabletting machine, and the tablets may be coated – sometimes in a fluidized bed coater.

Drying in a fluidized bed comes with a problem. Because a fluidized bed is such a good mixer, a simple design of a continuous fluidized bed dryer will naturally give a product with a wide range of moisture content as a result of the wide range of particle residence times characteristic of a well-mixed device. If this is not acceptable, then channels and weirs can be created in the bed in order to create more of a plug-flow characteristic in the fluidized bed. In this way, the range of moisture content in the product particles can be made much narrower.

Advanced materials

Fluidized beds are also increasingly used in the production of advanced materials by chemical vapour deposition (CVD) and related methods. For example, one of the two main methods for producing high-purity silicon for photovoltaic cells is a fluidized bed, as illustrated in Figure 6.39.

Silicon can be produced by decomposition at high temperatures (650–1100 °C) of the gases silane (SiH_4) or trichlorosilane ($SiHCl_3$) by the reactions:

$$SiH_4 \rightarrow Si + 2H_2$$

$$SiHCl_3 + H_2 \rightarrow Si + 3HCl$$

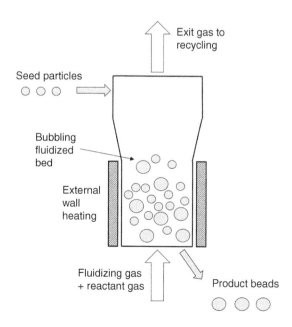

Figure 6.39 Schematic of a fluidized bed for production of silicon. Adapted from Filtvedt et al. (2010)

The bed consists of silicon particles, which are fluidized with hydrogen or helium, and it is usually heated to the decomposition temperature through the walls, taking advantage of the high heat transfer coefficients which are possible in fluidization. Silicon is deposited on the seed particles, which grow until they are ready to be removed. In principle, this kind of reactor can be operated continuously, as a batch process or in a discontinuous way with periodic removal of product and injection of new seed particles.

Transport

Fluidization is also used in the transport of particulate solids. Pneumatic transport or pneumatic conveying is the use of gas to transport solids within a plant or between plants. One simple use of fluidization for transport is the *air slide*. By fluidizing the powder, it can be made to flow down channels inclined at only a few degrees to the horizontal. In dense-phase pneumatic conveying, blow tanks are used. Solids are charged into the blow tank, which is then pressurized to force solids along pipelines. In some designs, the solids near to the exit must be fluidized in order to assist their flow into the pipeline.

Some designs of catalytic reactors in the oil and petrochemical industries use standpipes operating in fluidized bed flow in order to transport solids from regions of low pressure to regions of high pressure. Since the gas pressure increases from the top of the standpipe to the bottom, aeration gas must be added at locations along the standpipe in order to maintain the fluidized conditions. Refer to Chapter 7 for more detail of pneumatic transport.

As these examples illustrate, there is a surprisingly wide range of fluidized bed applications in the process industries and the fluidized bed is often at the heart of the process. The applications of fluidized bed technology are dealt with in more detail in Kunii and Levenspiel (1991) and Grace, Bi and Ellis (2020).

6.2.12   Worked Examples

WORKED EXAMPLE 6.2

3.6 kg of solid particles of density 2590 kg/m^3 and surface–volume mean size 748 μm form a packed bed of height 0.475 m in a circular vessel of diameter 0.0757 m. Water of density 1000 kg/m^3 and viscosity 0.001 Pa s is passed upwards through the bed. Calculate (a) the bed pressure drop at minimum fluidization, (b) the superficial liquid velocity at minimum fluidization, (c) the mean bed void fraction at a superficial liquid velocity of 1.0 cm/s, (d) the bed height at this velocity and (e) the pressure drop across the bed at this velocity.

Solution

(a) Applying Equation (6.36) to the packed bed, we find the packed bed void fraction:

$$\text{mass of solids} = 3.6 = (1 - \varepsilon) \times 2590 \times \frac{\pi (0.0757)^2}{4} \times 0.475$$

hence, $\varepsilon = 0.3498$

Frictional pressure drop across the bed when fluidized:

$$(-\Delta p) = \frac{weight\ of\ particles - buoyancy\ force}{cross - sectional\ area\ of\ bed}$$

$$(-\Delta p) = \frac{Mg - Mg\left(\rho_f / \rho_p\right)}{A}$$

(since buoyancy force = weight of fluid displaced by particles)

$$\text{Hence, } (-\Delta p) = \frac{Mg}{A}\left(1 - \frac{\rho_f}{\rho_p}\right) = \frac{3.6 \times 9.81}{4.50 \times 10^{-3}}\left(1 - \frac{1000}{2590}\right) = 4817\,\text{Pa}$$

(b) Assuming that the void fraction at the onset of fluidization is equal to the void fraction of the packed bed, we use the Ergun equation to express the relationship between packed pressure drop and superficial liquid velocity:

$$\frac{(-\Delta p)}{H} = 3.55 \times 10^7 U^2 + 2.648 \times 10^6 U$$

Equating this expression for pressure drop across the packed bed to the fluidized bed pressure drop, we determine the superficial fluid velocity at minimum fluidization, U_{mf}.

$$U_{mf} = 0.365\,\text{cm/s}$$

(c) The Richardson-Zaki equation [Equation (6.33)] allows us to estimate the expansion of a liquid fluidized bed.

$$U = U_T \varepsilon^n \tag{6.33}$$

Using the method given in Chapter 3, we determine the single particle terminal velocity, U_T.

$$Ar = 6527.9; \quad C_D Re_p^2 = 8704; \quad Re_p = 90; \quad U_T = 0.120\,\text{m/s}$$

Note that Re_p is calculated at U_T. At this value of Reynolds number, the flow is intermediate between viscous and inertial, and so in Equation (6.33) we use the correlation recommended by Khan and Richardson (1989) to determine the exponent n over the entire range of Reynolds number (see Chapter 3):

$$\frac{4.8 - n}{n - 2.4} = 0.043 Ar^{0.57}\left[1 - 2.4\left(\frac{x}{D}\right)^{0.27}\right] \tag{3.59}$$

where Ar is the Archimedes number $[x^3 \rho_f(\rho_p - \rho_f)g/\mu^2]$, x is the particle diameter and D is the vessel diameter. The most appropriate particle diameter to use here is the surface–volume mean.

$$\text{With } Ar = 6527.9, n = 3.202$$

Hence from Equation (6.33), $\varepsilon = 0.460$ when $U = 0.01\,\text{m/s}$.

Mean bed void fraction is 0.460 when the superficial liquid velocity is 1 cm/s.

(d) From Equation (6.37), we now determine the mean bed height at this velocity:

$$\text{Bed height (at } U = 0.01\,\text{m/s)} = \frac{(1 - 0.3498)}{(1 - 0.460)}0.475 = 0.572\,\text{m}$$

(e) The pressure drop across the bed remains essentially constant once the bed is fluidized. Hence at a superficial liquid velocity of 1 cm/s the pressure drop across the bed is 4817 Pa.

However, the measured pressure drop across the bed will include the hydrostatic head of the liquid in the bed. Applying the mechanical energy equation between the bottom (1) and the top (2) of the fluidized bed:

$$\frac{p_1 - p_2}{\rho_f g} + \frac{U_1^2 - U_2^2}{2g} + (z_1 - z_2) = \text{friction head loss} = \frac{4817}{\rho_f g}$$
$$U_1 = U_2; \quad z_1 - z_2 = -H = -0.572 \text{ m}$$

Hence, $p_1 - p_2 = 10\,428$ Pa.

WORKED EXAMPLE 6.3

A powder having the size distribution given below and a particle density of 2500 kg/m^3 is fed into a fluidized bed of cross-sectional area 4 m^2 at a rate of 1.0 kg/s.

Size range number (*i*)	Size range (μm)	Mass fraction in the feed
1	10–30	0.20
2	30–50	0.65
3	50–70	0.15

The bed is fluidized using air of density 1.2 kg/m^3 at a superficial velocity of 0.25 m/s. Processed solids are continuously withdrawn from the base of the fluidized bed in order to maintain a constant bed mass. Solids carried over with the gas leaving the vessel are collected by a bag filter operating at 100% total efficiency. None of the solids caught by the filter are returned to the bed. Assuming that the fluidized bed is well mixed and that the freeboard height is greater than the transport disengagement height under these conditions, calculate at equilibrium:

(a) The flow rate of solids entering the filter bag.

(b) The size distribution of the solids in the bed.

(c) The size distribution of the solids entering the filter bag.

(d) The rate of withdrawal of processed solids from the base of the bed.

(e) The solids loading in the gas entering the filter.

Solution

First calculate the elutriation rate constants for the three size ranges under these conditions from the Zenz and Weil correlation [Equation (6.58)]. The value of particle size x used in the correlation is the arithmetic mean of each size range:

$$x_1 = 20 \times 10^{-6} \text{ m}; \quad x_2 = 40 \times 10^{-6} \text{ m}; \quad x_3 = 60 \times 10^{-6} \text{ m}$$

With $U = 0:25\,\text{m/s}; \rho_p = 2500\,\text{kg/m}^3$ and $\rho_f = 1.2\,\text{kg/m}^3$

$$K_{1\infty}^* = 3.21 \times 10^{-2}\,\text{kg/m}^2\,\text{s}$$
$$K_{2\infty}^* = 8.74 \times 10^{-3}\,\text{kg/m}^2\,\text{s}$$
$$K_{3\infty}^* = 4.08 \times 10^{-3}\,\text{kg/m}^2\,\text{s}$$

Referring to Figure 6.W3.1 the overall and component material balances over the fluidized bed system are as follows:

$$\text{Overall balance:}\ F = Q + R \tag{6.W3.1}$$

$$\text{Component balance:}\ Fm_{F_i} = Qm_{Q_i} + Rm_{R_i} \tag{6.W3.2}$$

where F, Q and R are the mass flow rates of solids in the feed, withdrawal and filter discharge, respectively, and m_{F_i}, m_{Q_i} and m_{B_i} are the mass fractions of solids in size range i in the feed, withdrawal and filter discharge, respectively (Figure 6.W3.1).

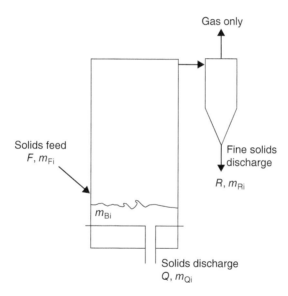

Figure 6.W3.1 Schematic diagram showing solids flows and size distributions for the fluidized bed.

From Equation (6.51) the entrainment rate of the size range i at the gas exit from the freeboard is given by:

$$R_i = Rm_{R_i} = K_{i\infty}^* Am_{B_i} \tag{6.W3.3}$$

and

$$R = \sum R_i = \sum Rm_{R_i} \tag{6.W3.4}$$

Combining these equations with the assumption that the bed is well mixed ($m_{Q_i} = m_{B_i}$),

$$m_{B_i} = \frac{Fm_{F_i}}{F - R + K_{i\infty}^* A} \tag{6.W3.5}$$

Now both m_{B_i} and R are unknown. However, noting that $\sum m_{B_i} = 1$; we have:

$$\frac{1.0 \times 0.2}{1.0 - R + (3.21 \times 10^{-2} \times 4)} + \frac{1.0 \times 0.65}{1.0 - R + (8.74 \times 10^{-3} \times 4)} + \frac{1.0 \times 0.15}{1.0 - R + (4.08 \times 10^{-3} \times 4)} = 1.0$$

Solving for R by trial and error, $R = 0.05\,\text{kg/s}$

(b) Substituting $R = 0.05\,\text{kg/s}$ in Equation (6W3.5), $m_{B_1} = 0.1855$, $m_{B_2} = 0.6599$ and $m_{B_3} = 0.1552$

Therefore, the size distribution of beds:

Size range number (i)	Size range (μm)	Mass fraction in bed
1	10–30	0.1855
2	30–50	0.6599
3	50–70	0.1552

(c) From Equation (6W3.3), knowing R and m_{B_i}, we can calculate m_{R_i}:

$$m_{R_i} = \frac{K_{1\infty}^* A m_{B_1}}{R} = \frac{3.21 \times 10^{-2} \times 4 \times 0.1855}{0.05} = 0.476$$

similarly, $m_{R_2} = 0.4614$, $m_{R_3} = 0.0506$

Therefore, size distribution of solids entering the filter:

Size range number (i)	Size range (μm)	Mass fraction entering filter
1	10–30	0.476
2	30–50	0.4614
3	50–70	0.0506

(d) From Equation (6.W3.1), the rate of withdrawal of solids from the bed, $Q = 0.95\,\text{kg/s}$.

(e) Solids loading for gas entering the filter:

$$\frac{\text{mass flow of solids}}{\text{volume flow of gas}} = \frac{R}{UA} = 0.05\,\text{kg/m}^3$$

WORKED EXAMPLE 6.4

A gas-phase catalytic reaction is performed in a fluidized bed operating at a superficial gas velocity of 0.3 m/s. For this reaction under these conditions, it is known that the reaction is first order in reactant A. The following information is given:

- Bed height at minimum fluidization = 1.5 m.

- Operating mean bed height = 1.65 m.

- Void fraction at minimum fluidization = 0.47.

- Reaction rate constant = 75.47 (per unit volume of solids).

- $U_{mf} = 0.033$ m/s;

- Mean bubble rise velocity = 0.111 m/s.

- Mass transfer coefficient between bubbles and emulsion = 0.1009 (based on unit bubble volume).

Use the reactor model of Orcutt et al. to determine:

(a) The conversion of reactant A.

(b) The effect on the conversion found in (a) of reducing the inventory of catalyst by one-half.

(c) The effect on the conversion found in (a) of halving the bubble size (assuming the interphase mass transfer coefficient is inversely proportional to the square root of the bubble diameter).

Discuss your answers to (b) and (c) and state which mechanism is controlling conversion in the reactor.

Solution

(a) From Section 6.2.8 the model of Orcutt et al. gives for a first-order reaction:

$$\text{conversion}, 1 - \frac{C_H}{C_0} = (1 - \beta e^{-\chi}) - \frac{(1 - \beta e^{-\chi})^2}{\dfrac{kH_{mf}(1 - \varepsilon_p)}{U} + (1 - \beta e^{-\chi})}$$

where

$$\chi = \frac{K_C H}{U_B} \quad \text{and} \quad \beta = (U - U_{mf})/U$$

From the information given:

$$K_C = 0.1009, U_B = 0.111 \text{ m/s}, U = 0.3 \text{ m/s}, U_{mf} = 0.033 \text{ m/s}$$
$$H = 1.65 \text{ m}, H_{mf} = 1.5 \text{ m}, k = 75.47$$

Hence, $\chi = 1.5$, $\beta = 0.89$ and $kH_{mf}(1 - \varepsilon_p)/U = 200$ (assuming $\varepsilon_p = \varepsilon_{mf}$)

So, from Equation (6.80), conversion = 0.798.

(b) If the inventory of solids in the bed is halved, both the operating bed height H and the height at minimum fluidization H_{mf} are halved. Thus, assuming all else remains constant, under the new conditions:

$$\chi = 0.75, \beta = 0.89 \text{ and } kH_{mf}\left(1 - \varepsilon_p\right)/U = 100$$

and so the new conversion = 0.576.

(c) If the bubble size is halved and K_c is proportional to $1/\sqrt{(\text{double diameter})}$,

$$\text{new } K_C = 1.414 \times 0.1009 = 0.1427$$

Hence, $\chi = 2.121$, giving conversion = 0.889.

(d) Comparing the conversion achieved in (c) with that achieved in (a), we see that improving interphase mass transfer has a significant effect on the conversion. We may also note that doubling the reaction rate (say by increasing the reactor temperature) and keeping everything else constant has a negligible effect on the conversion achieved in (a). We conclude, therefore, that under these conditions the transfer of gas between the bubble phase and the emulsion phase controls the conversion.

TEST YOURSELF – FLUIDIZATION

6.4 Write down the equation for the force balance across a fluidized bed and use it to come up with an expression for the pressure drop across a fluidized bed.

6.5 15 kg of particles of particle density 2000 kg/m^3 are fluidized in a vessel of cross-sectional area 0.03 m^2 by a fluid of density 900 kg/m^3. (a) What is the pressure drop across the bed? (b) If the bed height is 0.6 m, what is the bed void fraction?

6.6 Sketch a plot of pressure drop across a bed of powder versus the velocity of the fluid flowing upwards through it. Include packed bed and fluidized bed regions. Mark on the minimum fluidization velocity.

6.7 What are the chief behavioural characteristics of the four Geldart powder groups?

6.8 What differentiates a Geldart Group A powder from a Geldart Group B powder?

6.9 According to Richardson and Zaki, how does bed void fraction in a liquid-fluidized bed vary with fluidizing velocity at Reynolds numbers less than 0.3?

6.10 What is the basic assumption of the two-phase theory? Write down an equation that describes bed expansion as a function of superficial fluidizing velocity according to the two-phase theory.

6.11 Explain what is meant by particle convective heat transfer in a fluidized bed. In which Geldart group is particle convective heat transfer dominant?

6.12 Under what conditions does gas convective heat transfer play a significant role?

6.13 A fast gas-phase catalytic reaction is performed in a fluidized bed using a particulate catalyst. Would conversion be increased by improving conditions for mass transfer between the particulate phase and the bubble phase?

EXERCISES – FLUID FLOW THROUGH PACKED BEDS

6.1 A packed bed of solid particles of density $2500 \, kg/m^3$ occupies a depth of 1 m in a vessel of cross-sectional area $0.04 \, m^2$. The mass of solids in the bed is 50 kg and the surface–volume mean diameter of the particles is 1 mm. A liquid of density $800 \, kg/m^3$ and viscosity 0.002Pa s flows upwards through the bed, which is restrained at its upper surface.

    (a) Calculate the void fraction (volume fraction occupied by voids) of the bed.

    (b) Calculate the pressure drop across the bed when the volume flow rate of liquid is $1.44m^3/h$.

[Answer: (a) 0.50; (b) 6560 Pa (Ergun).]

6.2 A packed bed of solids of density $2000 \, kg/m^3$ occupies a depth of 0.6 m in a cylindrical vessel of inside diameter 0.1 m. The mass of solids in the bed is 5 kg and the surface–volume mean diameter of the particles is 300 μm. Water (density $1000 \, kg/m^3$ and viscosity 0.001 Pa s) flows upwards through the bed.

    (a) What is the void fraction of the packed bed?

    (b) Calculate the superficial liquid velocity at which the pressure drop across the bed is 4130 Pa.

[Answer: (a) 0.4692; (b) 1.5 mm/s (Ergun).]

6.3 A gas absorption tower of diameter 2 m contains ceramic Raschig rings randomly packed to a height of 5 m. Air containing a small proportion of sulphur dioxide passes upwards through the absorption tower at a flow rate of $6 \, m^3/s$. The viscosity and density of the gas may be taken as 1.80×10^{-5} Pa s and $1.2 \, kg/m^3$, respectively. Details of the packing are given below:

Ceramic Raschig rings
surface area per unit volume of packed bed, $S_B = 190 \, m^2/m^3$
the void fraction of a randomly packed bed = 0.71

    (a) Calculate the diameter, d_{sv}, of a sphere with the same surface–volume ratio as the Raschig rings.

    (b) Calculate the frictional pressure drop across the packing in the tower.

    (c) *Discuss* how this pressure drop will vary with the flow rate of the gas within ±10% of the quoted flow rate.

    (d) *Discuss* how the pressure drop across the packing would vary with gas pressure and temperature.

[Answer: (a) 9.16 mm; (b) 3460 Pa; for (c), (d) use the hint that turbulence dominates.]

6.4 A solution of density 1100 kg/m^3 and viscosity 2×10^{-3} Pa s is flowing under gravity at a rate of 0.24 kg/s through a bed of catalyst particles. The bed diameter is 0.2 m and the depth is 0.5 m. The particles are cylindrical, with a diameter of 1 mm and a length of 2 mm. They are loosely packed to give a void fraction of 0.3. Calculate the depth of liquid above the top of the bed. (*Hint:* apply the mechanical energy equation between the bottom of the bed and the surface of the liquid.)

[Answer: 0.716 m.]

6.5 In the regeneration of an ion exchange resin, hydrochloric acid of density 1200 kg/m^3 and viscosity 2×10^{-3} Pa s flows upwards through a bed of resin particles of density 2500 kg/m^3 resting on a porous support in a tube 4 cm in diameter. The particles are spherical, have a diameter of 0.2 mm and form a bed of void fraction 0.5. The bed is 60 cm deep and is unrestrained at its upper surface. Plot the frictional pressure drop across the bed as a function of acid flow rate up to a value of 0.1 l/minute.

[Answer: Pressure drop increases linearly up to a value of 3826 Pa beyond which point the bed will fluidize and maintain this pressure drop (see Section 6.2.1).]

6.6 The reactor of a catalytic reformer contains spherical catalyst particles of diameter 1.46 mm. The packed volume of the reactor is to be 3.4 m^3 and the void fraction is 0.45. The reactor feed is a gas of density 30 kg/m^3 and viscosity 2×10^{-5} Pa s flowing at a rate of 11 320 m^3/h. The gas properties may be assumed constant. The pressure loss through the reactor is restricted to 68.95 kPa. Calculate the cross-sectional area for flow and the bed depth required.

[Answer: area = 4.78 m^2; depth = 0.711 m.]

EXERCISES – FLUIDIZATION

6.7 A packed bed of solid particles of density 2500 kg/m^3, occupies a depth of 1 m in a vessel of cross-sectional area 0.04 m^2. The mass of solids in the bed is 50 kg and the surface–volume mean diameter of the particles is 1 mm. A liquid of density 800 kg/m^2 and viscosity 0.002 Pa s flows upwards through the bed.

(a) Calculate the void fraction (volume fraction occupied by voids) of the bed.

(b) Calculate the pressure drop across the bed when the volume flow rate of the liquid is 1.44 m^3/h.

(c) Calculate the pressure drop across, the bed when it becomes fluidized.

[Answer: (a) 0.5; (b) 6560 Pa; (c) 8338 Pa.]

6.8 130 kg of uniform spherical particles with a diameter of 50 μm and particle density 1500 kg/m^3 are fluidized by water (density 1000 kg/m^3, viscosity 0.001 Pa s) in a circular bed of cross-sectional area 0.2 m^2. The single particle terminal velocity of the particles is 0.68 mm/s and the void fraction at minimum fluidization is known to be 0.47.

(a) Calculate the bed height at minimum fluidization.

(b) Calculate the mean bed void fraction when the liquid flow rate is 2×10^{-5} m^3/s.

[Answer: (a) 0.818 m; (b) 0.6622.]

6.9 A packed bed of solid particles of density 2500 kg/m^3, occupies a depth of 1 m in a vessel of cross-sectional area 0.04 m^2. The mass of solids in the bed is 59 kg and the surface–volume mean diameter of the particles is 1 mm. A liquid of density 800 kg/m^3 and viscosity 0.002 Pa s flows upwards through the bed.

(a) Calculate the void fraction (volume fraction occupied by voids) of the bed.

(b) Calculate the pressure drop across the bed when the volume flow rate of the liquid is 0.72 m^3/h.

(c) Calculate the pressure drop across the bed when it becomes fluidized.

[Answer: (a) 0.41; (b) 7876 Pa; (c) 9839 Pa.]

6.10 12 kg of spherical resin particles of density 1200 kg/m^3 and uniform diameter 70 μm are fluidized by water (density 1000 kg/m^3 and viscosity 0.001 Pa s) in a vessel of diameter 0.3 m and form an expanded bed of height 0.25 m.

(a) Calculate the difference in pressure between the base and the top of the bed.

(b) If the flow rate of water is increased to 7 cm^3/s, what will be the resultant bed height and bed void fraction (liquid volume fraction)?

State and justify the major assumptions.

[Answer: (a) Frictional pressure drop = 277.5 Pa, pressure difference = 2730 Pa; (b) height = 0.465 m; void fraction = 0.696.]

6.11 A packed bed of solids of density 2000 kg/m^3 occupies a depth of 0.6 m in a cylindrical vessel of inside diameter 0.1 m. The mass of solids in the bed is 5 kg and the surface–volume mean diameter of the particles is 300 μm. Water (density 1000 kg/m^3 and viscosity 0.001 Pa s) flows upwards through the bed.

(a) What is the void fraction of the packed bed?

(b) Use a force balance over the bed to determine the bed pressure drop when fluidized.

(c) Hence, assuming laminar flow and that the void fraction at minimum fluidization is the same as the packed bed void fraction, determine the minimum fluidization velocity. Verify the assumption of laminar flow.

[Answer: (a) 0.4692; (b) 3124 Pa; (c) 1.145 mm/s.]

6.12 A packed bed of solids of density 2000 kg/m^3 occupies a depth of 0.5 m in a cylindrical vessel of inside diameter 0.1 m. The mass of solids in the bed is 4 kg and the surface–volume mean diameter of the particles is 400 μm. Water (density 1000 kg and viscosity 0.001 Pa s) flows upwards through the bed.

(a) What is the void fraction of the packed bed?

(b) Use a force balance over the bed to determine the bed pressure drop when fluidized.

(c) Hence, assuming laminar flow and that the void fraction at minimum fluidization is the same as the packed bed void fraction, determine the minimum fluidization velocity. Verify the assumption of laminar flow.

[Answer: (a) 0.4907; (b) 2498 Pa; (c) 2.43 mm/s.]

6.13 By applying a force balance, calculate the minimum fluidizing velocity for a system with particles of particle density 5000 kg/m^3 and mean volume diameter 100 μm and a fluid of density 1.2 kg/m^3 and viscosity 1.8×10^{-5} Pa s. Assume that the void fraction at minimum fluidization is 0.5.

If in the above example the particle size is changed to 2 mm, what is U_{mf}?

[Answer: 0.045 m/s; 2.26 m/s.]

6.14 A powder of mean sieve size 60 μm and particle density 1800 kg/m^3 is fluidized by air of density 1.2 kg/m^3 and viscosity 1.84×10^{-5} Pa s in a circular vessel of diameter 0.5 m. The mass of powder charged to the bed is 240 kg and the volume flow rate of air to the bed is 140 m^3/h. It is known that the average bed void fraction at minimum fluidization is 0.45 and correlation reveals that the average bubble rise velocity under the conditions in question is 0.8 m/s. Estimate:

(a) The minimum fluidization velocity, U_{mf}.

(b) The bed height at minimum fluidization.

(c) The visible bubble flow rate.

(d) The bubble fraction.

(e) The particulate phase void fraction.

(f) The mean bed height.

(g) The mean bed void fraction.

[Answer: (a) Baeyens and Geldart correlation [Equation (6.27)], 0.0027 m/s; (b) 1.24 m; (c) 0.038 m^3/s (assumes $U_{mf} \cong U_{mb}$); (d) 0.245; (e) 0.45; (f) 1.64 m; (g) 0.585.]

6.15 A batch fluidized bed process has an initial charge of 2000 kg of solids of particle density 1800 kg/m^3 and with the size distribution shown below:

Size range number (i)	Size range (μm)	Mass fraction in the feed
1	15–30	0.10
2	30–50	0.20
3	50–70	0.30
4	70–100	0.40

The bed is fluidized by a gas of density 1.2 kg/m^3 Pa s at a superficial gas velocity of 0.4 m/s. The fluid bed vessel has a cross-sectional area of 1 m^2.

Using a discrete time interval calculation with a time increment of 5 minutes, calculate:

(a) The size distribution of the bed after 50 minutes.

(b) The total mass of solids lost from the bed in that time.

(c) The maximum solids loading at the process exit.

(d) The entrainment flux above the transport disengagement height of solids in size range 1 (15–30 µm) after 50 minutes.

Assume that the process exit is positioned above TDH and that none of the entrained solids are returned to the bed.

[Answer: (a) (range 1) 0.029, (2) 0.165, (3) 0.324, (4) 0.482; (b) 527 kg; (c) 0.514 kg/m^3 s; (d) 0.024 kg/m^2 s.]

6.16 A powder has a particle density of 1800 kg/m^3 and the following size distribution:

Size range number (i)	Size range (µm)	Mass fraction in the feed
1	20–40	0.10
2	40–60	0.35
3	60–80	0.40
4	80–100	0.15

It is fed into a fluidized bed 2 m in diameter at a rate of 0.2 kg/s. The cyclone inlet is 4 m above the distributor and the mass of solids in the bed is held constant at 4000 kg by withdrawing solids continuously from the bed. The bed is fluidized using dry air at 700 K (density 0.504 kg/m^3 and viscosity 3.33 × 10^{-5} Pa s) giving a superficial gas velocity of 0.3 m/s. Under these conditions the mean bed void fraction is 0.55 and the mean bubble size at the bed surface is 5 cm. For this powder, under these conditions, $U_{mb} = 0.155$ cm/s.
Assuming that none of the entrained solids are returned to the bed, estimate:

(a) The flow rate and size distribution of the entrained solids entering the cyclone.

(b) The equilibrium size distribution of solids in the bed.

(c) The solids loading of the gas entering the cyclone.

(d) The rate at which solids are withdrawn from the bed.

[Answer: (a) 0.0485 kg/s, (range 1) 0.213, (2) 0.420, (3) 0.295, (4) 0.074; (b) (range 1) 0.0638, (2) 0.328, (3) 0.433, (4) 0.174; (c) 51.5 g/m^3; (d) 0.152 kg/s.]

6.17 A gas-phase catalytic reaction is performed in a fluidized bed operating at a superficial gas velocity equivalent to 10 × U_{mf}. For this reaction under these conditions, it is known that the reaction is first order in reactant A. Given the following information:

$$kH_{mf}(1 - \varepsilon_p)/U = 100; \chi = \frac{K_C H}{U_B} = 1.0$$

use the reactor model of Orcutt et al. to determine:

(a) The conversion of reactant A.

(b) The effect on the conversion found in (a) of doubling the inventory of catalyst.

(c) The effect on the conversion found in (a) of halving the bubble size by using suitable baffles (assuming the interphase mass transfer coefficient is inversely proportional to the bubble diameter).

If the reaction rate were two orders of magnitude smaller, comment on the wisdom of installing baffles in the bed with a view to improving conversion.

[Answer: (a) 0.6645; (b) 0.8744; (c) 0.8706.]

7

Pneumatic Transport and Standpipes

In this chapter we deal with two examples of the transport of particulate solids in the presence of a gas. The first example is *pneumatic transport* (sometimes referred to as *pneumatic conveying*), which is the use of a gas to transport a particulate solid through a pipeline. The second example is the *standpipe*, which has been used for many years, particularly on refineries, and has become a standard method for transferring solids downwards from a vessel at low pressure to a vessel at a higher pressure.

7.1 PNEUMATIC TRANSPORT

For many years gases have been used successfully in industry to transport a wide range of particulate solids – from wheat flour to wheat grain and plastic chips to minerals. Early applications of pneumatic transport were designed to operate in dilute suspension using large volumes of air at high velocity. A more modern development is the so-called 'dense phase' mode of transport in which the solid particles are not fully suspended. The attractions of dense phase transport lie in its low air requirements. Thus, in dense phase transport, a minimum amount of air (or other suspension gas) is delivered to the process with the solids (a particular attraction in feeding solids into fluidized bed reactors, for example). A low air requirement also generally means a lower energy requirement (despite the higher pressures needed). The resulting low solids velocities mean that in dense phase transport, product degradation by attrition and pipeline erosion are not the major problems they can be in dilute phase pneumatic transport.

In this section we will look at the distinguishing characteristics of dense and dilute phase transport and the types of equipment and systems used with each. The design of dilute phase systems is dealt with in detail and the approach to design of dense phase systems is summarized.

Introduction to Particle Technology, Third Edition. Martin Rhodes and Jonathan Seville.
© 2024 John Wiley & Sons Ltd. Published 2024 by John Wiley & Sons Ltd.
Website: www.wiley.com/go/rhodes/particle3e

7.1.1   Dilute Phase and Dense Phase Transport

The pneumatic transport of particulate solids is broadly classified into two flow regimes: dilute (or lean) phase flow and dense phase flow. Dilute phase flow in its most recognizable form is characterized by high gas velocities (greater than 20 m/s), low solids concentrations (less than 1% by volume) and low-pressure drops per unit length of transport line (typically less than 5 mbar/m). Dilute phase pneumatic transport is limited to short route, continuous transport of solids at rates of less than 10 t/h and is the only system capable of operation under negative pressure, which can have advantages in some applications. Under these dilute flow conditions, the solid particles behave as individuals, fully suspended in the gas, and fluid-particle forces dominate. At the opposite end of the scale is dense phase flow, characterized by low gas velocities (1–5 m/s, high solids concentrations (greater than 30% by volume) and high-pressure drops per unit length of pipe (typically greater than 20 mbar/m). In dense phase transport, particles are not fully suspended and there is much interaction between them.

The boundary between dilute phase flow and dense phase flow, however, is not clear cut and there are as yet no universally accepted definitions of dense phase and dilute phase transport.

Konrad (1986) lists four alternative means of distinguishing dense phase flow from dilute phase flow:

(a) On the basis of solids/air mass flow rates.

(b) On the basis of solids concentration.

(c) Dense phase flow exists where the solids completely fill the cross section of the pipe at some point.

(d) Dense phase flow exists when, for horizontal flow, the gas velocity is insufficient to support all particles in suspension, and, for vertical flow, where reverse flow of solids occurs.

In this chapter, the *choking* and *saltation* velocities will be used to mark the boundaries between dilute phase transport and dense phase transport in vertical and horizontal pipelines, respectively. These terms are defined below in considering the relationships between gas velocity, solids mass flow rate and pressure drop per unit length of transport line in both horizontal and vertical transport.

7.1.2   The Choking Velocity in Vertical Transport

We will see in Section 7.1.4 that the pressure drop across a length of transport line has in general six components:

- Pressure drop due to gas acceleration.

- Pressure drop due to particle acceleration.

- Pressure drop due to gas–pipe friction.

- Pressure drop related to solid–pipe friction.

- Pressure drop due to the static head of the solids.

- Pressure drop due to the static head of the gas.

The general relationship between gas velocity and pressure gradient $\Delta p/\Delta L$ for a vertical transport line is shown in Figure 7.1. Line AB represents the frictional pressure loss due to gas only in a vertical transport line. Curve CDE is for a solids flux of G_1 and curve FG is for a higher feed rate G_2. At point C the gas velocity is high, the concentration is low, and frictional resistance between gas and pipe wall predominates. As the gas velocity is decreased the frictional resistance decreases, but, since the concentration of the suspension increases, the static head required to support these solids increases. If the gas velocity is decreased below point D then the increase in static head outweighs the decrease in frictional resistance and $\Delta p/\Delta L$ rises again. In the region DE the decreasing velocity causes a rapid increase in solids concentration and a point is reached when the gas can no longer entrain all the solids. At this point a flowing, slugging fluidized bed (see Chapter 6) is formed in the transport line. This phenomenon is known as *choking* and is usually accompanied by large pressure fluctuations. The *choking velocity*, U_{CH}, is the lowest velocity at which this dilute phase transport line can be operated at the solids feed rate G_1. At the higher solids feed rate, G_2, the choking velocity is higher. The choking velocity marks the boundary between dilute phase and dense phase vertical pneumatic transport. Note that choking can be reached by decreasing the gas velocity at a constant solids flow rate, or by increasing the solids flow rate at a constant gas velocity.

It is not possible to predict theoretically the conditions for choking to occur. However, many correlations for predicting choking velocities are available in the literature.

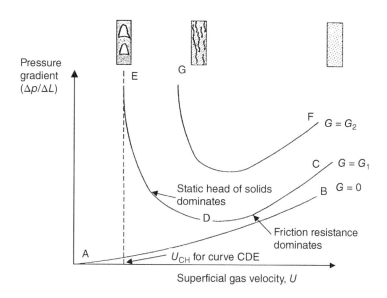

Figure 7.1 Phase diagram for dilute phase vertical pneumatic transport

Knowlton (1986) recommends the correlation of Punwani *et al.* (1976), which takes account of the considerable effect of gas density. This correlation is presented below:

$$\frac{U_{CH}}{\varepsilon_{CH}} - U_T = \frac{G}{\rho_p(1 - \varepsilon_{CH})} \tag{7.1}$$

$$\rho_f^{0.77} = \frac{2250D(\varepsilon_{CH}^{-4.7} - 1)}{\left(\frac{U_{CH}}{\varepsilon_{CH}} - U_T\right)^2} \tag{7.2}$$

where ε_{CH} is the void fraction in the pipe at the choking velocity U_{CH}, ρ_p is the particle density, ρ_f is the gas density, G is the mass flux of solids ($=M_p/A$) and U_T is the free fall, or terminal velocity, of a single particle in the gas. (Note that the constant is dimensional and that SI units must be used.)

Equation (7.1) represents the solids velocity at choking and includes the assumption that the slip velocity U_{slip} is equal to U_T (see Section 7.1.4 for a definition of slip velocity). Equations (7.1) and (7.2) must be solved simultaneously by trial and error to give ε_{CH} and U_{CH}.

7.1.3 The Saltation Velocity in Horizontal Transport

The general relationship between gas velocity and pressure gradient $\Delta p/\Delta L$ for a horizontal transport line is shown in Figure 7.2 and is in many ways similar to that for a vertical transport line. Line *AB* represents the curve obtained for gas only in the line, *CDEF* for a solids flux, G_1, and curve *GH* for a higher solids feed rate, G_2. At point *C*, the gas velocity is sufficiently high to carry all the solids in very dilute suspension. The solid particles are prevented from settling to the walls of the pipe by the turbulent eddies generated in the flowing gas. If the gas velocity is reduced whilst solids feed rate is kept constant, the frictional resistance and $\Delta p/\Delta L$ decrease. The solids move more slowly and the solids concentration increases. At point *D* the gas velocity is insufficient to maintain the solids in suspension and the solids begin to settle out in the bottom of the pipe. The gas velocity at which this occurs is termed the *saltation velocity*. Further decrease in

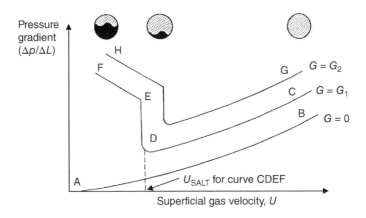

Figure 7.2 Phase diagram for dilute phase horizontal pneumatic transport

gas velocity results in rapid *salting out* of solids and rapid increase in $\Delta p / \Delta L$ as the area available for flow of gas is restricted by settled solids.

In regions *E* and *F* some solids may move in dense phase flow along the bottom of the pipe whilst others travel in dilute phase flow in the gas in the upper part of the pipe. The saltation velocity marks the boundary between dilute phase flow and dense phase flow in horizontal pneumatic transport.

Once again, it is not possible to predict theoretically the conditions under which saltation will occur. However, many correlations for predicting saltation velocity are available in the literature. The correlation of Rizk (1973), based on a semi-theoretical approach, is quite simple to use and is most unambiguously expressed as:

$$\frac{M_p}{\rho_f U_{salt} A} = \left(\frac{1}{10^{(1440x + 1.96)}} \right) \left(\frac{U_{salt}}{\sqrt{gD}} \right)^{(1100x + 2.5)} \tag{7.3}$$

where $\dfrac{M_P}{\rho_f U_{salt} A}$ is the solids loading $\left(\dfrac{\text{mass flow rate of solids}}{\text{mass flow rate of gas}} \right)$

$\dfrac{U_{salt}}{\sqrt{gD}}$ is the Froude number at saltation

U_{salt} is the superficial gas velocity (see Section 7.1.4 for the definition of superficial velocity) at saltation when the mass flow rate of solids is M_p, the pipe diameter is D and the particle size is x. (The units are SI.) The error range for this correlation is around $\pm 50\%$, which is typical of many of the correlations for saltation velocity found in the literature.

7.1.4 Fundamentals

In this section we generate some basic relationships governing the flow of gas and particles in a pipe.

Gas and particle velocities

We have to be careful in the definition of gas and particle velocities and in the relative velocity between them, the slip velocity. The terms are often used loosely in the literature and are defined below.

The term *superficial velocity* is commonly used, as in Chapter 6. Here we define separate superficial gas and solids (particles) velocities as:

$$\text{superficial gas velocity } U_{fs} = \frac{\text{volume flow of gas}}{\text{cross-sectional area of pipe}} = \frac{Q_f}{A} \tag{7.4}$$

$$\text{superficial solids velocity } U_{ps} = \frac{\text{volume flow of solids}}{\text{cross-sectional area of pipe}} = \frac{Q_p}{A} \tag{7.5}$$

where subscript 's' denotes superficial and subscripts 'f' and 'p' refer to the fluid and particles, respectively.

The fraction of pipe cross-sectional area available for the flow of gas is usually assumed to be equal to the volume fraction occupied by gas, i.e. the void fraction ε. The fraction of pipe area available for the flow of solids is therefore $(1 - \varepsilon)$.

And so, actual gas velocity,

$$U_f = \frac{Q_f}{A\varepsilon} \tag{7.6}$$

and actual particle velocity,

$$U_p = \frac{Q_p}{A(1-\varepsilon)} \tag{7.7}$$

Note that *actual* velocities are averages of distributions of velocities over time and space.

Thus, superficial velocities are related to actual velocities by the equations:

$$U_f = \frac{U_{fs}}{\varepsilon} \tag{7.8}$$

$$U_p = \frac{U_{ps}}{1-\varepsilon} \tag{7.9}$$

It is common practice in dealing with fluidization and pneumatic transport simply to use the symbol U to denote superficial fluid velocity. This practice will be followed in this chapter. Also, in line with common practice, the symbol G will be used to denote the mass flux of solids, i.e. $G = M_P/A$, where M_P is the mass flow rate of solids.

The relative velocity between particle and fluid U_{rel} is defined as:

$$U_{rel} = U_f - U_p \tag{7.10}$$

This velocity is often also referred to as the *slip velocity* U_{slip}.

It is often assumed that in vertical dilute phase flow the slip velocity is equal to the single particle terminal velocity U_T. This becomes less justifiable as the concentration increases and particle clustering can occur.

Continuity

Consider a length of transport pipe into which are fed particles and gas at mass flow rates of M_P and M_f, respectively. The continuity equations for particles and gas are as follows:

for the particles:

$$M_p = AU_p(1-\varepsilon)\rho_p \tag{7.11}$$

for the gas:

$$M_f = AU_f\varepsilon\rho_f \tag{7.12}$$

Combining these continuity equations gives an expression for the ratio of mass flow rates. This ratio is known as the solids loading:

$$\text{Solids loading,} \quad \frac{M_p}{M_f} = \frac{U_p(1-\varepsilon)\rho_p}{U_f\varepsilon\rho_f} \tag{7.13}$$

This shows us that the average void fraction ε, at a particular position along the length of the pipe, is a function of the solids loading and the magnitudes of the gas and solids velocities for given gas and particle density.

Pressure drop

In order to obtain an expression for the total pressure drop along a section of transport line we will write down the momentum equation for a section of pipe. Consider a section of pipe of cross-sectional area A and length δL inclined to the horizontal at an angle θ and carrying a suspension of void fraction ε (see Figure 7.3).

The momentum balance equation is:

$$\left(\begin{array}{c} \text{net force acting} \\ \text{on pipe contents} \end{array} \right) = \left(\begin{array}{c} \text{rate of increase in} \\ \text{momentum of contents} \end{array} \right)$$

Therefore,

$$\left(\begin{array}{c} \text{pressure} \\ \text{force} \end{array} \right) - \left(\begin{array}{c} \text{gas-wall} \\ \text{friction force} \end{array} \right) - \left(\begin{array}{c} \text{solids-wall} \\ \text{friction force} \end{array} \right) - \left(\begin{array}{c} \text{gravitational} \\ \text{force} \end{array} \right)$$

$$= \left(\begin{array}{c} \text{rate of increase} \\ \text{in momentum} \\ \text{of the gas} \end{array} \right) + \left(\begin{array}{c} \text{rate of increase} \\ \text{in momentum} \\ \text{of the solids} \end{array} \right)$$

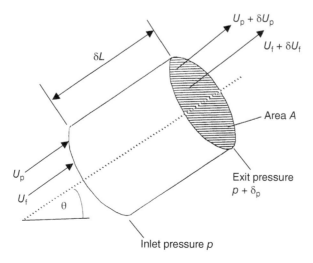

Figure 7.3   Section of conveying pipe: basis for momentum equation

or

$$- A\delta p - F_{fw}A\delta L - F_{pw}A\delta L - \left[A(1-\varepsilon)\rho_p\delta L\right]g\sin\theta - (A\varepsilon\rho_f\delta L)g\sin\theta$$
$$= \rho_f A\varepsilon U_f\delta U_f + \rho_p A(1-\varepsilon)U_p\delta U_p \tag{7.14}$$

where F_{fw} and F_{pw} are the gas–wall friction force and solids–wall friction force per unit volume of pipe, respectively.

Rearranging Equation (7.14) and integrating assuming constant gas density and void fraction:

$$p_1 - p_2 = \underset{(1)}{\frac{1}{2}\varepsilon\rho_f U_f^2} + \underset{(2)}{\frac{1}{2}(1-\varepsilon)\rho_p U_p^2} + \underset{(3)}{F_{fw}\,g\,L} + \underset{(4)}{F_{pw}L}$$
$$+ \underset{(5)}{\rho_p L(1-\varepsilon)g\sin\theta} + \underset{(6)}{\rho_f L\varepsilon\,g\,\sin\theta} \tag{7.15}$$

Readers should note that Equations (7.14) and (7.15) apply in general to the flow of any gas–particle mixture in a pipe. No assumption has been made as to whether the particles are transported in dilute phase or dense phase.

Equation (7.15) indicates that the total pressure drop along a straight length of pipe carrying solids in dilute phase transport is made up of a number of terms:

(1) Pressure drop due to gas acceleration.

(2) Pressure drop due to particle acceleration.

(3) Pressure drop due to gas–wall friction.

(4) Pressure drop related to solids–wall friction.

(5) Pressure drop due to the static head of the solids.

(6) Pressure drop due to the static head of the gas.

Some of these terms may be ignored depending on circumstance. If the gas and the solids are already accelerated in the line, then the first two terms should be omitted from the calculation of the pressure drop; if the pipe is horizontal, terms (5) and (6) can be omitted. The main difficulties are in knowing what the solids–wall friction is, and whether the gas–wall friction can be assumed independent of the presence of the solids; these will be covered in Section 7.1.5.

7.1.5 Design for Dilute Phase Transport

Design of a dilute phase transport system involves selection of a combination of pipe size and gas velocity to ensure dilute flow, calculation of the resulting pipeline pressure drop and selection of appropriate equipment for moving the gas and separating the solids from the gas at the end of the line.

Gas velocity

In both horizontal and vertical dilute phase transport it is desirable to operate at the lowest possible velocity in order to minimize frictional pressure loss, reduce attrition and reduce running costs. For a particular pipe size and solids flow rate, the saltation velocity is always higher than the choking velocity. Therefore, in a transport system comprising both vertical and horizontal lines, the gas velocity must be selected to avoid saltation. In this way choking will also be avoided. These systems would ideally operate at a gas velocity slightly to the right of point D in Figure 7.2. In practice, however, U_{salt} is not known with great confidence and so conservative design leads to operation well to the right of point D, with the consequent increase in frictional losses. Another factor encouraging caution in selecting the design velocity is the fact that the region near to point D is unstable; slight perturbations in the system may bring about saltation.

If the system consists only of a lift line, then the choking velocity becomes the important criterion. Here again, since U_{CH} cannot be predicted with confidence, conservative design is necessary. In systems using a centrifugal blower, characterized by reduced capacity at increased pressure, choking can almost be self-induced. For example, if a small perturbation in the system gives rise to an increase in solids feed rate, the pressure gradient in the vertical line will increase (Figure 7.1). This results in a higher back pressure at the blower giving rise to reduced volume flow of gas. Less gas means higher pressure gradient and the system soon reaches the condition of choking. The system fills with solids and can only be restarted by draining off the solids.

Bearing in mind the uncertainty in the correlations for predicting choking and saltation velocities, safety margins of 50% and greater are recommended when selecting the operating gas velocity.

Pipeline pressure drop

Equation (7.15) applies in general to the flow of any gas–particle mixture in a pipe. In order to make the equation specific to dilute phase transport, we must find expressions for terms 3 (gas–wall friction) and 4 (solids–wall friction).

In dilute transport the gas–wall friction is often assumed independent of the presence of the solids and so the friction factor for the gas may be used (e.g. Fanning friction factor – see worked example on dilute pneumatic transport in Section 7.3).

Several approaches to estimating solids–wall friction are presented in the literature. Here we will use the modified Konno and Saito (1969) correlation for estimating the pressure loss due to solid–wall friction in vertical transport and the Hinkle (1953) correlation for estimating this pressure loss in horizontal transport. Thus, for vertical transport (Konno and Saito, 1969):

$$F_{pw}L = 0.057GL\sqrt{\frac{g}{D}} \tag{7.16}$$

and for horizontal transport:

$$F_{pw}L = \frac{2f_p(1-\varepsilon)\rho_p U_p^2 L}{D} \tag{7.17a}$$

or

$$F_{pw}L = \frac{2f_p GU_p L}{D} \tag{7.17b}$$

where

$$U_p = U\left(1 - 0.0638x^{0.3}\rho_p^{0.5}\right) \tag{7.18}$$

and (Hinkle, 1953)

$$f_p = \frac{3}{8}\frac{\rho_f}{\rho_p}\frac{D}{x}C_D\left(\frac{U_f - U_p}{U_p}\right)^2 \tag{7.19}$$

where C_D is the drag coefficient between the particle and gas (see Chapter 3).

Note that Hinkle's analysis assumes that particles lose momentum by collision with the pipe walls. The pressure loss due to solids–wall friction is the gas pressure loss as a result of re-accelerating the solids. Thus, from Chapter 3, the drag force on a single particle is given by:

$$F_D = \frac{\pi x^2}{4}\rho_f C_D\frac{\left(U_f - U_p\right)^2}{2} \tag{7.20}$$

If the void fraction is ε, then the number of particles per unit volume of pipe N_v is:

$$N_v = \frac{(1 - \varepsilon)}{\pi x^3/6} \tag{7.21}$$

Therefore, the force exerted by the gas on the particles in unit volume of pipe F_v is:

$$F_v = F_D\frac{(1 - \varepsilon)}{\pi x^3/6} \tag{7.22}$$

Based on Hinkle's assumption, this is equal to the solids-wall friction force per unit volume of pipe F_{pw}. Hence,

$$F_{pw}L = \frac{3}{4}\rho_f C_D\frac{L}{x}(1 - \varepsilon)\left(U_f - U_p\right)^2 \tag{7.23}$$

Expressing this in terms of a friction factor f_p we obtain Equations (7.17a) and (7.17b) and (7.19).

Equation (7.15) relates to pressure losses along lengths of straight pipe. Pressure losses are also associated with bends in pipelines and estimations of the value of these losses will be covered in the next section.

Bends

Bends complicate the design of pneumatic dilute phase transport systems and when designing a transport system, it is best to use as few bends as possible. Bends increase the pressure drop in a line, and also are the points of most serious erosion and particle attrition. Solids normally in suspension in straight, horizontal or vertical pipes tend to salt out at bends due to the centrifugal force encountered while travelling around the bend. Particles which salt out must then be this re-entrained and re-accelerated after it, resulting in the higher pressure drops associated with bends. There is a greater tendency for particles to salt out in a horizontal pipe which is preceded by a downflowing vertical to horizontal bend than in any other configuration. If this type of bend is present in a system, it is possible for solids to remain on the bottom of the pipe for very long distances following the bend before they redisperse. Therefore, it is recommended that downflowing vertical to horizontal bends be avoided, if at all possible, in dilute phase pneumatic transport systems. In the past, designers of dilute phase pneumatic transport systems have made use of gradual, long radius elbows on the assumption that this would reduce the erosion and increase bend service life relative to sharp 90° elbows. Zenz (1964), however, recommended that blinded tees (Figure 7.4) be used in place of elbows in pneumatic transport systems. The theory behind the use of the blinded tee is that a cushion of stagnant particles collects in the blinded or unused branch of the tee, and the conveyed particles then impinge upon the stagnant particles in the tee rather than on the metal surface, as they would do in a long radius or short radius elbow. In the case of the blind tee the particles must be re-accelerated effectively from a standstill, giving rise to a significantly higher pressure drop than for long-radius bends. So, the increase in service life for a blind tee bend is gained at the expense of increased bend pressure loss during operation. Blind tees are unlikely to be used in applications where the solids holdup in the system cannot be tolerated (in the food and pharmaceutical industries for example).

Despite a considerable amount of research into bend pressure drop, there is no widely accepted method of predicting accurate bend pressure drops other than by experiment for the actual conditions expected. Tripathi and co-workers (2015) tested seven models for bend pressure loss against experimental data for fly ash in test rigs of different configurations over a range of conditions covering dilute phase and dense phase. They recommended using the models of Chambers and Marcus (1986), Pan (1992), and Pan and Wypych (1998) for estimation of bend losses of fly ash conveying systems from dense

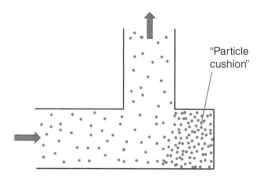

Figure 7.4   Blinded tee bend

phase to dilute phase. The model of Chambers and Marcus (1986) is now used quite widely and is shown below:

$$\Delta p_{\text{bends}} = N_{\text{bends}} B \left(1 + \frac{M_{\text{p}}}{M_{\text{f}}} \right) \frac{\rho U^2}{2} \tag{7.24}$$

where N_{bends} is the number of bends in the system, B is the bend factor or coefficient (accounting for bend radius), $\frac{M_{\text{p}}}{M_{\text{f}}}$ is the solids loading (kg/kg), ρ is the gas density at bend exit conditions (kg/m^3), U is the gas velocity at bend exit conditions (m/s) A value of $B = 0.5$ is recommended for bends with a 1 m radius of curvature).

Equipment

Any form of pneumatic conveying requires a device to feed the solids into the air stream. In dilute phase conveying, it is common for solids to be fed from a hopper at a controlled rate into the air stream. The system may be at positive pressure relative to atmospheric, at negative pressure or employ a combination of both. Positive pressure systems are usually limited to a maximum pressure of 1 bar gauge and negative pressure systems to a vacuum of about 0.4 bar below atmospheric pressure by the types of blowers and exhausters used.

Typical dilute phase systems are shown in Figures 7.5 and 7.6. Blowers are normally of the positive displacement type which may or may not have speed control in order to vary volume flow rate. Rotary airlocks enable solids to be fed at a controlled rate into the air stream against the air pressure. Alternatively, screw feeders are frequently used to transfer solids. Cyclone separators (see Chapter 8) are used to recover the solids from the gas stream at the receiving end of the transport line. Filters of various types and with various

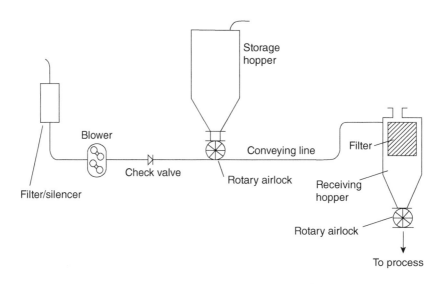

Figure 7.5   Dilute phase transport: positive pressure system

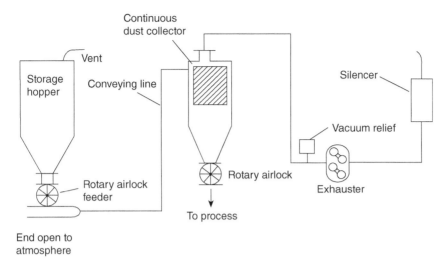

Figure 7.6 Dilute phase transport: negative pressure system

methods of solids recovery are used to clean up the transport gas before discharge or recycle.

In some circumstances it may not be desirable to use once-through air as the transport gas, so the gas is recirculated in a closed loop (e.g. to avoid the risk of contamination of the factory with toxic or radioactive substances; where an inert gas is used to avoid the risk of explosion; in order to control humidity when the solids are moisture sensitive). If a rotary positive displacement blower is used then the solids must be separated from the gas by a cyclone separator and by an in-line fabric filter. If lower system pressures are acceptable (0.2 bar gauge) then a centrifugal blower may be used in conjunction with only a cyclone separator. The centrifugal fan is able to operate without damage when small quantities of solids are present, whereas the positive displacement blower will not tolerate dust.

7.1.6 Dense Phase Transport

Flow patterns

As pointed out in the introduction to this chapter, there are many different definitions of dense-phase transport and of the transition point between dilute phase transport and dense phase transport. For the purpose of this section dense phase transport is described as the condition in which solids are conveyed such that they are not entirely suspended in the gas. Thus, the transition point between dilute phase transport and dense phase transport is saltation for horizontal transport and choking for vertical transport.

However, even within the dense phase regime several different flow patterns occur in both horizontal and vertical transport. Each of these flow patterns has particular characteristics giving rise to particular relationships between gas velocity, solids flow rate and pipeline pressure drop. In Figure 7.7, for example, five different flow patterns are identified within the dense phase regime for horizontal transport.

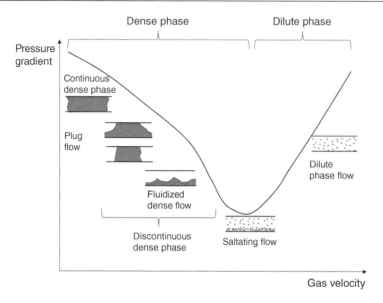

Figure 7.7   Flow patterns in horizontal pneumatic conveying

The continuous dense phase flow pattern, in which the solids occupy the entire pipe, is virtually the same as *extrusion*. Transport in this form requires very high gas pressures and is limited to short straight pipe lengths and larger particles (which have a high permeability).

Discontinuous dense phase flow can be divided into three fairly distinct flow patterns: (i) *discrete plug flow* in which discrete plugs of solids occupy the full pipe cross section; (ii) *dune flow*, often called *fluidized dense flow* in which a layer of solids settled at the bottom of the pipe moves along in the form of rolling dunes; and (iii) a pattern in which the rolling dunes completely fill the pipe cross section but in which there are no discrete plugs (also known as *plug flow*).

Saltating flow is encountered at gas velocities just below the saltation velocity. Particles are conveyed in suspension above a layer of settled solids. Particles may be deposited and re-entrained from this layer. As the gas velocity is decreased, the thickness of the layer of settled solids increases and eventually results in dune flow.

It should be noted, first, that not all powders exhibit all these flow patterns and, secondly, that within any transport line it is possible to encounter more than one regime.

The main advantages of dense phase transport arise from the low gas requirements and low solids velocities. Low gas volume requirements generally mean low energy requirements per kilogram of product conveyed and mean that smaller pipelines and solids–gas separation are required. Indeed, in some cases, since the solids are not suspended in the transport gas, it may be possible to operate without a filter at the receiving end of the pipeline. Low solids velocities mean that abrasive and friable materials may be conveyed without major pipeline erosion or product degradation.

It is interesting to look at the characteristics of the different dense phase flow patterns with a view to selecting the optimum for a dense phase transport system. The *continuous dense phase flow* pattern is the most attractive from the point of view of low gas requirements and solid velocities but has the serious drawback that it is limited to use in the transport of larger particles along short straight pipes and requires very high pressures. *Saltating flow* occurs at a velocity too close to the saltation velocity and is therefore unstable. In addition, this flow pattern offers little advantage from the point of view of gas and solids velocity. We are then left with the so-called *discontinuous dense phase flow* pattern including *plug flow* and *dune flow*.

Operation in *plug flow* can be unpredictable and may require high pressures or give rise to complete pipeline blockages. To see why this might be so, we will consider how the pressure drop across a plug of solids depends on its length. Large cohesionless particles [typically Geldart Group D particles (Geldart's classification of powders – see the Fluidization section in Chapter 6)] give rise to a permeable plug permitting the passage of a significant gas flow at low-pressure drops. In this case the stress developed in the plug would be low and a linear dependence of pressure drop on plug length would result. Plugs of fine cohesive particles (typically Geldart Group C) would be virtually impermeable to gas flow at the pressures usually encountered. In this case, the plug moves as a piston in a cylinder by purely mechanical means. The stress developed within the plug is high. The high normal stress results in a high wall shear stress which gives rise to an exponential increase in pressure drop with plug length. There is an analogy here with the stress developed in storage vessels (refer to Chapter 9). Thus, it is the degree of permeability of the plug which determines the relationship between plug length and pressure drop: the pressure drop across a plug can vary between a linear and an exponential function of the plug length depending on the permeability of the plug.

Large cohesionless particles form permeable plugs, as referred to above, and these are therefore suitable for plug flow. In other materials, where interaction under the action of stress and interparticle forces gives rise to low permeability plugs, plug flow is only possible if some mechanism is used to limit plug length, avoiding blockages. Some of these mechanisms are discussed in the next section. Note, however, that in practice, transport in *plug flow* makes up only a small fraction of industrial applications of dense phase flow and is mostly limited to large cohesionless particles. Most industrial dense phase transport applications use *fluidized dense flow* (Figure 7.7).

Equipment

All commercial dense phase transport systems employ a blow tank which may be fitted with a fluidizing element (Figure 7.8) or without (Figure 7.9). The blow tank is automatically taken through repeated cycles of filling, pressurizing and discharging. Since typically one third of the cycle time is used for filling the blow tank, a system required to give a mean delivery rate of 20 t/h must be able to deliver a peak rate of over 30 t/h. Dense phase transport is thus a batch operation because of the high pressures involved, whereas dilute phase transport can be continuous because of the relatively low pressures and the use of rotary valves. The dense phase system can be made to operate in semi-continuous mode by using two blow tanks in parallel.

Some commercial systems aim to prevent the formation of long plugs by either (a) using a bypass system in which the pressure build-up behind a plug causes more air to flow around the bypass line and break up the plug from its front end (Figure 7.10) or (b) detecting the pressure build-up using pressure actuated valves which divert auxiliary air to break up the plugs into smaller lengths (Figure 7.11).

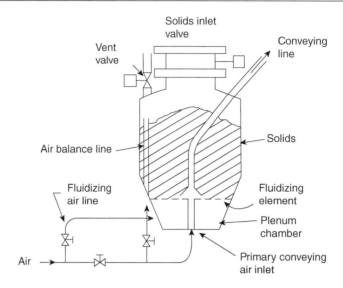

Figure 7.8   Dense phase transport blow tank with fluidizing element

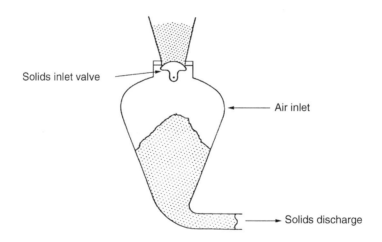

Figure 7.9   Blow tank without fluidizing element

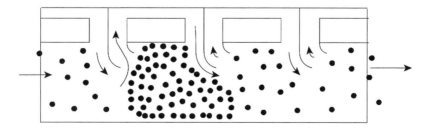

Figure 7.10   Dense phase conveying system using a bypass line to break up plugs of solids

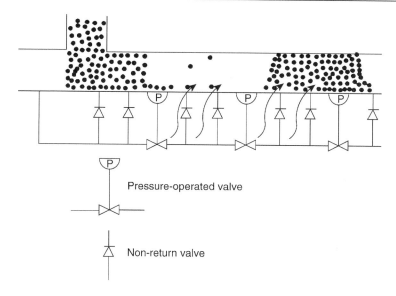

Pressure-operated valve

Non-return valve

Figure 7.11 Dense phase conveying system using pressure-actuated valves to direct gas

Other systems attempt to form stable plugs. Cohesionless granular materials do form plugs naturally under certain conditions. However, to form stable plugs of manageable length of other materials, it is generally necessary to induce them artificially, for example by using an air knife (Figure 7.12) or an alternating valves system (Figure 7.13) to chop up solids fed in continuous dense phase flow from a blow tank.

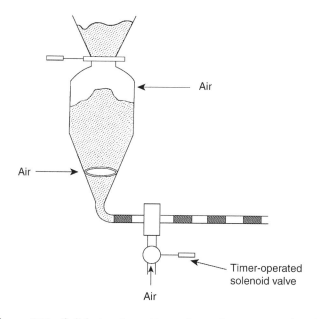

Figure 7.12 Solid plug formation using a timer-operated air knife

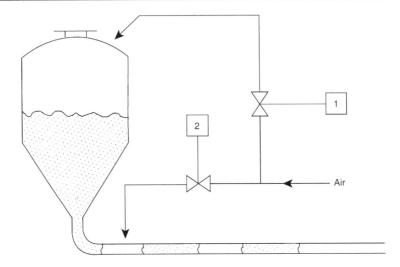

Figure 7.13 Solid plug formation using alternating air valves (valves 1 and 2 open and close alternately to create plugs of solids in the discharge pipe)

Design for dense phase transport

Whereas dilute phase transport systems can be designed, albeit with a large safety margin, from first principles together with the help of some empirical correlations, the design of commercial dense phase systems is largely empirical. Although in theory the equation for pressure drop in two-phase flow developed earlier in this chapter [Equation (7.15)] may be applied to dense phase flow, in practice it is of little use. Generally, a test facility which can be made to simulate most transport situations is used to monitor the important transport parameters during tests on a particular material. From these results, details of the dense phase transport characteristics of the material can be built up and the optimum conditions of pipe size, air flow rate, and type of dense phase system can be determined. Commercial dense phase systems are designed on the basis of past experience together with the results of tests such as these. Details of how this is done may be found in Mills (1990).

7.1.7   Matching the System to the Powder

Generally speaking, it is possible to convey any powder in the dilute phase mode, but because of the attractions of dense phase transport, there is great interest in assessing the suitability of a powder for transport in this mode. The most commonly used procedure is to undertake a series of tests on a sample of the powder in a pilot plant. This is obviously expensive. An alternative approach may be taken to reduce the range and cost of the pilot plant tests. This approach involves predicting which mode of dense phase conveying (if any) is suitable for a particular powder based on measured properties of that powder. Jones and Williams (2008) reviewed the available prediction methods and their findings are summarized here. These authors used a considerable body of published data on a range of powders for which the pneumatic conveying mode of flow was known, in order to test the predictive capability of the available methods. The methods were classified into two types: those based primarily on basic powder properties (mean particle size and particle density – Geldart, 1973; Molerus, 1982; Dixon, 1979; Pan, 1999) and the so-called air-particle parameter methods based on bulk powder properties (for example, de-aeration rate and permeability – Mainwaring and Reed, 1987; Jones, 1988; Chambers *et al.*, 1998; Sanchez *et al.*, 2003). Here we limit ourselves to one example from each type of prediction method.

Jones and Williams (2008) found that Geldart's classification of powders (see Section 6.2.4), using the basic powder properties of mean particle size and particle density and developed for application to fluidized beds rather than pneumatic transport (see Figure 6.8) gives a reasonable prediction of the mode of flow for conveying – powders in Groups A and C were likely to be suitable for fluidized dense phase conveying, whereas Group D powders (particularly those of lower density) were capable of plug flow conveying. The mode of flow suitable for Group B powders was less clear. The predictive capability of the Geldart classification was improved if it was represented in terms of the loose poured bulk density of the powder rather than the particle density. The modified Geldart classification predicts that Group B powders with a loose poured bulk density greater than $1000 \, \text{kg/m}^3$ may be conveyed only in dilute phase.

The so-called air-particle parameter methods were designed specifically for predicting pneumatic conveying mode. These methods recognize the fact that basic powder properties, such as mean particle size and particle density, do not fully predict the bulk properties of the powder and that the bulk properties relevant to conveying are better determined experimentally. Jones (1988) used experimentally derived permeability and de-aeration rate data to classify materials into the three modes of flow – fluidized dense flow, dilute phase and plug flow. This approach gives reasonable predictions and clearly identifies conditions where only dilute phase conveying is possible.

7.2 STANDPIPES

Standpipes are used for transferring solids downwards from a region of low pressure to a region of higher pressure. The overview of standpipe operation given here is based largely on the work of Knowlton (1997).

Typical overflow and underflow standpipes are shown in Figure 7.14, where they are used to transfer solids continuously from an upper fluidized bed to a lower fluidized bed.

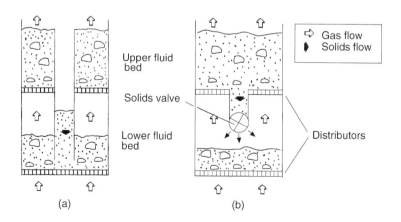

Figure 7.14 (a) Overflow- and (b) underflow-type standpipes transporting solids from a low-pressure fluidized bed to a bed at higher pressure

For solids to be transferred downwards against the pressure gradient, gas must flow upwards relative to the solids. The friction losses developed by the flow of the gas through the packed or fluidized bed of solids in the standpipe generate the required pressure gradient. If the gas must flow upwards relative to the downflowing solids there are two possible cases: (i) the gas flows upwards relative to both the standpipe wall and the solids and (ii) the gas flows downwards relative to the standpipe wall, but at a lower velocity than the solids, so the direction of flow is upwards relative to the solids.

A standpipe may operate in two basic flow regimes depending on the relative velocity of the gas to the solids: packed bed flow and fluidized bed flow.

7.2.1 Standpipes in Packed Bed Flow

If the relative upward velocity of the gas $(U_f - U_p)$ is less than the relative velocity at incipient fluidization $(U_f - U_p)_{mf}$, then packed bed flow results and the relationship between gas velocity and pressure gradient is in general determined by the Ergun equation [see Equation (6.11)].

The Ergun equation is usually expressed in terms of the superficial gas velocity through the packed bed. However, for the purposes of standpipe calculations it is useful to write the Ergun equation in terms of the magnitude of the velocity of the gas relative to the velocity of the solids $|U_{rel}| (= |U_f - U_p|)$. (Refer to Section 7.1.4 for clarification of relationships between superficial and actual velocities.)

$$\text{Superficial gas velocity, } U = \varepsilon |U_{rel}| \tag{7.25}$$

And so, in terms of $|U_{rel}|$ the Ergun equation becomes:

$$\frac{(-\Delta p)}{H} = \left[150 \frac{\mu}{x_{sv}^2} \frac{(1-\varepsilon)^2}{\varepsilon^2}\right] |U_{rel}| + \left[1.75 \frac{\rho_f}{x_{sv}} \frac{(1-\varepsilon)}{\varepsilon}\right] |U_{rel}|^2 \tag{7.26}$$

This equation allows us to calculate the value of $|U_{rel}|$ required to give a particular pressure gradient. We now adopt a sign convention for velocities. For standpipes it is convenient to take downward velocities as positive. In order to create the pressure gradient in the required direction (higher pressure at the lower end of the standpipe), the gas must flow upwards relative to the solids. Hence, U_{rel} should always be negative in normal operation. Solids flow is downwards, so U_P, the actual velocity of the solids (relative to the pipe wall), is always positive.

Knowing the magnitude and direction of U_P and U_{rel}, the magnitude and direction of the actual gas velocity (relative to the pipe wall) may be found from $U_{rel} = U_f - U_p$. In this way the quantity of gas passing up or down the standpipe may be estimated.

7.2.2 Standpipes in Fluidized Bed Flow

If the relative upward velocity of the gas $(U_f - U_p)$ is greater than the relative velocity at incipient fluidization $(U_f - U_p)_{mf}$, then fluidized bed flow will result. In fluidized bed flow the pressure gradient is independent of relative gas velocity. Assuming that in fluidized bed flow the entire apparent weight of the particles is supported by the gas flow, then the pressure gradient is given by (see Chapter 7):

$$\frac{(-\Delta p)}{H} = (1 - \varepsilon)\left(\rho_p - \rho_f\right)g \qquad (7.27)$$

where $(-\Delta p)$ is the pressure drop across a height H of solids in the standpipe, ε is the void fraction and ρ_p is the particle density.

Fluidized bed flow may be non-bubbling flow or bubbling flow. Non-bubbling flow occurs only with Geldart Group A solids (described in Chapter 6) when the relative gas velocity lies between the relative velocity for incipient fluidization and the relative velocity for minimum bubbling $(U_f - U_p)_{mb}$. For Geldart Group B materials (Chapter 6) with $(U_f - U_p) > (U_f - U_p)_{mf}$ and for Group A solids with $(U_f - U_p) > (U_f - U_p)_{mb}$ bubbling fluidized flow results.

Four types of bubbling fluidized bed flow in standpipes are possible depending on the direction of motion of the gas in the bubble phase and particulate phases relative to the standpipe walls. These are depicted in Figure 7.15. In practice, bubbles are undesirable in a standpipe. The presence of rising bubbles hinders the flow of solids and reduces the pressure gradient developed in the standpipe. If the bubble rise velocity is greater than

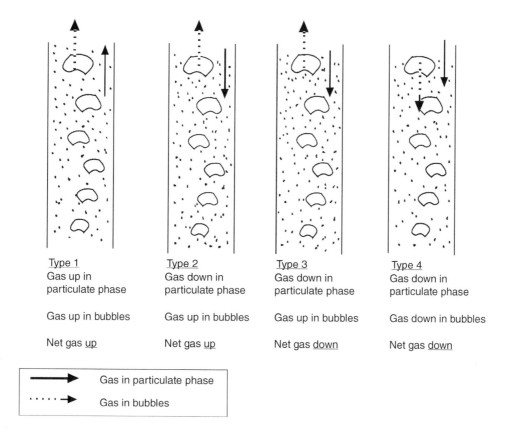

Type 1
Gas up in particulate phase

Gas up in bubbles

Net gas up

Type 2
Gas down in particulate phase

Gas up in bubbles

Net gas up

Type 3
Gas down in particulate phase

Gas up in bubbles

Net gas down

Type 4
Gas down in particulate phase

Gas down in bubbles

Net gas down

→ Gas in particulate phase

· · · · → Gas in bubbles

Figure 7.15 Types of fluidized flow in a standpipe

the solids velocity, then the bubbles will rise and grow by coalescence. Larger standpipes are easier to operate since they can tolerate larger bubbles than smaller standpipes. For optimum standpipe operation when using Group B solids, the relative gas velocity should be slightly greater than the relative velocity for minimum fluidization. For Group A solids, the relative gas velocity should lie between $(U_f - U_p)_{mf}$ and $(U_f - U_p)_{mb}$.

In practice, additional gas (*aeration*) is often added along the length of a standpipe in order to maintain the solids in a fluidized state just above minimum fluidization velocity. If this were not done then, with a constant mass flow of gas, relative velocities would decrease towards the high-pressure end of the standpipe. The lower velocities would result in lower mean void fractions and the possibility of an unfluidized region at the bottom of the standpipe. Aeration gas is added in stages up the length of the standpipe and only the minimum requirement is added at any level. If too much is added, bubbles are created, which may hinder solids flow. The analysis below, based on that of Kunii and Levenspiel (1991), enables calculation of the position and quantity of aeration gas to be added.

The starting point is Equation (7.13), the equation derived from the continuity equations for gas and solids flow in a pipe. For fine Group A solids, the relative velocity between gas and particles will be very small in comparison with the actual velocities, and so we can assume with little error that $U_P = U_f$. Hence, from Equation (7.13):

$$\frac{M_p}{M_f} = \frac{(1-\varepsilon)}{\varepsilon} \frac{\rho_p}{\rho_f} \tag{7.28}$$

Using subscripts 1 and 2 to refer to the upper (low pressure) and lower (high pressure) levels in the standpipe, since M_p, M_f and ρ_p are constant:

$$\frac{(1-\varepsilon_1)}{\varepsilon_1} \frac{1}{\rho_{f_1}} = \frac{(1-\varepsilon_2)}{\varepsilon_2} \frac{1}{\rho_{f_2}} \tag{7.29}$$

And so, since the pressure ratio $p_2/p_1 = \rho_{f_2}/\rho_{f_1}$, then

$$\frac{p_2}{p_1} = \frac{(1-\varepsilon_2)}{\varepsilon_2} \frac{\varepsilon_1}{(1-\varepsilon_1)} \tag{7.30}$$

Let us assume that the void fraction ε_2 is the lowest void fraction acceptable for maintaining fluidized standpipe flow. Equation (7.30) allows calculation of the equivalent maximum pressure ratio, and hence the pressure drop between levels 1 and 2. Assuming the solids are fully supported, this pressure difference will be equal to the apparent weight per unit cross-sectional area of the standpipe [Equation (7.27)].

$$(p_2 - p_1) = \left(\rho_p - \rho_f\right)(1 - \varepsilon_a)Hg \tag{7.31}$$

where ε_a is the average void fraction over the section between levels 1 and 2, H is the distance between the levels and g is the acceleration due to gravity. In practice, ρ_f is usually negligible by comparison with ρ_p.

If ε_1 and ε_2 are known, H may be calculated from Equation (7.31).

The objective of adding aeration gas is to raise the void fraction at the lower level to equal that at the upper level. Applying Equation (7.28),

$$\frac{(1-\varepsilon_2)}{\varepsilon_2} = \frac{M_p}{(M_f + M_{f_2})} \frac{\rho_{f_2}}{\rho_p} = \frac{M_p}{M_f} \frac{\rho_{f_1}}{\rho_p} \qquad (7.32)$$

where M_{f_2} is the mass flow of aeration air added at level 2.

Then rearranging,

$$M_{f_2} = M_f \left(\frac{\rho_{f_2}}{\rho_{f_1}} - 1\right) \qquad (7.33)$$

and, since from Equation (7.28), $M_f = M_p \dfrac{\varepsilon_1}{(1-\varepsilon_1)} \dfrac{\rho_{f_1}}{\rho_p}$

$$M_{f_2} = M_p \frac{\varepsilon_1}{(1-\varepsilon_1)} \frac{\rho_{f_1}}{\rho_p} \left(\frac{\rho_{f_2}}{\rho_{f_1}} - 1\right) \qquad (7.34)$$

and so mass flow of aeration air to be added,

$$M_{f_2} = \frac{\varepsilon_1}{(1-\varepsilon_1)} \frac{M_p}{\rho_p} \left(\rho_{f_2} - \rho_{f_1}\right) \qquad (7.35)$$

from which it can also be shown that:

$$Q_{f_2} = Q_p \frac{\varepsilon_1}{(1-\varepsilon_1)} \left(1 - \frac{\rho_{f_1}}{\rho_{f_2}}\right) \qquad (7.36)$$

where Q_{f_2} is the volume flow rate of gas to be added at pressure p_2 and Q_p is the volume flow rate of solids down the standpipe.

For long standpipes, aeration gas will need to be added at several levels in order to keep the void fraction within the required range (see the worked example on standpipe aeration).

7.2.3 Pressure Balance During Standpipe Operation

As an example of the operation of a standpipe, we will consider how an overflow standpipe operating in fluidized bed flow reacts to a change in gas flow rate. Figure 7.16(a) shows the pressure profile over such a system. The pressure balance equation over this system is:

$$\Delta p_{SP} = \Delta p_{LB} + \Delta p_{UB} + \Delta p_d \qquad (7.37)$$

where Δp_{SP}, Δp_{LB}, Δp_{UB} and Δp_d are the pressure drops across the standpipe, the lower fluidized bed, the upper fluidized bed and the distributor of the upper fluidized bed, respectively.

Let us consider a disturbance in the system such that the gas flow through the fluidized beds increases [Figure 7.16(b)]. If the gas flow through the lower bed increases, although the pressure drops across the lower and upper beds will remain constant, the pressure drop across the upper distributor will increase to $\Delta p_{d(new)}$. To match this increase, the pressure across the standpipe must rise to $\Delta p_{SP(new)}$ [Figure 7.16(b)]. In the case of an overflow standpipe operating in fluidized flow the increase in standpipe pressure drop results from a rise in the height of solids in the standpipe to $H_{SP(new)}$.

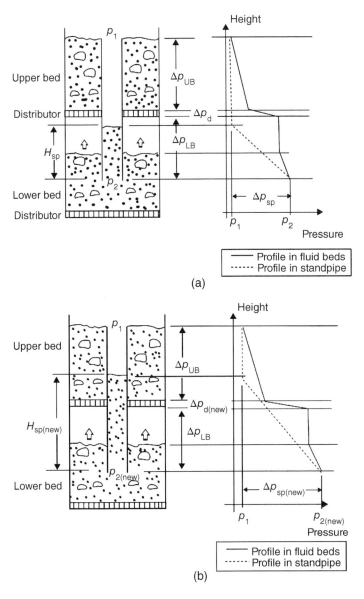

Figure 7.16 Operation of an overflow standpipe: (a) before increasing gas flow; (b) change in pressure profile due to increased gas flow through the fluid beds

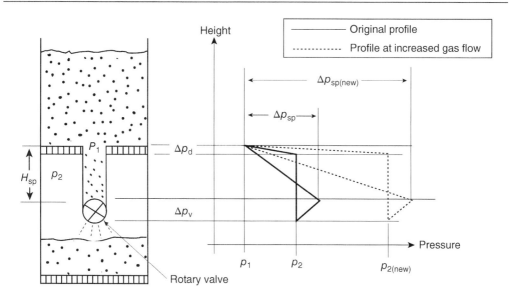

Figure 7.17 Pressure balance during operation of an underflow standpipe: effect of increasing gas flow through fluid beds

Consider now the case of an underflow standpipe operating in packed bed flow (Figure 7.17). The pressure balance across the system is given by:

$$\Delta p_{SP} = \Delta p_d + \Delta p_V \tag{7.38}$$

where Δp_{SP}, Δp_d and Δp_V are the pressure drops across the standpipe, the distributor of the upper fluidized bed and the standpipe valve (a rotary valve in this case), respectively.

If the gas flow from the lower bed increases, the pressure drop across the upper bed distributor increases to $\Delta p_{d(new)}$. The pressure balance then calls for an increase in stand-pipe pressure drop. Since in this case the standpipe length is fixed, in packed bed flow this increase in pressure drop is achieved by an increase in the magnitude of the relative veloc-ity $|U_{rel}|$. The standpipe pressure drop will increase to $\Delta p_{SP(new)}$ and the valve pressure drop, which depends on the solids flow, will remain essentially constant. Once the stand-pipe pressure gradient reaches that required for fluidized bed flow, its total pressure drop will remain constant so it will not be able to adjust to system changes.

A standpipe commonly used in industry is the underflow vertical standpipe with slide valve at the lower end. In this case the standpipe generates more head than is required and the excess is used across the slide valve in controlling the solids flow. Such a stand-pipe is used in the fluid catalytic cracking (FCC) unit to transfer solids from the reactor to the regenerator.

7.2.4 Further Reading

Readers wishing to learn more about solids circulation systems, standpipe flow and non-mechanical valves are referred to Kunii and Levenspiel (1991) or the chapter by Knowlton in Grace *et al.* (1997).

7.3   WORKED EXAMPLES

WORKED EXAMPLE 7.1

Design a positive pressure dilute phase pneumatic transport system to transport 900 kg/h of sand of particle density 2500 kg/m^3 and mean particle size 100 μm between two points in a plant separated by 10 m vertical distance and 30 m horizontal distance using ambient air. Assume that six 90° bends are required and that the allowable pressure loss is 0.50 bar.

Solution

In this case, to design the system means to determine the pipe size and air flow rate which would give a total system pressure loss near to the allowable pressure loss.

The design procedure requires trial and error calculations. Pipes are available in fixed sizes and so the procedure adopted here is to select a pipe size and determine the saltation velocity from Equation (7.3). The system pressure loss is then calculated at a superficial gas velocity equal to 1.5 times the saltation velocity [this gives a reasonable safety margin bearing in mind the accuracy of the correlation in Equation (7.3)]. The calculated system pressure loss is then compared with the allowable pressure loss. The pipe size selected may then be altered and the above procedure repeated until the calculated pressure loss matches that allowed.

Step 1. Selection of pipe size

Select 78 mm inside diameter pipe

Step 2. Determine gas velocity

Use the Rizk correlation of Equation (7.3) to estimate the saltation velocity, U_{salt}. Equation (7.3) rearranged becomes:

$$U_{\text{salt}} = \left(\frac{4 M_{\text{p}} 10^{\alpha} g^{\beta/2} D^{(\beta/2)-2}}{\pi \rho_{\text{f}}} \right)^{1/(\beta+1)}$$

where $\alpha = 1440x + 1.96$ and $\beta = 1100 + 2.5$.

In the present case $\alpha = 2.104$, $\beta = 2.61$ and $U_{\text{salt}} = 9.88$.

Therefore, superficial gas velocity $U = 1.5 \times 9.88 \text{ m/s} = 14.82 \text{ m/s}$.

Step 3. Pressure loss calculations

(a) *Horizontal sections.* Starting with Equation (7.15) an expression for the total pressure loss in the horizontal sections of the transport line may be generated. We will assume that all the initial acceleration of the solids and the gas takes place in the horizontal sections and so terms (1) and (2) are required. For term (3) the Fanning friction equation is used, which effectively assumes that the pressure loss due to gas–wall friction is independent of the presence of solids. For term (4) we employ the Hinkle correlation [Equations (7.17a) and (7.17b)]. Terms (5) and (6) become zero as θ = 0 for horizontal pipes. Thus, the pressure loss Δp_{H}, in the horizontal sections of the transport line is given by:

$$\Delta p_{\text{H}} = \frac{\rho_{\text{f}} \varepsilon_{\text{H}} U_{\text{fH}}^2}{2} + \frac{\rho_{\text{p}}(1-\varepsilon_{\text{H}}) U_{\text{pH}}^2}{2} + \frac{2 f_{\text{g}} \rho_{\text{f}} U^2 L_{\text{H}}}{D} + \frac{2 f_{\text{p}} \rho_{\text{p}}(1-\varepsilon_{\text{H}}) U_{\text{pH}}^2 L_{\text{H}}}{D}$$

where the subscript H refers to the values specific to the horizontal sections.

To use this equation, we need to know ε_H, U_{f_H} and U_{PH}. Hinkle's correlation gives us U_{P_H}:

$$U_{pH} = U\left(1 - 0.0638x^{0.3}\rho_p^{0.5}\right) = 11.84\,\text{m/s}$$

From continuity, $G = \rho_p(1 - \varepsilon_H)U_{pH}$

$$\text{thus } \varepsilon_H = 1 - \frac{G}{\rho_p U_{pH}} = 0.9982$$
$$\text{and } U_{fH} = \frac{U}{\varepsilon_H} = \frac{14.82}{0.9982} = 14.85\,\text{m/s}$$

Friction factor f_p is found from Equation (7.19) with C_D estimated at the relative velocity $(U_f - U_P)$, using the approximate correlations given below [or by using an appropriate C_D versus Re chart (see Chapter 2)]:

$$Re_p < 1 \quad C_D = 24/Re_p$$
$$1 < Re_p < 500 \quad C_D = 18.5\,Re_p^{-0.6}$$
$$500 < Re_p < 2 \times 10^5 \quad C_D = 0.44$$

Thus, for flow in the horizontal sections,

$$Re_p = \frac{\rho_f\left(U_{fH} - U_{pH}\right)x}{\mu}$$

for ambient air $\rho_f = 1.2\,\text{kg/m}^3$ and $\mu = 18.4 \times 10^{-6}\,\text{Pa s}$, giving:

$$Re_p = 19.63$$

and so, using the approximate correlations above,

$$C_D = 18.5\,Re_p^{-0.6} = 3.1$$

Substituting $C_D = 3.1$ into Equation (7.19) we have:

$$f_p = \frac{3}{8} \times \frac{1.2}{2500} \times 3.1 \times \frac{0.078}{100 \times 10^{-6}}\left(\frac{14.85 - 11.84}{11.84}\right)^2$$

The gas friction factor[1] is taken as $f_g = 0.005$. This gives $\Delta p_H = 14.864\,\text{Pa}$.

(b) *Vertical sections.* Starting again with Equation (7.15), the general pressure loss equation, an expression for the total pressure loss in the vertical section may be derived. Since the initial acceleration of solids and gas was assumed to take place in the horizontal sections, terms (1) and (2) become zero. The Fanning friction equation is used to estimate the pressure loss due to gas–wall friction [term (3)] assuming solids have a negligible effect on this pressure loss. For term (4) the modified Konno and Saito correlation [Equation (7.16)] is used. For vertical transport θ becomes equal to $90°$ in terms (5) and (6).

---

[1] One could use the Blasius correlation ($f_g = 0.079\,Re^{-0.25}$) for the Fanning gas-to-wall friction factor for smooth pipes for Re in the range $2100-10^5$, which gives $f_g = 0.00477$ in this case.

Thus, the pressure loss Δp_v, in the vertical sections of the transport line is given by:

$$\Delta p_v = \frac{2f_g \rho_f U^2 L_v}{D} + 0.057 G L_v \sqrt{\frac{g}{D}} + \rho_p(1 - \varepsilon_v)gL_v + \rho_f \varepsilon_v g L_v$$

where subscript v refers to values specific to the vertical sections.
To use this equation, we need to calculate the void fraction of the suspension in the vertical pipeline ε_v.

Assuming particles behave as individuals, then slip velocity is equal to single particle terminal velocity, U_T (also noting that the superficial gas velocity in both horizontal and vertical sections is the same and equal to U), i.e.:

$$U_{pv} = \frac{U}{\varepsilon_v} - U_T$$

continuity gives particle mass flux, $G = \rho_p(1 - \varepsilon_v)U_{pv}$.

Combining these equations gives a quadratic in ε_v which has only one physically possible root.

$$\varepsilon_v^2 U_T - \left(U_T + U + \frac{G}{\rho_p} \right) \varepsilon_v + U = 0$$

The single particle terminal velocity U_T may be estimated as shown in Chapter 3, giving $U_T = 0.52$ m/s assuming the particles are spherical.

And so, solving the quadratic equation, $\varepsilon_v = 0.9985$ and thus $\Delta p_v = 1148$ Pa.

(a) *Bends*. The pressure loss across the six 90° bend in the system will be calculated using Equation (7.24):

$$\Delta p_{bends} = N_{bends} B \left(1 + \frac{M_p}{M_f} \right) \frac{\rho U^2}{2} \tag{7.24}$$

$N_{bends} = 6$; $B = 0.5$ (for 90° bends of radius 1 m); $\rho = 1.2$ kg/m^3; $U = 14.82$ m/s

$M_p = 900$ kg/h $= 0.25$ kg/s

$$M_f = \pi \left(\frac{0.078^2}{4} \right) \times 14.82 \times 1.2 = 0.085 \frac{\text{kg}}{\text{s}}$$

Hence, solids loading $\frac{M_p}{M_f} = 2.941$ kg/kg

And so, pressure loss due to system bends = 1558 Pa

Therefore:

$$\begin{pmatrix} \text{total pressure} \\ \text{loss} \end{pmatrix} = \begin{pmatrix} \text{loss across} \\ \text{vertical sections} \end{pmatrix} + \begin{pmatrix} \text{loss across} \\ \text{horizontal} \\ \text{sections} \end{pmatrix} + \begin{pmatrix} \text{loss across} \\ \text{bends} \end{pmatrix}$$

$$= 1148 + 14\,864 + 1558 \text{ Pa}$$
$$= 0.176 \text{ bar}$$

Step 4. Compare calculated and allowable pressure losses

The allowable system pressure loss is 0.50 bar and so we may select a smaller pipe size and repeat the above calculation procedure. The table below gives the results for a range of pipe sizes.

Pipe inside diameter (mm)	Total system pressure loss (bar)
78	0.176
63	0.270
50	0.433
40	0.687

In this case we would select 50 mm pipe which gives a total system pressure loss of 0.433 bar. (An economic option could be found if capital and running costs were incorporated.) The design details for this selection are given below:

$$\text{pipe size} = 50\,\text{mm inside diameter}$$
$$\text{air flow rate} = 0.0317\,\text{m}^3/\text{s}$$
$$\text{air superficial velocity} = 16.15\,\text{m/s}$$
$$\text{saltation velocity} = 10.77\,\text{m/s}$$
$$\text{solids loading} = 6.57\,\text{kg solid/kg air}$$
$$\text{total system pressure loss} = 0.433\,\text{bar}$$

WORKED EXAMPLE 7.2

A 20 m long standpipe carrying a Geldart group A solid at a rate of 80 kg/s is to be aerated in order to maintain fluidized flow with a void fraction in the range 0.5–0.53. Solids enter the top of the standpipe at a void fraction of 0.53. The pressure and gas density at the top of the standpipe are 1.3 bar (abs) and 1.0 kg/m^3, respectively.

The particle density of the solids is 1200 kg/m^3.

Determine the aeration positions and rates.

Solution

From Equation (7.30), pressure ratio,

$$\frac{p_2}{p_1} = \frac{(1-0.50)}{0.50}\frac{0.53}{(1-0.53)} = 1.128$$

Therefore, $p_2 = 1.466\,\text{bar(abs)}$

Pressure difference $p_2 - p_1 = 0.166 \times 10^5\,\text{Pa}$.

Hence, from Equation (7.31) [with $\varepsilon_a = (0.5 + 0.53)/2 = 0.515$],

$$\text{length to first aeration point}, H = \frac{0.166 \times 10^5}{1200 \times (1 - 0.515) \times 9.81} = 2.91 \text{ m}$$

Assuming ideal gas behaviour, density at level 2, $\rho_{f_2} = \rho_{f_1} \left(\frac{p_2}{p_1} \right) = 1.128 \text{ kg/m}^3$

Applying Equation (7.35), aeration gas mass flow at the first aeration point,

$$M_{f_2} = \frac{0.53}{(1 - 0.53)} \frac{80}{1200} (1.128 - 1.0) = 0.0096 \text{ kg/s}$$

The above calculation is repeated in order to determine the position and rates of subsequent aeration points. The results are summarized below:

	First point	Second point	Third point	Fourth point	Fifth point
Distance from top of standpipe (m)	2.91	6.18	9.88	14.04	18.75
Aeration rate (kg/s)	0.0096	0.0108	0.0122	0.0138	0.0155
Pressure at aeration point (bar)	1.47	1.65	1.86	2.10	2.37

WORKED EXAMPLE 7.3

A 10 m long vertical standpipe of inside diameter 0.1 m transports solids at a flux of 100 kg/m²s from an upper vessel which is held at a pressure of 1.0 bar to a lower vessel held at 1.5 bar. The particle density of the solids is 2500 kg/m³ and the surface-volume mean particle size is 250 µm. Assuming that the void fraction is constant along the standpipe and equal to 0.50, and that the effect of pressure change may be ignored, determine the direction and flow rate of gas passing between the vessels. (Properties of gas in the system: density, 1 kg/m³; viscosity, 2×10^{-5} Pa s.)

Solution

First check that the solids are moving in packed bed flow. We do this by comparing the actual pressure gradient with the pressure gradient for fluidization.

Assuming that in fluidized flow the apparent weight of the solids will be supported by the gas flow, Equation (7.27) gives the pressure gradient for fluidized bed flow:

$$\frac{(-\Delta p)}{H} = (1 - 0.5) \times (2500 - 1) \times 9.81 = 12258 \text{ Pa/m}$$

$$\text{Actual pressure gradient} = \frac{(1.5 - 1.0) \times 10^5}{10} = 5000 \text{ Pa/m}$$

Since the actual pressure gradient is well below that for fluidized flow, the standpipe is operating in packed bed flow.

The pressure gradient in packed bed flow is generated by the upward flow of gas relative to the solids in the standpipe. The Ergun equation [Equation (7.26)] provides the relationship between gas flow and pressure gradient in a packed bed.

Knowing the required pressure gradient, the packed bed void fraction and the particle and gas properties, Equation (7.26) can be solved for $|U_{rel}|$, the magnitude of the relative gas velocity:

Ignoring the negative root of the quadratic,

$$|U_{rel}| = 0.1026 \, \text{m/s}$$

We now adopt a sign convention for velocities. For standpipes it is convenient to take downward velocities as positive. In order to create the pressure gradient in the required direction, the gas must flow upwards relative to the solids. Hence, U_{rel} is negative:

$$U_{rel} = -0.1026 \, \text{m/s}$$

From the continuity for the solids [Equation (7.11)],

$$\text{solids flux,} \; \frac{M_P}{A} = U_P (1 - \varepsilon) \rho_P$$

The solids flux is given as $100 \, \text{kg/m}^2\text{s}$ and so

$$U_P = \frac{100}{(1 - 0.5) \times 2500} = 0.08 \, \text{m/s}$$

Solids flow is downwards, so $U_P = +0.08 \, \text{m/s}$.

The relative velocity $U_{rel} = U_f - U_P$

hence, actual gas velocity $U_f = -0.1026 + 0.08 = -0.0226 \, \text{m/s}$ (upwards)

Therefore the gas flows *upwards* at a velocity of $0.0226 \, \text{m/s}$ relative to the standpipe walls. The superficial gas velocity is therefore:

$$U = \varepsilon U_f = -0.0113 \, \text{m/s}$$

From the continuity for the gas [Equation (7.12)] mass flow rate of gas,

$$M_f = \varepsilon U_f \rho_f A$$
$$= -8.9 \times 10^{-5} \, \text{kg/s}$$

So for the standpipe to operate as required, $8.9 \times 10^{-5} \, \text{kg/s}$ of gas must flow from the lower vessel to the upper vessel.

TEST YOURSELF

7.1 In horizontal pneumatic transport of particulate solids, what is meant by the term saltation velocity?

7.2 In vertical pneumatic transport of particulate solids, what is meant by the term choking velocity?

7.3 In horizontal pneumatic transport of particulate solids, why is there a minimum in the pressure drop versus gas velocity plot?

7.4 In vertical pneumatic transport of particulate solids, why is there a minimum in the pressure drop versus gas velocity plot?

7.5 There are six components in the equation describing the pressure drop across a pipe carrying solids by pneumatic transport. Write down these six components, in words.

7.6 In a dilute phase pneumatic transport system, what are the two main reasons for using a rotary airlock?

7.7 In a dense phase pneumatic transport system, why is it necessary to limit plug length in some cases? Describe three ways in which the plug length might be limited in practice?

7.8 How do we determine whether a standpipe is operating in packed bed flow or fluidized bed flow?

7.9 For a standpipe operating in packed bed flow, how do we determine the quantity of gas flow and whether the gas is flowing upwards or downwards?

7.10 For a standpipe operating in fluidized bed flow, why is it often necessary to add aeration gas at several points along the standpipe? What approach is taken to calculation of the quantities of gas to be added and the positions of the aeration points?

EXERCISES

7.1 Design a positive pressure dilute phase pneumatic transport system to carry 500 kg/h of a powder of particle density 1800 kg/m^3 and mean particle size 150 μm across a horizontal distance of 100 m and a vertical distance of 20 m using ambient air. Assume that the pipe is smooth, that four 90° bends (radius of curvature 1 m) are required and that the allowable pressure loss is 0.7 bar. See below for Blasius correlation for the gas-wall friction factor for smooth pipes.

[Answer: A 50 mm diameter pipe gives total pressure drop of 0.504 bar; superficial gas velocity 13.8 m/s.]

7.2 It is required to use an existing 50 mm inside diameter vertical smooth pipe as lift line to transfer 2000 kg/h of sand of mean particle size 270 μm and particle density 2500 kg/m^3 to a process 50 m above the solids feed point. A blower is available which is capable of delivering 60 m^3/h of ambient air at a pressure of 0.3 bar. Will the system operate as required?

[Answer: Using a superficial gas velocity of 8.49 m/s (= 1.55 × U_{CH}) the total pressure drop is 0.344 bar. System will not operate as required since allowable Δp = 0.3 bar]

7.3 Design a negative pressure dilute phase pneumatic transport system to carry 700 kg/h of plastic spheres of particle density 1000 kg/m^3 and mean particle size 1 mm between two points in a factory separated by a vertical distance of 15 m and a horizontal distance of 80 m using ambient air. Assume that the pipe is smooth, that five 90° bends (radius of curvature 1 m) are required and that the allowable pressure loss is 0.3 bar. See below for Blasius correlation for the gas-wall friction factor for smooth pipes.

[Answer: Using a superficial gas velocity of 16.4 m/s in a pipe of inside diameter 40 mm, the total pressure drop is 0.28 bar.]

7.4 A 25 m long standpipe carrying Group A solids at a rate of 75 kg/s is to be aerated in order to maintain fluidized flow with a void fraction in the range of 0.50–0.55. Solids enter the top of the standpipe at a void fraction of 0.55. The pressure and gas density at the top of the standpipe are 1.4 bar (abs) and 1.1 kg/m^3, respectively. The particle density of the solids is 1050 kg/m^3. Determine the aeration positions and rates.

[Answer: Positions: 6.36 m, 14.13 m, 23.6 m. Rates: 0.0213 kg/s, 0.0261 kg/s, 0.0319 kg/s.]

7.5 A 15 m long standpipe carrying Group A solids at a rate of 120 kg/s is to be aerated in order to maintain fluidized flow with a void fraction in the range of 0.50–0.54. Solids enter the top of the standpipe at a void fraction of 0.54. The pressure and gas density at the top of the standpipe are 1.2 bar (abs) and 0.9 kg/m^3, respectively. The particle density of the solids is 1100 kg/m^3. Determine the aeration positions and rates. What is the pressure at the lowest aeration point?

[Answer: Positions: 4.03 m, 6.76 m, 14.3 m. Rates: 0.0200 kg/s, 0.0235 kg/s, 0.0276 kg/s. Pressure: 1.94 bar.]

7.6 A 5 m long vertical standpipe of inside diameter 0.3 m transports solids at flux of 500 kg/m^{2}s from an upper vessel which is held at a pressure of 1.25 bar to a lower vessel held at 1.6 bar. The particle density of the solids is 1800 kg/m^3 and the surface-volume mean particle size is 200 μm. Assuming that the void fraction is 0.48 and is constant along the standpipe, determine the direction and flow rate of gas passing between the vessels. (Properties of gas in the system: density, 1.5 kg/m^3; viscosity, 1.9×10^{-5} Pa s.)

[Answer: 0.023 kg/s downwards]

7.7 A vertical standpipe of inside diameter 0.3 m transports solids at a flux of 300 kg/m^2 s from an upper vessel which is held at a pressure of 2.0 bar to a lower vessel held at 2.72 bar. The particle density of the solids is 2000 kg/m^3 and the surface-volume mean particle size is 220 μm. The density and viscosity of the gas in the system are 2.0 kg/m^3 and 2×10^{-5} Pa s, respectively. Assuming that the void fraction is 0.47 and is constant along the standpipe:

(a) Determine the minimum standpipe length required to avoid fluidized flow.

(b) Given that the actual standpipe is 8 m long, determine the direction and flow rate of gas passing between the vessels.

[Answer: (a) 6.92 m; (b) 0.0114 kg/s downwards]

Blasius correlation for the Fanning gas-wall friction factor for smooth pipes for Re in the range 2100 to 10^5: $f_g = 0.079 \, \text{Re}^{-0.25}$

8

Separation of Particles from a Gas

There are many cases during the processing and handling of particulate solids when particles are required to be separated from suspension in a gas. We saw in Chapter 6 that in fluidized beds the passage of gas through the bed entrains fine particles. These particles must be removed from the gas and returned to the bed before the gas can be discharged or sent to the next stage in the process. Keeping the very small particles in the fluidized bed may be crucial to the successful and safe operation of the process, as is the case in drying of pharmaceutical powders, for example.

In Chapter 7, we saw how a gas may be used to transport powders within a process. The efficient separation of the product from the gas at the end of the transport line plays an important part in the successful application of this method of powder transportation. In cases where gases are expanded to generate power, for example, turbines must be protected from particles in the gas. The same is true of air intakes to many types of machinery.

In any application, the size of the particles to be removed from the gas determines, to a large extent, the method to be used for their separation, as introduced in Section 5.6. Generally speaking, particles larger than about 100 μm can be separated easily by gravity settling. For particles less than 10 μm, more energy intensive methods such as *filtration*, *wet scrubbing* and *electrostatic precipitation* must be used.

Figure 8.1 shows typical *grade efficiency* curves for gas–particle separation devices. The grade efficiency curve describes how the separation efficiency of the device varies with particle size. The primary focus of this chapter is on the device known as the cyclone separator or *cyclone*. Gas cyclones[1] are generally not suitable for separation involving suspensions with a large proportion of particles less than 10 μm. They are best suited as primary separation devices and for relatively coarse particles, with an electrostatic precipitator or filter being used downstream to remove very fine particles. The second part of this chapter includes an introduction to filtration.

---

[1] Cyclones for solid–liquid separation, known as *hydrocyclones*, also exist but are not considered here.

Introduction to Particle Technology, Third Edition. Martin Rhodes and Jonathan Seville.
© 2024 John Wiley & Sons Ltd. Published 2024 by John Wiley & Sons Ltd.
Website: www.wiley.com/go/rhodes/particle3e

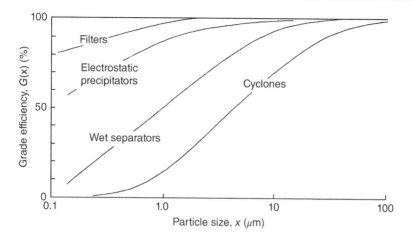

Figure 8.1 Typical grade efficiency curves for gas-particle separators

Readers wishing to know more about other methods of gas–particle separation and about the choice between methods are referred to Svarovsky (1981), Perry's Chemical Engineering Handbook, 9th Edition (Green and Southard, 2018) and Seville (1997).

8.1 GAS CYCLONES

The most common type of cyclone is known as the reverse flow type (Figure 8.2). Inlet gas is brought tangentially into the cylindrical section and a strong vortex is thus created inside the cyclone body. Particles in the gas are subjected to centrifugal forces which move them radially outwards, against the inward flow of gas and towards the inside surface of the cyclone on which the solids separate. The direction of flow of the vortex reverses near the bottom of the cyclone and the gas leaves via the outlet in the vertical

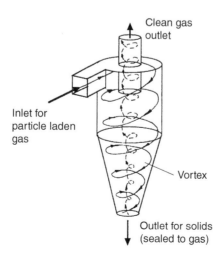

Figure 8.2 Schematic diagram of a reverse flow cyclone separator

pipe at the top, which is called the *vortex finder*. The solids at the wall of the cyclone are pushed downwards by the outer vortex and out of the solids exit, which is sealed to gas flow. In a cyclone, centrifugal forces on the particles greatly exceed forces due to gravity, so the orientation of the cyclone is not critical, except in ensuring effective discharge of the solids into the collection vessel.

8.1.1   Flow Characteristics of Cyclones

Rotational flow in the forced vortex within the cyclone body gives rise to a radial pressure gradient. This pressure gradient, combined with the frictional pressure losses at the gas inlet and outlet and losses due to changes in flow direction, make up the total pressure drop. This pressure drop, measured between the inlet and the gas outlet, is usually proportional to the square of gas flow rate through the cyclone. A resistance coefficient, the *Euler number Eu*, relates the cyclone pressure drop Δp to a characteristic velocity v:

$$Eu = \Delta p / \left(\rho_f v^2 / 2 \right) \tag{8.1}$$

where ρ_f is the gas density.

The characteristic velocity v can be defined for gas cyclones in various ways but the simplest and most appropriate definition is based on the cross-section of the cylindrical body of the cyclone, so that:

$$v = 4q / \left(\pi D^2 \right) \tag{8.2}$$

where q is the gas flow rate and D is the cyclone inside diameter.

The Euler number represents the ratio of pressure forces to the inertial forces acting on a fluid element. Its value is practically constant for a given cyclone geometry, independent of the cyclone body diameter (see Section 8.1.3).

8.1.2   Efficiency of Separation of Cyclones

Total efficiency and grade efficiency

Consider a cyclone to which the solids mass flow rate is M, the mass flow rate discharged from the solids exit is M_c (known as the coarse product) and the solids mass flow rate leaving with the gas is M_f (known as the fine product). The total material balance on the solids over this cyclone may be written:

$$\text{Total:}\ M = M_f + M_c \tag{8.3}$$

and the 'component' material balance for each particle size x (assuming no breakage or agglomeration of particles within the cyclone) is:

$$\text{Component}: M(\mathrm{d}F/\mathrm{d}x) = M_f(\mathrm{d}F_f/\mathrm{d}x) + M_c(\mathrm{d}F_c/\mathrm{d}x) \tag{8.4}$$

where, dF/dx, dF_f/dx and dF_c/dx are the differential frequency size distributions by mass (i.e. mass fraction of size x) for the feed, fine product and coarse product respectively. F, F_f and F_c are the cumulative frequency size distributions by mass (mass fraction less than size x) for the feed, fine product and coarse product, respectively. Refer to Chapter 1 for further details on representations of particle size distributions.

The total efficiency of separation of particles from the gas E_T, is defined as the fraction of the total feed which appears in the coarse product collected, i.e.:

$$E_T = M_c/M \tag{8.5}$$

The efficiency with which the cyclone collects particles of a certain size is described by the *grade efficiency* $G(x)$, which is defined as:

$$G(x) = \frac{\text{mass of solids of size } x \text{ in coarse product}}{\text{mass of solids of size } x \text{ in feed}} \tag{8.6}$$

Using the notation for size distribution described above:

$$G(x) = \frac{M_c(dF_c/dx)}{M(dF/dx)} \tag{8.7}$$

Combining with Equation (8.5), we find an expression linking grade efficiency with total efficiency of separation:

$$G(x) = E_T \frac{(dF_c/dx)}{(dF/dx)} \tag{8.8}$$

From Equations (8.3) to (8.5), we have:

$$(dF/dx) = E_T(dF_c/dx) + (1 - E_T)(dF_f/dx) \tag{8.9}$$

Equation (8.9) relates the size distributions of the feed (no subscript), the coarse product (subscript c) and the fine product (subscript f). In cumulative form this becomes:

$$F = E_T F_c + (1 - E_T)F_f \tag{8.10}$$

Simple theoretical analysis for the gas cyclone separator

Referring to Figure 8.3, consider a reverse flow cyclone with a cylindrical section of radius R. Particles entering the cyclone with the gas stream are forced into circular motion. The net flow of gas is radially inwards towards the central gas outlet. The forces acting on a particle following a circular path are drag, buoyancy and centrifugal force. The balance between these forces determines the equilibrium orbit adopted by the particle. The drag force is caused by the inward flow of gas past the particle and acts radially inwards.

Consider a particle of diameter x and density ρ_p following an orbit of radius r in a gas of density ρ_f and viscosity μ. Let the tangential velocity of the particle be U_θ and the radial

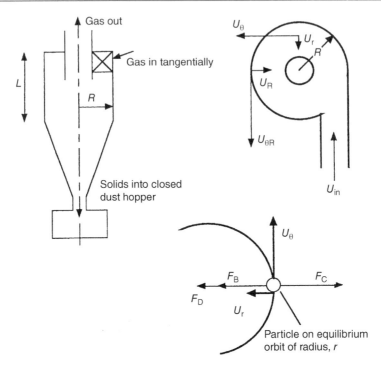

Figure 8.3 Reverse flow cyclone – a simple theory for separation efficiency

inward velocity of the gas be U_r. If we assume that Stokes' law (see Section 3.1) applies under these conditions, then the drag force is given by:

$$F_D = 3\pi x \mu U_r \tag{8.11}$$

The centrifugal and buoyancy forces acting on the particle moving with a tangential velocity component U_θ at radius r are, respectively:

$$F_C = \frac{\pi x^3}{6} \rho_p \frac{U_\theta^2}{r} \tag{8.12}$$

$$F_B = \frac{\pi x^3}{6} \rho_f \frac{U_\theta^2}{r} \tag{8.13}$$

Under the action of these forces the particle moves inwards or outwards until the forces are balanced and the particle assumes its equilibrium orbit. In this idealized approach, therefore, a particle in an equilibrium orbit will remain there for ever if this is within the cyclone; the particle will not be collected and will not leave with the exit gas. At this point,

$$F_C = F_D + F_B \tag{8.14}$$

and so

$$x^2 = \frac{18\mu}{\left(\rho_p - \rho_f\right)} \left(\frac{r}{U_\theta^2}\right) U_r \tag{8.15}$$

To go any further, we need a relationship between U_θ and the radius r for the vortex in a cyclone. Now for a rotating solid body, $U_\theta = r\omega$, where ω is the angular velocity, and for a free vortex $U_\theta r$ = constant. For the confined vortex inside the cyclone body, it has been found experimentally that the following holds approximately:

$$U_\theta r^{1/2} = \text{constant}$$

hence

$$U_\theta r^{1/2} = U_{\theta R} R^{1/2} \tag{8.16}$$

If we also assume uniform flow of gas towards the central outlet, then we are able to derive the radial variation in the radial component of gas velocity U_r:

$$\text{Gas flow rate}, q = 2\pi r L U_r = 2\pi R L U_R \tag{8.17}$$

hence

$$U_R = U_r(r/R) \tag{8.18}$$

Combining Equations (8.16) and (8.18) with Equation (8.15), we find:

$$x^2 = \frac{18\mu}{\left(\rho_p - \rho_f\right)} \frac{U_R}{U_{\theta R}^2} r \tag{8.19}$$

where r is the radius of the equilibrium orbit for a particle of diameter x.

If we assume that all particles with an equilibrium orbit radius greater than or equal to the cyclone body radius will be collected, then substituting $r = R$ in Equation (8.19) we derive the expression below for the critical particle diameter for separation, x_{crit}:

$$x_{crit}^2 = \frac{18\mu}{\left(\rho_p - \rho_f\right)} \frac{U_R}{U_{\theta R}^2} R \tag{8.20}$$

The values of the radial and tangential velocity components at the cyclone wall, U_R and $U_{\theta R}$, in Equation (8.20) may be found from a knowledge of the cyclone geometry and the gas flow rate.

This analysis predicts an ideal grade efficiency curve shown in Figure 8.4. All particles of diameter x_{crit} and greater are collected and all particles of size less than x_{crit} are not collected.

Cyclone grade efficiency in practice

In practice, gas velocity fluctuations and particle–particle interactions result in some particles larger than x_{crit} being lost and some particles smaller than x_{crit} being collected. Consequently, in practice the cyclone does not achieve such a sharp cut-off as predicted by the

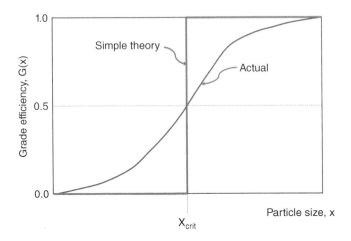

Figure 8.4   Theoretical and actual grade efficiency curves

theoretical analysis above. In common with other separation devices in which body forces are opposed by drag forces, the grade efficiency curve for gas cyclones is usually S-shaped.

For such a curve, the particle size for which the grade efficiency is 50%, x_{50}, is often used as a single number measurement of the efficiency of the cyclone. x_{50} is also known as the equiprobable size since it is that size of particle which has a 50% probability of appearing in the coarse product. This also means that, in a large population of particles, 50% of the particles of this size will appear in the coarse product. x_{50} is sometimes simply referred to as the *cut size* of the cyclone (or other separation device).

The concept of x_{50} cut size is useful where the efficiency of a cyclone is to be expressed as a single number independent of the feed solid size distribution, such as in scale-up calculations.

8.1.3   Scale-up of Cyclones

The scale-up of cyclones is based on a dimensionless group, the *cyclone Stokes number*, Stk_{50} which characterizes the separation performance of a family of geometrically similar cyclones:

$$Stk_{50} = \frac{x_{50}^2 \rho_p v}{18 \mu D} \tag{8.21}$$

where μ is gas viscosity, ρ_p is solids density, v is the characteristic velocity defined by Equation (8.2) and D is the diameter of the cyclone body. The physical significance of the Stokes number is that it is a ratio of the centrifugal force (less buoyancy) to the drag force, both acting on a particle of size x_{50}. Readers will note the similarity between our theoretical expression of Equation (8.20) and the Stokes number of Equation (8.21). There is therefore some theoretical justification for the use of the Stokes number in scale-up.

We will also meet the Stokes number in Chapter 14, when we consider the capture of particles in the respiratory airways. Analysis shows that for a gas carrying particles in a duct, the Stokes number is the dimensionless ratio of the force required to cause a particle to change direction and the drag force available to bring about that change. The greater the value of Stokes number is above unity, the greater is the tendency for particles to impact with the airway walls and so be captured. There are obvious similarities between the conditions required for collection of particles in the gas cyclone and those required for deposition of particles by inertial impaction on the walls of the lungs and on fibres within filters (see Section 8.2). In each case, for the particle not to be captured when the gas changes direction, the available drag force must be sufficient to bring about the change of direction of the particle.

For large industrial cyclones, the Stokes number, like the Euler number defined previously, is approximately independent of Reynolds number. For suspensions of concentration less than about 5 g/m^3, the Stokes and Euler numbers are usually constant for a given cyclone geometry (i.e. a set of geometric proportions relative to cyclone diameter D). The geometries and values of Eu and Stk_{50} for two common industrial cyclones, the Stairmand high efficiency (HE) and the Stairmand high rate (HR) are included here in Figure 8.5 as examples of the available standard cyclone designs. Note that here, as in all cyclones, the gas outlet pipe projects into the body of the cyclone by a distance J. As indicated earlier, this projection is known as the vortex finder and is crucial to the design as it stops gas by-passing the main cyclone body and going straight to the exit.

The use of the two dimensionless groups Eu and Stk_{50} in cyclone scale-up and design is demonstrated in the worked examples at the end of this chapter.

As can be seen from Equation (8.21), the separation efficiency is described only by the cut size x_{50} and no regard is given to the shape of the grade efficiency curve. If the whole grade efficiency curve is required in performance calculations, it may be generated around the given cut size using plots or analytical functions of a generalized grade efficiency function available from the literature or from previously measured data. For example, Green and Southard (2018) give the grade efficiency expression:

$$\text{Grade efficiency} = \frac{(x/x_{50})^2}{\left[1 + (x/x_{50})^2\right]} \tag{8.22}$$

for a reverse flow cyclone with the geometry:

A	B	C	E	J	K	N
4.0	2.0	2.0	0.25	0.625	0.5	0.5

[Letters refer to the cyclone geometry diagram shown in Figure 8.5(a)].

This expression gives rise to the grade efficiency curve shown in Figure 8.6 for an x_{50} cut size of 5 μm. It should be noted, however, that the shape of the grade efficiency curve can be affected by operating conditions and the cyclone size or design, particularly the configuration of the solids and gas exit arrangements.

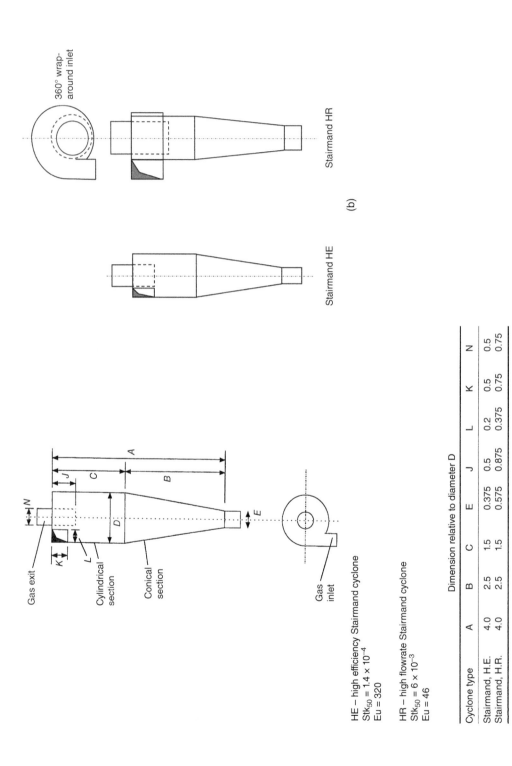

HE – high efficiency Stairmand cyclone
$Stk_{50} = 1.4 \times 10^{-4}$
$Eu = 320$

HR – high flowrate Stairmand cyclone
$Stk_{50} = 6 \times 10^{-3}$
$Eu = 46$

Cyclone type	Dimension relative to diameter D							
	A	B	C	E	J	L	K	N
Stairmand, H.E.	4.0	2.5	1.5	0.5	0.375	0.2	0.5	0.5
Stairmand, H.R.	4.0	2.5	1.5	0.875	0.575	0.375	0.75	0.75

(a)

Stairmand HE

360° wrap-around inlet

Stairmand HR

(b)

Figure 8.5 (a) Geometries and Euler and Stokes numbers for Stairmand HE and HR cyclones. (b) Comparison of the geometries of Stairmand HE and HR cyclones

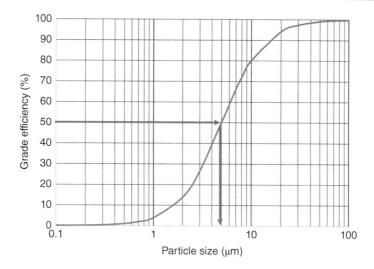

Figure 8.6 Grade efficiency curve described by Equation (8.22) for a cut size $x_{50} = 5\,\mu m$

8.1.4 Range of Operation of Cyclones

One of the most important characteristics of gas cyclones is the way in which their efficiency is affected by flow rate. For a particular cyclone and inlet particle concentration, total efficiency of separation and pressure drop vary with gas flow rate as shown in Figure 8.7. Theory predicts that efficiency increases with increasing gas flow rate, because of the resulting increase in centrifugal force. However, in practice, the total efficiency curve falls away at high flow rates because re-entrainment of separated solids increases

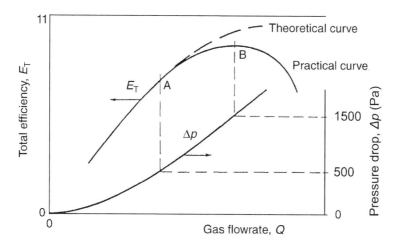

Figure 8.7 Total separation efficiency and pressure drop versus gas flow rate through a typical reverse flow cyclone

with increased turbulence at high velocities. Optimum operation is achieved somewhere between points A and B, where maximum total separation efficiency is achieved with reasonable pressure loss (and hence reasonable power consumption). The position of point B changes only slightly for different dusts. Correctly designed and operated cyclones should operate at pressure drops within a recommended range; and this, for most cyclone designs operated at ambient conditions, is between 500 and 1500 Pa. Within this range, the total separation efficiency E_T increases with applied pressure drop, in accordance with the inertial separation theory shown above.

Above the top limit of this pressure range the total efficiency no longer increases with increasing pressure drop and it may actually decline due to re-entrainment of dust from the solids outlet. It is, therefore, wasteful of energy to operate cyclones above the limit. At pressure drops below the bottom limit of the range, the cyclone represents little more than a settling chamber, giving low efficiency due to low velocities within it which may not be capable of generating a stable vortex.

8.1.5   Some Practical Design and Operation Details of Cyclones

The following practical considerations for design and operation of reverse flow gas cyclones are among those listed by Svarovsky (1986).

Effect of dust loading on efficiency

One of the important operating variables affecting total efficiency is the concentration of particles in the suspension (known as the dust loading). Generally, high dust loadings (above about $5 \, \text{g/m}^3$) lead to higher total separation efficiencies due to particle enlargement through clustering and agglomeration (caused, for example, by the effect of humidity).

Cyclone types

The many reverse flow cyclone designs available today may be divided into two main groups: high efficiency designs (e.g. Stairmand HE) and the high rate designs (e.g. Stairmand HR). High efficiency cyclones give high recoveries and are characterized by relatively small inlets and gas outlets. The high rate designs have lower total efficiencies but offer low resistance to flow so that a unit of a given size will give much higher gas capacity than a high efficiency design of the same body diameter. The high rate cyclones have large inlets and gas outlets and are usually shorter. The geometries and values of Eu and Stk_{50} for two common cyclones, the Stairmand HE and the Stairmand HR, are given in Figure 8.5. The standard entry velocity for both Stairmand types is 15.2 m/s, which gives a pressure drop in the desired range. The recommended volumetric throughput, which is equivalent to about 15.2 m/s, is $1.5 \, D^2 \, \text{m}^3/\text{s}$ for the HE and $4.3 \, D^2 \, \text{m}^3/\text{s}$ for the HR. Readers may like to check this using Equations (8.1) and (8.2) and the cyclone geometries given in Figure 8.5.

For well-designed cyclones there is a direct correlation between Eu and Stk_{50}. High values of the resistance coefficient usually lead to low values of Stk_{50} (therefore low cut sizes and high efficiencies), and vice versa. The general trend can be described by the following *approximate* empirical correlation:

$$Eu = \sqrt{\frac{12}{Stk_{50}}}$$

(8.23)

Abrasion

Abrasion in gas cyclones is an important aspect of cyclone performance and it is affected by the way cyclones are installed and operated as much as by the material construction and design. Materials of construction are usually steels of different grades, sometimes lined with rubber, refractory or other material. Within the cyclone body there are two critical zones for abrasion: in the cylindrical part just beyond the inlet opening and in the conical part near the dust discharge.

Attrition of solids

Attrition or break-up of solids is known to take place in gas cyclones, but little is known about how it is related to particle properties, although large particles are more likely to be affected by attrition than finer fractions. Attrition is most detectable in recirculating systems such as fluidized beds where cyclones are used to return the carry-over material back to the bed (see Chapters 6 and 11). The complete inventory of the bed may pass through the cyclones many times per hour and the effect of attrition is thus increased through repeated stress on the particles.

Blockages

Blockages, usually caused by overloading of the solids outlet, are one of the most common causes of failure in cyclone operation. If the solids outlet becomes blocked, the cylindrical section of the cyclone rapidly fills up with dust, the pressure drop increases and efficiency falls dramatically. Blockages can arise due to mechanical defects in the cyclone body (bumps on the cyclone cone, protruding welds or gaskets) and/or changes in chemical or physical properties of the solids (e.g. condensation of water vapour from the gas onto the surface of particles).

Discharge hoppers and diplegs

The design of the solids discharge is important for correct functioning of a gas cyclone. If the cyclone operates under vacuum, any inward leakages of air at the discharge end cause particles to be re-entrained and this leads to a sharp decrease in separation efficiency. If the cyclone is under pressure, outward leakages may cause a slight increase in separation efficiency, but also result in loss of product and pollution of the local environment. It is therefore best to keep the solids discharge as gas-tight as possible.

The strong vortex inside a cyclone can reach into the space underneath the solids outlet and it is important that no powder is allowed to build up within at least one cyclone diameter below the underflow orifice. A conical vortex breaker positioned just under the dust discharge orifice may be used to prevent the vortex from intruding into the discharge hopper below. Some cyclone manufacturers use a 'stepped' cone to counter the effects of re-entrainment and abrasion.

In fluidized beds with internal cyclones, *diplegs* are used to return the collected entrained particles into the fluidized bed. Diplegs are vertical pipes connected directly to the solids discharge of the cyclone and extending down to below the fluidized bed surface. Particles discharged from the cyclone collect as a moving settled suspension in the lower part of the dipleg before it enters the bed. The level of the settled suspension in the dipleg is always higher than the fluidized bed surface and it provides a necessary resistance to minimize the counter-current flow of gas up the dipleg which could otherwise re-entrain particles and reduce the cyclone efficiency.

Cyclones in series

Connecting cyclones in series is often done in practice to increase recovery. Usually, the primary cyclone would be of medium or low efficiency design and the secondary and subsequent cyclones of progressively more efficient design or smaller diameter (Figure 8.8).

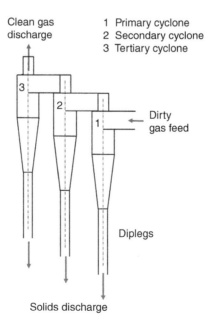

Figure 8.8   Cyclones in series

Cyclones in parallel

The x_{50} cut size achievable for a given cyclone geometry and gas throughput decreases with decreasing cyclone size [see Equation (8.21)]. The size of a single cyclone for treating a given volumetric flow rate of gas is determined by that gas flow rate through Equations (8.1) and (8.2). For large gas flow rates, the resulting cyclone may be so large that the x_{50} cut size is unacceptably high. (It is also the case that very large cyclones are difficult and expensive to manufacture.) The solution is to split the gas flow into several smaller cyclones operating in parallel. In this way, both the operating pressure drop and x_{50} cut size requirements can be achieved. The worked examples at the end of the chapter demonstrate how the number and diameter of cyclones in parallel are estimated.

8.2 FILTRATION

Filtration is a common method of particle removal from a gas. Filters are generally more efficient than cyclones, but at the cost of greater pressure drop and increased maintenance. Filters can be used alone or mounted downstream of a cyclone and used to remove particles which are too small for the cyclone to collect. Gas filtration refers to the process by which the dust-laden gas passes through a permeable material, commonly termed the filter *medium*. Two basic types of filtration are *depth filtration* and *barrier* or *surface filtration*, according to where particle collection occurs (Figure 8.9). It is almost always the case in filtration that the particles to be collected are much smaller than the elements of the filter medium and the spaces between them, so particles are initially able to penetrate deeply into the structure. This is depth filtration, and some filters operate only in this mode; they are generally designed for one-off use and subsequent disposal. This is typical of contamination-critical applications in the pharmaceutical and nuclear industries, for example. Their design relies on the principles introduced in Section 5.6: inertial deviation from the gas streamlines results in a high collection efficiency for larger particles and diffusional collection is strong for smaller ones. The overall effect is that depth filters generally show a *most penetrating size* at around 1 μm.

In general process industry applications, surface filters are more common and have the advantage of being cleanable. In a perfect surface filter, particles are not able to penetrate into the medium and are separated at the surface. In practice, however, it is almost always the case for a new filter that an initial period of depth filtration occurs – termed the

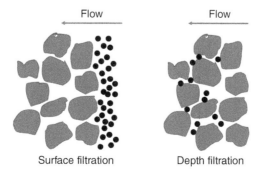

Figure 8.9 Comparison of surface filtration and depth filtration

conditioning period – during which a thin layer of particles is formed on the filter surface, and it is this which then acts to separate subsequent particles from the flow. Some filter media are designed with a finer surface layer to enhance this effect.

Design of a surface filtration system involves the selection of a medium suitable for the particles to be collected and choice of an operating *face velocity*, from which the required filter area is easily calculated. The face velocity is the volumetric flow of gas divided by the superficial area of the filter medium available for flow. The choice of face velocity is very system-specific: it is dependent on the nature of the particles to be separated, their size distribution and concentration, the gas temperature, the method of filter cleaning and the type of filter medium (Seville, 1997). For many ambient temperature applications, face velocities lie in the range 0.5–5 cm/s. Low face velocity means large filter area and high capital cost. Too high a face velocity not only results in an initially high pressure drop but may make the filter inoperable because of penetration of particles into the filter structure, inadequate cleaning and uncontrolled subsequent pressure rise.

Common filter media include woven and non-woven fabrics, ceramics, sintered metals and paper. A much-used arrangement is for the medium to be formed into cylinders or bags with wire supports in the case of fabrics and paper (Figure 8.10). The cylinders are hung inside a housing (known as a baghouse). Gas flows from the outside inwards through the medium and into a plenum chamber at the top. The dust which builds up on the outside of the bags is periodically removed by discharging a pulse of compressed air through a venturi nozzle and into the bag or by mechanically shaking the bag. As the filter cake builds up, the pressure drop across the cake and medium increases. The cleaning action, which may be triggered by a pressure drop signal or by a timer, removes most of the cake and the pressure drop falls back but usually not to its original value (Figure 8.11). The original *clean medium* pressure drop is not achieved due to dust particles remaining both on the medium surface and within its depth.

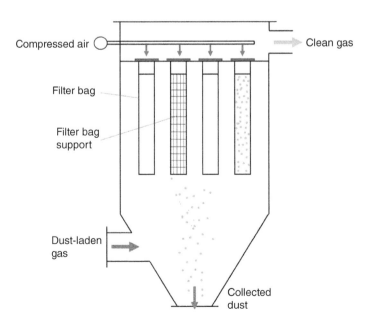

Figure 8.10   Bag filter arrangement for gas cleaning

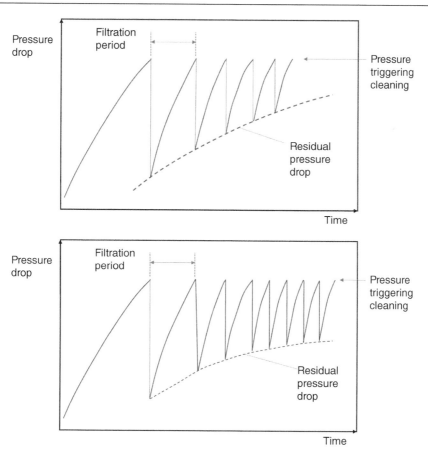

Figure 8.11 Examples of pressure drop changes over several periods of filtration and cleaning during conditioning of a new filter medium (top) uncontrolled pressure rise, leading to inoperability, (bottom) controlled pressure rise typical of successful conditioning

Note (a) that the pressure drop at which cleaning is triggered and the average pressure drop are both much larger than the initial pressure drop of the clean medium alone, and (b) that the cycle time reduces as the medium becomes conditioned and reaches steady operation. In practice, for large filter installations cleaning of rows of elements occurs in sequence so as to average out perturbations to the flow and the overall pressure drop and avoid overloading the solids discharge arrangement.

For typical dust sizes and face velocities, the Reynolds number for flow through the dust cake is low (less than 10) and the Carman–Kozeny equation (Equation 6.9) describes the flow-pressure drop relationship well.

$$\frac{(-\Delta p)}{H} = 180 \frac{\mu U}{x^2} \frac{(1-\varepsilon)^2}{\varepsilon^3} \tag{6.9}$$

The Carmen–Kozeny equation, in its original form [Equation (6.8)], may be used to describe the pressure drop across the filter medium:

$$\frac{(-\Delta p)}{H} = K_3 \frac{(1-\varepsilon)^2}{\varepsilon^3} \mu U S_v^2 \tag{6.8}$$

where S_v is the specific surface area of the material of the medium, ε is the void fraction of the medium and K_3 is a constant depending on its geometrical structure.

8.3   WORKED EXAMPLES

WORKED EXAMPLE 8.1 – DESIGN OF A CYCLONE

Determine the diameter and number of gas cyclones required to treat $2\,\mathrm{m}^3/\mathrm{s}$ of ambient air (viscosity, $18.25 \times 10^{-6}\,\mathrm{Pas}$; density, $1.2\,\mathrm{kg/m}^3$) laden with solids of density $1000\,\mathrm{kg/m}^3$ at a suitable pressure drop and with a cut size of $4\,\mu\mathrm{m}$. Use a Stairmand HE (high efficiency) cyclone for which $Eu = 320$ and $Stk_{50} = 1.4 \times 10^{-4}$.

$$\text{Optimum pressure drop} = 100\,\mathrm{m\ gas}$$
$$= 100 \times 1.2 \times 9.81\,\mathrm{Pa}$$
$$= 1177\,\mathrm{Pa}$$

Solution

From Equation (8.1), characteristic velocity, $v = 2.476\,\mathrm{m/s}$.

Hence, from Equation (8.2), diameter of cyclone $D = 1.014\,\mathrm{m}$.

With this cyclone, using Equation (8.21), cut size, $x_{50} = 4.34\,\mu\mathrm{m}$.

This is too high, and we must therefore opt for passing the gas through several smaller cyclones in parallel.

Assuming that n cyclones in parallel are required and that the total flow is equally divided, then for each cyclone the flow rate will be $q = 2/n$.

Therefore, from Equations (8.1) and (8.2), the new cyclone diameter, $D = 1.014/n^{0.5}$. Substituting in Equation (8.21) for D, the required cut size and v (2.476 m/s, as originally calculated, since this is determined solely by the pressure drop requirement), we find that

$$n = 1.386$$

We will therefore need two cyclones. Now with $n = 2$, we recalculate the cyclone diameter from $D = 1.014/n^{0.5}$ and the actual achieved cut size from Equation (8.21).

Thus, $D = 0.717\,\mathrm{m}$, and using this value for D in Equation (8.21) together with the required cut size and $v = 2.476\,\mathrm{m/s}$, we find that the actual cut size is $3.65\,\mu\mathrm{m}$.

Therefore, two 0.717 m diameter Stairmand HE cyclones in parallel will give a cut size of $3.65\,\mu\mathrm{m}$ using a pressure drop of 1177 Pa.

WORKED EXAMPLE 8.2

Tests on a reverse flow gas cyclone give the results shown in the table below:

Size range (µm)	0–5	5–10	10–15	15–20	20–25	25–30
Feed size analysis m (g)	10	15	25	30	15	5
Course product size analysis m_c (g)	0.1	3.53	18.0	27.3	14.63	5.0

(a) From these results determine the total efficiency of the cyclone.

(b) Plot the grade efficiency curve and hence show that the x_{50} cut size is 10 µm.

(c) The dimensionless constants describing this cyclone are: $Eu = 384$ and $Stk_{50} = 1 \times 10^{-3}$. Determine the diameter and number of cyclones to be operated in parallel to achieve this cut size when handling 10 m³/s of a gas of density 1.2 kg/m³ and viscosity 18.4×10^{-6} Pa s, laden with dust of particle density 2500 kg/m³. The available pressure drop is 1200 Pa.

(d) What is the actual cut size of your design?

Solution

(a) From the test results:

Mass of feed, $M = 10 + 15 + 25 + 30 + 15 + 5 = 100$ g

Mass of coarse product, $M_c = 0.1 + 3.53 + 18.0 + 27.3 + 14.63 + 5.0 = 68.56$ g

Therefore, from Equation (8.5), total efficiency,

$$E_T = \frac{M_c}{M} = 0.6856 \text{ (or } 68.56\%)$$

(b) From Equation (8.7), grade efficiency,

$$G(x) = \frac{M_c}{M} \frac{dF_c dx}{dF/dx} = E_T \frac{dF_c/dx}{dF/dx}$$

In this case, $G(x)$ may be obtained directly from the results table as:

$$G(x) = \frac{m_c}{m}$$

And so, the grade efficiency curve data becomes:

Size range (µm)	0–5	5–10	10–15	15–20	20–25	25–30
$G(x)$	0.01	0.235	0.721	0.909	0.975	1.00

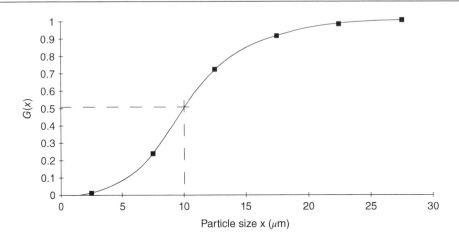

Figure 8.W2.1   Grade efficiency curve

Plotting these data gives $x_{50} = 10$ μm, as may be seen in Figure 8.W2.1.

For interest, we can calculate the size distributions of the feed dF/dx and the coarse product dF_c/dx:

Size range (pm)	0–5	5–10	10–15	15–20	20–25	25–30
dF_c/dx	0.00146	0.0515	0.263	0.398	0.2134	0.0729
dF/dx	0.1	0.15	0.25	0.30	0.15	0.05

We can then verify the calculated $G(x)$ values. For example, in the size range 10–15:

$$G(x) = E_T \frac{dF_c/dx}{dF/dx} = 0.6856 \times \frac{0.263}{0.25} = 0.721$$

(c) Using Equation (8.1), noting that the allowable pressure is 1200 Pa, we calculate the characteristic velocity v:

$$v = \sqrt{\frac{2\Delta p}{Eu\rho_f}} = \sqrt{\frac{2 \times 1200}{384 \times 1.2}} = 2.282 \,\text{m/s}$$

If we have n cyclones in parallel then assuming even distribution of the gas between the cyclones, flow rate to each cyclone $q = Q/n$ and from Equation (8.2),

$$D = \sqrt{\frac{4Q}{n\pi v}} = \sqrt{\frac{4 \times 10}{n\pi \times 2.282}} = \frac{2.362}{\sqrt{n}}$$

Now substitute this expression for D and the required cut size x_{50} in Equation (8.21) for Stk_{50}:

$$Stk_{50} = \frac{x_{50}^2 \rho_p v}{18\mu D}$$

$$1 \times 10^{-3} = \frac{\left(10 \times 10^{-6}\right)^2 \times 2500 \times 2.282}{18 \times 18.4 \times 10^{-6} \times (2.362/\sqrt{n})}$$

giving $n = 1.88$.

We therefore require two cyclones. With two cyclones, using all of the allowable pressure drop the characteristic velocity will be the same (2.282 m/s) and the required cyclone diameter may be calculated from the expression derived above:

$$D = \frac{2.362}{\sqrt{n}}$$

giving $D = 1.67$ m.

(d) The actual cut size achieved with two cyclones is calculated from Equation (8.21) with $D = 1.67$ m and $v = 2.282$ m/s:

$$\text{Actual cut size, } x_{50} = \sqrt{\frac{1 \times 10^{-3} \times 18 \times 18.4 \times 10^{-6} \times 1.67}{2500 \times 2.282}} = 9.85 \times 10^{-6} \text{ m}$$

Summary. Two cyclones (described by $Eu = 384$ and $Stk_{50} = 1 \times 10^{-3}$) of diameter 1.67 m and operating at a pressure drop of 1200 Pa will achieve an equiprobable cut size of 9.85 μm.

TEST YOURSELF

8.1 Typically, in what particle size range are industrial cyclone separators useful?

8.2 With the aid of a sketch, describe the operation of a reverse flow cyclone separator.

8.3 What forces act on a particle inside a cyclone separator? What factors govern the magnitudes of each of these forces?

8.4 For a gas–particle separation device, define *total efficiency* and *grade efficiency*. Using these definitions and the mass balance, derive an expression relating the size distributions of the feed, coarse product and fine product for a gas–particle separation device.

8.5 What is meant by the x_{50} cut size?

8.6 Define the two dimensionless numbers which are used in the scale-up of cyclone separators.

8.7 Theory suggests that the total efficiency of a cyclone separator will increase with increasing gas flow rate. Explain why, in practice, cyclone separators are operated within a certain range of pressure drops.

8.8 Under what conditions might we choose to operate cyclone separators in parallel?

8.9 In a given application, why would you choose to use (a) a cyclone, (b) a filter, (c) a cyclone followed by a filter?

8.10 In a particular application, a filter is used to collect a dust of size distribution 1–50 μm. A cyclone with a cut size of 10 μm is now installed before the filter in order to collect the coarse fraction of the dust. (The gas flow rate is unchanged.) What happens to the pressure drop across the filter?

8.11 Which takes up most space: a cyclone or a filter, assuming that both are capable of removing the dust to the required efficiency? (Take a gas flow of 1 m^3/s, a cyclone entry velocity of 15 m/s and a filter face velocity of 3 cm/s. Think about how you are going to arrange the filter bags and allow some space for the plenum chamber and cleaning arrangement.)

EXERCISES

8.1 A gas–particle separation device is tested and gives the results shown in the table below:

Size range (μm)	0–10	10–20	20–30	30–40	40–50
Range mean (μm)	5	15	25	35	45
Feed mass (kg)	45	69	120	45	21
Coarse product mass (kg)	1.35	19.32	99.0	44.33	21.0

(a) Find the total efficiency of the device.

(b) Produce a plot of the grade efficiency for this device and determine the equiprobable cut size.

[Answer: (a) 61.7%; (b) 19.4 μm.]

8.2 A gas–particle separation device is tested and gives the results shown in the table below:

Size range (μm)	6.6–9.4	9.4–13.3	13.3–18.7	18.7–27.0	27.0–37.0	37.0–53.0
Feed size distribution	0.05	0.2	0.35	0.25	0.1	0.05
Coarse product size distribution	0.016	0.139	0.366	0.30	0.12	0.06

Given that the total mass of feed is 200 kg and the total mass of coarse product collected is 166.5 kg:

(a) Find the total efficiency of the device.

(b) Determine the size distribution of the fine product.

(c) Plot the grade efficiency curve for this device and determine the equiprobable size.

(d) If this same device were fed with a material with the size distribution below, what would be the resulting coarse product size distribution?

Size range (pm)	6.6–9.4	9.4–13.3	13.3–18.7	18.7–27.0	27.0–37.0	37.0–53.0
Feed size distribution	0.08	0.13	0.27	0.36	0.14	0.02

[Answer: (a) 83.25%; (b) 0.219, 0.503, 0.271, 0.0015, 0.0006, 0.0003; (c) 10.5 μm; (d) 0.025, 0.089, 0.276, 0.422, 0.165, 0.024].

8.3 (a) Explain what a 'grade efficiency curve' is with reference to a gas–solids separation device and sketch an example of such a curve for a gas cyclone separator.

(b) Determine the diameter and number of Stairmand HR gas cyclones to be operated in parallel to treat $3\,\text{m}^3/\text{s}$ of gas of density $0.5\,\text{kg/m}^3$ and viscosity 2×10^{-5} Pa s carrying a dust of density $2000\,\text{kg/m}^3$. A x_{50} cut size of at most 7 μm is to be achieved at a pressure drop of 1200 Pa.

(For a Stairmand HR cyclone: $Eu = 46$ and $Stk_{50} = 6 \times 10^{-3}$.)

(c) Give the actual cut size achieved by your design.

(d) A change in process conditions requirements necessitates a 50% drop in gas flow rate. What effect will this have on the cut size achieved by your design?

[Answer: (a) Two cyclones 0.43 m in diameter; (b) $x_{50} = 6.8\,\mu m$; (c) new $x_{50} = 9.6\,\mu m$.]

8.4 (a) Determine the diameter and number of Stairmand HE gas cyclones to be operated in parallel to treat $1\,m^3/s$ of gas of density $1.2\,kg/m^3$ and viscosity 18.5×10^{-6} Pas carrying a dust of density $1000\,kg/m^3$. An x_{50} cut size of at most $5\,\mu m$ is to be achieved at a pressure drop of 1200 Pa.

(For a Stairmand HE cyclone: $Eu = 320$ and $Stk_{50} = 1.4 \times 10^{-4}$.)

(b) Give the actual cut size achieved by your design.

[Answer: (a) One cyclone, 0.714 m in diameter; (b) $x_{50} = 3.6\,\mu m$.]

8.5 Stairmand HR cyclones are to be used to clean up $2.5\,m^3/s$ of ambient air (density, $1.2\,kg/m^3$; viscosity, 18.5×10^{-6} Pa s) laden with dust of particle density $2600\,kg/m^3$. The available pressure drop is 1200 Pa and the required cut size is to be not more than $6\,\mu m$.

(a) What size of cyclones is required?

(b) How many cyclones are needed and in what arrangement?

(c) What is the actual cut size achieved?

[Answer: (a) Diameter = 0.311 m; (b) five cyclones in parallel; (c) actual cut size = $6\,\mu m$.]

9

Storage and Flow of Powders – Hopper Design

The short-term storage of raw materials, intermediates and products in the form of particulate solids in process plants presents problems which are often underestimated and which, as was pointed out in the introduction of this book, may frequently be responsible for production stoppages.

One common problem in such plants is the interruption of flow from the hopper, or converging section beneath a storage vessel for powders. However, a technology is available which will allow us to design such storage vessels to ensure flow of the powders when desired. Within the bounds of a single chapter it is not possible to cover all aspects of the gravity flow of unaerated powders, and so here we will confine ourselves to a study of the design philosophy to ensure flow from conical hoppers when required. The approach used is that first proposed by Jenike (1964).

9.1 MASS FLOW AND CORE FLOW

Mass flow: In perfect mass flow, all the powder in a silo is in motion whenever any of it is drawn from the outlet as shown in Figure 9.1(a). The flowing channel coincides with the walls of the silo. Mass flow hoppers have smooth and steep walls. Figure 9.2(a–d) shows sketches taken from a sequence of photographs of a hopper operating in mass flow. The use of alternate layers of coloured powder in this sequence clearly shows the key features of the flow pattern. Note how the powder surface remains level until it reaches the sloping section.

Core flow: This occurs when the powder flows towards the outlet of a silo in a channel formed within the powder itself [Figure 9.1(b)]. Figure 9.3(a–d) shows images taken from a sequence of photographs of a hopper operating in core flow. Note the regions of powder lower down in the hopper are stagnant until the hopper is almost empty. The inclined surface of the powder gives rise to size segregation (see Chapter 10).

Introduction to Particle Technology, Third Edition. Martin Rhodes and Jonathan Seville.
© 2024 John Wiley & Sons Ltd. Published 2024 by John Wiley & Sons Ltd.
Website: www.wiley.com/go/rhodes/particle3e

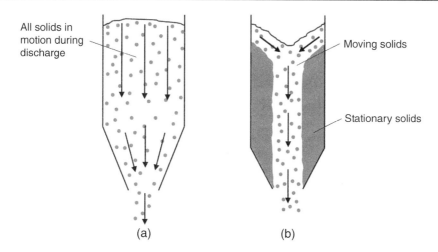

Figure 9.1   Mass flow and core flow in hoppers: (a) mass flow; (b) core flow

Mass flow has many advantages over core flow. In mass flow, the motion of the powder is uniform and steady state can be closely approximated over most of the discharge
period. The bulk density of the discharged powder is constant and practically independent of the height of material in the silo. In mass flow, stresses are generally low throughout the mass of solids, giving low compaction of the powder. There are no stagnant
regions in the mass flow hopper. Thus, the risk of product degradation is small compared
with the case of the core flow hopper. The first-in–first-out flow pattern of the mass flow
hopper ensures a narrow range of residence times for solids. Also, segregation of particles
according to size is far less of a problem in mass flow than in core flow. Mass flow has one
disadvantage which may be overriding in certain cases. Friction between the moving
solids and the silo and hopper walls results in erosion, which gives rise to contamination

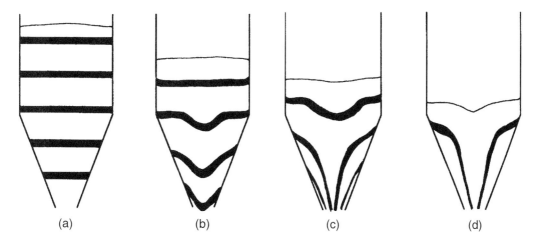

Figure 9.2   Sequence of images taken from photographs showing a mass flow pattern as a hopper
discharges. Order of the images is (a), (b), (c), (d). (The black bands are layers of coloured tracer
particles.)

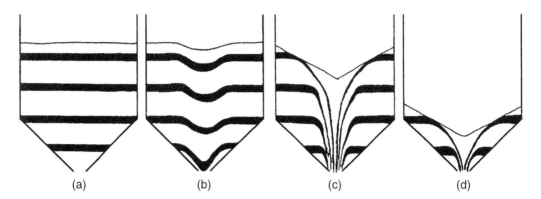

Figure 9.3 Sequence of images taken from photographs showing a core flow pattern as a hopper discharges. . Order of the images is (a), (b), (c), (d). (The black bands are layers of coloured tracer particles.)

of the solids by the material of the hopper wall. If either contamination of the solids or serious erosion of the wall material is unacceptable, then a core flow hopper should be considered. However, for the reasons given above, core flow is generally to be avoided and design for core flow is not considered here.

For conical hoppers the slope angle required to ensure mass flow depends on the powder/powder friction and the powder/wall friction, which were introduced in Section 2.4. Later we will see how these are quantified and how it is possible to determine the conditions which give rise to mass flow. Note that there is no such thing as a hopper which always delivers mass flow; a hopper which gives mass flow with one powder may give core flow with another.

9.2   THE DESIGN PHILOSOPHY

We will consider the blockage or obstruction to flow called *arching*; if arching does not occur then flow will take place (Figure 9.4). In general, powders develop strength under the action of compacting stresses. The greater the compacting stress, the greater the strength developed (Figure 9.5). (Free-flowing solids such as dry coarse sand do not develop strength as the result of compacting stresses and will always flow.)

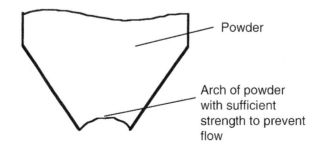

Figure 9.4   Arching in the flow of powder from a hopper

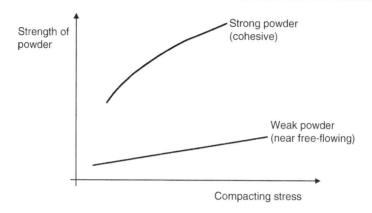

Figure 9.5   Variation of strength of powder with compacting stress for cohesive and free-flowing powders

9.2.1   Flow – No Flow Criterion

Gravity flow of a solid in a channel will take place provided the strength developed by the solids under the action of compacting stresses is insufficient to support an obstruction to flow. An arch occurs when the strength developed by the solids is greater than the stresses acting within the surface of the arch.

9.2.2   The Hopper Flow Factor *ff*

The *hopper flow factor ff*, relates the stress developed in a particulate solid to the compacting stress acting in a particular hopper. The hopper flow factor is defined as:

$$ff = \frac{\sigma_C}{\sigma_D} = \frac{\text{compacting stress in the hopper}}{\text{stress developed in the powder}} \qquad (9.1)$$

A high value of *ff* means low *flowability* since high σ_C means greater compaction, and a low value of σ_D means more chance of an arch forming.

The hopper-flow factor depends on:

• the nature of the solid;

• the nature of the wall material;

• the slope of the hopper wall.

These relationships will be quantified later.

9.2.3   Unconfined Yield Stress σ_y

We are interested in the strength developed by the powder in the arch surface. Suppose that the yield stress (i.e. the stress which causes flow) of the powder in the exposed surface of the arch is σ_y. (Note that this is similar to, but different from, the yield stress introduced in Chapter 2, which is a material property. The yield stress of a powder is a variable which depends on the state of the powder, as explained below.) The stress σ_y is known as the *unconfined yield stress* of the powder. Then if the stresses developed in the powder forming the arch are greater than the unconfined yield stress of the powder in the arch, flow will occur. That is, for flow:

$$\sigma_D > \sigma_y \tag{9.2}$$

Incorporating Equation (9.1), this criterion may be rewritten as:

$$\frac{\sigma_C}{ff} > \sigma_y \tag{9.3}$$

9.2.4 Powder Flow Function

Obviously, the unconfined yield stress, σ_y, of the solids varies with compacting stress, σ_c:

$$\sigma_y = \mathrm{fn}(\sigma_C)$$

This relationship is determined experimentally and is usually presented graphically (Figure 9.6). This relationship has several different names, some of which are misleading. Here we will call it the *powder flow function*. Note that it is a function *only* of the powder properties.

9.2.5 Critical Conditions for Flow

From Equation (9.3), the limiting condition for flow is:

$$\frac{\sigma_C}{ff} = \sigma_y$$

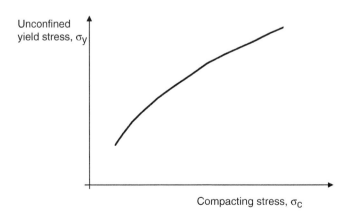

Figure 9.6 Powder flow function (a property of the solids only)

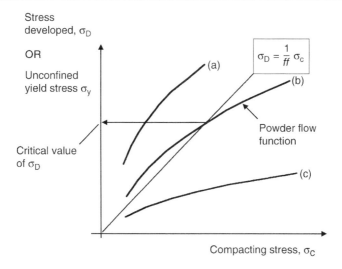

Figure 9.7   Determination of critical conditions for flow

This may be plotted on the same axes as the powder flow function (unconfined yield stress σ_y and compacting stress σ_C) in order to reveal the conditions under which flow will occur for this powder in the hopper. The limiting condition gives a straight line of slope $1/ff$. Figure 9.7 shows such a plot.

Where the powder has a yield stress greater than σ_C/ff, no flow occurs [powder flow function (a)]. Where the powder has a yield stress less than σ_C/ff, flow occurs [powder flow function (c)]. For powder flow function (b) there is a critical condition where unconfined yield stress, σ_y, is equal to stress developed in the powder, σ_C/ff. This gives rise to a critical value of stress, σ_{crit}, which is the critical stress developed in the surface of the arch:

$$\text{If actual stress developed} < \sigma_{crit} \Rightarrow \text{no flow}$$
$$\text{If actual stress developed} > \sigma_{crit} \Rightarrow \text{flow}$$

9.2.6   Critical Outlet Dimension

Intuitively, for a given hopper geometry, one would expect the stress developed in the arch to increase with the span of the arch and the weight of solids in the arch. In practice this is the case and the stress developed in the arch is related to the size of the hopper outlet B, and the bulk density, ρ_B, of the material by the relationship:

$$\text{Minimum outlet dimension } B = \frac{H(\theta)\sigma_{crit}}{\rho_B g} \tag{9.4}$$

where $H(\theta)$ is a factor determined by θ, the angle of the hopper wall from the vertical, and g is the acceleration due to gravity. An approximate expression for $H(\theta)$ for conical hoppers is:

$$H(\theta) = 2.0 + \frac{\theta}{60} \tag{9.5}$$

where θ is in degrees.

9.2.7   Summary

From the above discussion of the design philosophy for ensuring mass flow from a conical hopper, we see that the following are required:

(a) the relationship between the strength of the powder in the arch σ_y (unconfined yield stress) and the compacting stress acting on the powder σ_C;

(b) the variation of hopper flow factor ff, with:

   (a) the nature of the powder (characterized by the effective angle of internal friction δ);

   (b) the nature of the hopper wall (characterized by the angle of wall friction Φ_w);

   (c) the slope of the hopper wall (characterized by θ, the semi-included angle of the conical section, i.e. the angle between the sloping hopper wall and the vertical).

Knowing δ, Φ_w, and θ, the hopper flow factor ff can be fixed. The hopper flow factor is therefore a function both of powder properties and of the hopper properties (geometry and the material of construction of the hopper walls).

Knowing the hopper flow factor and the powder flow function (σ_y versus σ_C), the critical stress in the arch can be determined and the minimum size of outlet found corresponding to this stress.

9.3   SHEAR CELL TEST

The data listed above can be found by performing shear cell tests on the powder.

The *Jenike shear cell* (Figure 9.8) allows powders to be compacted to any degree and sheared under controlled load conditions. At the same time the shear force (and hence stress) can be measured.

Generally, powders change bulk density under shear. Under the action of shear, for a specific normal load:

• a loosely packed powder would contract (increase bulk density);

• a very tightly packed powder would expand (decrease bulk density);

• a critically packed powder would not change in volume.

For a particular bulk density there is a critical normal load which gives failure (yield) without volume change. A powder flowing in a hopper is in this critical condition. Yield without volume change is therefore of particular interest to us in design.

Using a standardized test procedure, five or six samples of powder are prepared all having the same bulk density. Referring to the diagram of the Jenike shear cell shown in Figure 9.8(a), a normal load is applied to the lid of the cell and the horizontal force applied to the sample via the bracket and loading pin is recorded. That horizontal force necessary to initiate shear or failure of the powder sample is noted. This procedure is

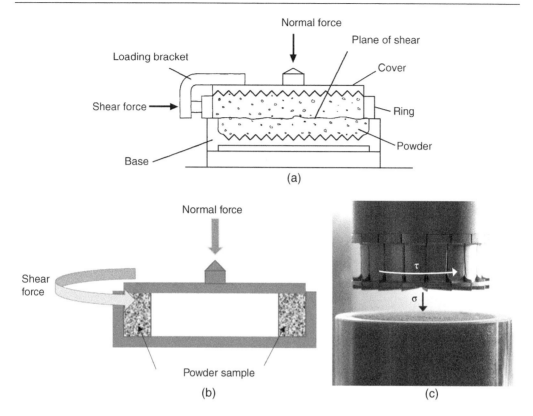

Figure 9.8 (a) Jenike shear cell. (b) Ring shear cell. (c) Powder rheometer (Freeman Technology).

repeated for each identical powder sample but with a reduced normal load applied to the lid each time. This test thus generates a set of five or six pairs of values for normal load and shear force and, hence pairs of values of compacting stress and shear stress for a powder of a particular bulk density. The pairs of values are plotted to give a *yield locus* (Figure 9.9). The end point of the yield locus corresponds to critical flow conditions where

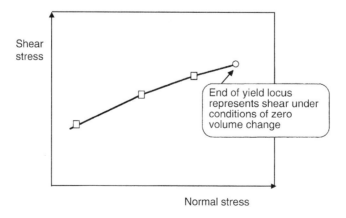

Figure 9.9 A single yield locus

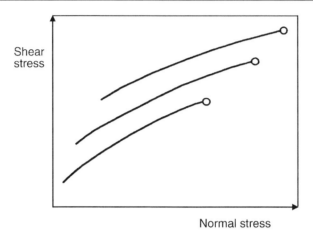

Figure 9.10   A family of yield loci

initiation of flow is not accompanied by a change in bulk density. Experience with the procedure permits the operator to select combinations of normal and shear forces which achieve the critical conditions. This entire test procedure is repeated two or three times with samples prepared to different bulk densities. In this way a family of yield loci is generated (Figure 9.10).

The Jenike apparatus is known as a direct shear cell, in which the shear force is applied in a linear direction. There are other devices in which a rotational shear force is applied to the powder for example, the ring shear cell [Figure 9.8(b)]. Multifunctional powder-testing instruments are also available, sometimes called powder rheometers, which may include a shear cell attachment such as the one shown in Figure 9.8(c). The procedure for these devices is essentially the same as that used with the Jenike shear cell.

These yield loci characterize the flow properties of the unaerated powder. The following section deals with the generation of the powder flow function from this family of yield loci.

9.4   ANALYSIS OF SHEAR CELL TEST RESULTS

The mathematical stress analysis of the flow of unaerated powders in a hopper requires the use of *principal stresses*. We need to use a geometric construction known as *Mohr's stress circle* in order to determine principal stresses from the results of the shear tests.

9.4.1   Mohr's Circle – in Brief

In any stress system there are two planes at right angles to each other in which the shear stresses are zero. The normal stresses acting on these planes are called the principal stresses.

The Mohr's circle represents the possible combinations of normal and shear stresses acting on any plane in a body (or powder) under stress. Figure 9.11 shows how the

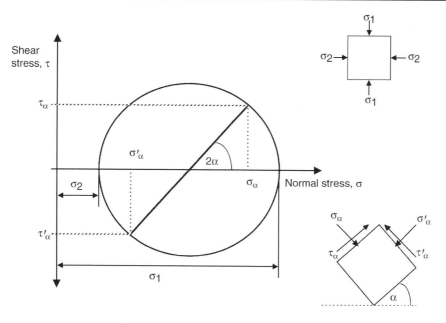

Figure 9.11 Mohr's circle construction

Mohr's circle relates to the stress system. Further information on the background to the development and use of Mohr's circles may be found in Nedderman (1992) or in most texts dealing with the strength of materials and the analysis of stress and strain in solids.

9.4.2 Application of Mohr's Circle to the Analysis of the Yield Locus

Each point on a yield locus represents that point on a particular Mohr's circle for which failure or yield of the powder occurs. A yield locus is then a tangent to all the Mohr's circles representing stress systems under which the powder will fail (flow). For example, in Figure 9.12 Mohr's circles (a) and (b) represent stress systems under which the powder

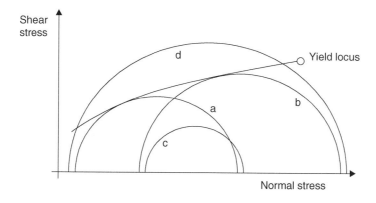

Figure 9.12 Identification of the applicable Mohr's circle

would fail. In circle (c) the stresses are insufficient to cause flow. Circle (d) is not relevant since the system under consideration cannot support stress combinations above the yield locus. It is therefore Mohr's circles which are tangential to yield loci that are important to our analysis.

9.4.3 Determination of σ_y and σ_c

Two tangential Mohr's circles are of particular interest. Referring to Figure 9.13, the smaller Mohr's circle represents conditions at the free surface of the arch: this free surface is a plane in which there is zero shear and zero normal stress and so the Mohr's circle which represents flow (failure) under these conditions must pass through the origin of the shear stress versus normal stress plot. This Mohr's circle gives the (major principal) unconfined yield stress, and this is the value we use for σ_y. The end point of the yield locus is tangent to the larger Mohr's circle and therefore this Mohr's circle represents conditions for critical failure. The major principal stress from this Mohr's circle is taken as our value of compacting stress σ_C.

Pairs of values of σ_y and σ_C are found from each yield locus and plotted against each other to give the powder flow function (Figure 9.6).

9.4.4 Determination of δ from Shear Cell Tests

Experiments carried out on hundreds of bulk solids (Jenike, 1964) have demonstrated that for an element of powder flowing in a hopper:

$$\frac{\sigma_1}{\sigma_2} = \frac{\text{major principal stress on the element}}{\text{minor principal stress on the element}} = \text{a constant}$$

This property of bulk solids is expressed by the relationship:

$$\frac{\sigma_1}{\sigma_2} = \frac{1 + \sin \delta}{1 - \sin \delta} \tag{9.6}$$

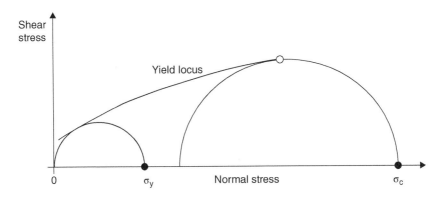

Figure 9.13 Determination of unconfined yield stress σ_y and compacting stress σ_C

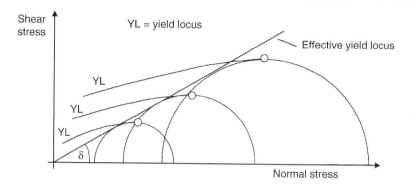

Figure 9.14   Definition of effective yield locus and effective angle of internal friction δ

where δ is the effective angle of internal friction of the solid. In terms of the Mohr's stress circle this means that a straight line through the origin, of slope tan δ, is tangential to all of Mohr's circles for critical failure (Figure 9.14).

This straight line is called the *effective yield locus* of the powder. By drawing in this line, the angle δ can be determined. Note that δ is not a real physical angle within the powder; it is the tangent of the ratio of shear stress to normal stress. Note also that for a free-flowing solid, which does not gain strength under compaction, there is only one yield locus and this locus coincides with the effective yield locus (Figure 9.15). (This type of relationship between normal stress and shear stress is known as *Coulomb friction*; see Section 2.4.)

9.4.5   The Kinematic Angle of Friction between Powder and Hopper Wall, Φ_W

The kinematic angle of friction between powder and hopper wall is otherwise known as the *angle of wall friction* Φ_W. This gives us the relationship between normal stress acting between powder and wall and the shear stress under flow conditions. To determine Φ_W it is necessary to first construct the wall yield locus from shear cell tests. The wall yield locus is determined by shearing the powder against a sample of the wall material under various

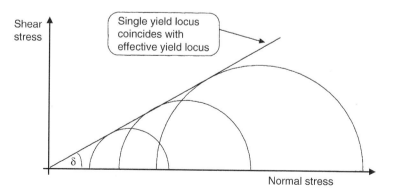

Figure 9.15   Yield locus for a free-flowing powder

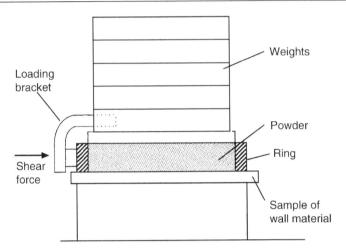

Figure 9.16 Apparatus for the measurement of kinematic angle of wall friction Φ_w

normal loads. The apparatus used is shown in Figure 9.16, and a typical wall yield locus is shown in Figure 9.17.

The kinematic angle of wall friction is given by the gradient of the wall yield locus (Figure 9.17), i.e.:

$$\tan \Phi_w = \frac{\text{shear stress at the wall}}{\text{normal stress at the wall}}$$

9.4.6 Determination of the Hopper Flow Factor, *ff*

The hopper flow factor *ff*, is a function of δ, Φ_w, and θ and can be calculated. However, Jenike (1964) obtained values for a conical hopper and for a wedge-shaped hopper with a slot outlet for values of δ of 30°, 40°, 50°, 60° and 70°. Examples of the 'flow factor charts'

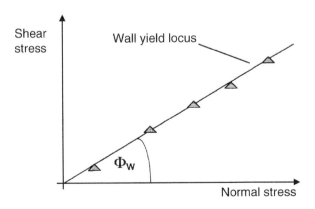

Figure 9.17 Kinematic angle of wall friction Φ_w

for conical hoppers are shown in Figure 9.18. It will be noticed that values of flow factor exist only in a triangular region; this defines the conditions under which mass flow is possible.

The following is an example of the use of these flow factor charts. Suppose that shear cell tests have given us δ and Φ_w equal to 30° and 19°, respectively; then entering the chart for conical hoppers with effective angle of friction δ = 30°, we find that the limiting value

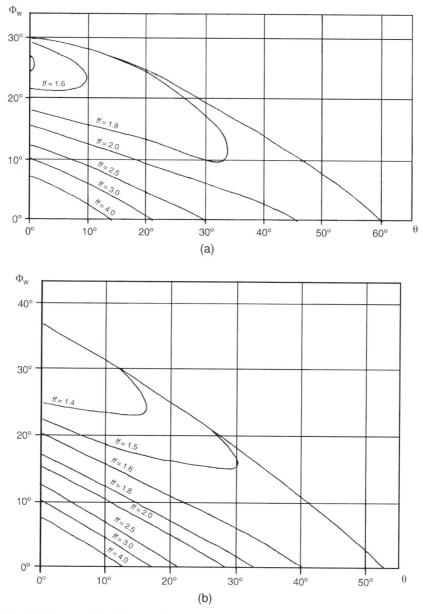

Figure 9.18 (a) Hopper flow factor values for conical channels, δ = 30°. (b) Hopper flow factor values for conical channels, δ = 40°. (c) Hopper flow factor values for conical channels, δ = 50°. (d) Hopper flowfactor values for conical channels, δ = 60°

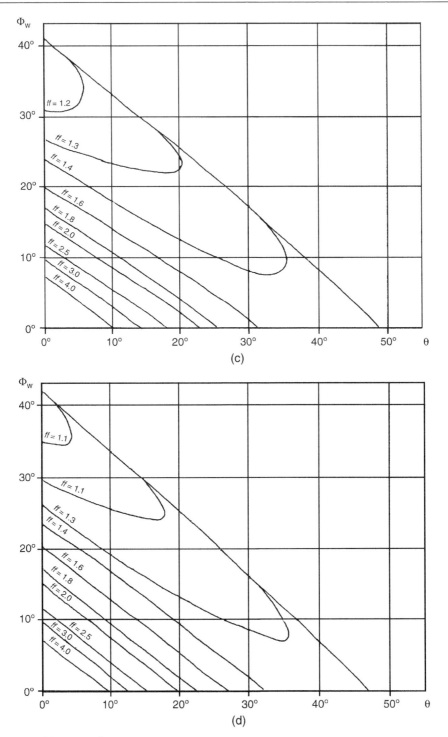

Figure 9.18   (Continued)

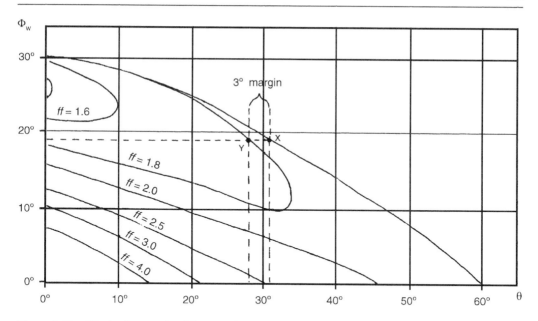

Figure 9.19 Worked example of the use of hopper flow factor charts. Hopper flow factor values for conical channels, $\delta = 30°$

of wall slope θ, to ensure mass flow is 30.5° (point X in Figure 9.19). In practice it is usual to allow a safety margin of 3°, and so, in this case the semi-included angle of the conical hopper θ would be chosen as 27.5°, giving a hopper flow factor $ff = 1.8$ (point Y, Figure 9.19).

The calculation of hopper flow factor ff and semi-included angle θ from first principles is beyond the scope of this text. However, Oko and co-workers (2010) used curve fitting of the hopper flow factor charts to develop correlations for semi-included angle θ, and flow factor ff, as functions of the angle of wall friction Φ_w for specified values of the effective angle of internal friction δ. Their correlations are shown in Tables 9.1 and 9.2.

Table 9.1 Correlations for ff in terms of δ and Φ_w

δ (degrees)	Correlation (θ [degrees])
30	$= -0.0331\Phi_w^2 - 0.6781\Phi_w + 52.663$
40	$= -0.0122\Phi_w^2 - 0.9024\Phi_w + 47.814$
50	$= -0.0027\Phi_w^2 - 1.0962\Phi_w + 46.10$
60	$= -0.0033\Phi_w^2 - 0.9695\Phi_w + 43.343$

Table 9.2 Correlations for ff in terms of δ and Φ_w

δ (degrees)	Correlation (ff [−])
30	$= -0.00031\Phi_w^2 - 0.0065\Phi_w + 2.0707$
40	$= -0.0001\Phi_w^2 - 0.0050\Phi_w + 1.6251$
50	$= -0.00004\Phi_w^2 - 0.0065\Phi_w + 1.4573$
60	$= -0.00003\Phi_w^2 - 0.0056\Phi_w + 1.3474$

These correlations may be used instead of hopper flow factor charts (Figure 9.18(a–d). With appropriate interpolation one can determine θ and ff for intermediate values of δ. See Worked Example 9.1.

9.5 SUMMARY OF DESIGN PROCEDURE

The following is a summary of the procedure for the design of conical hoppers for mass flows:

(i) Shear cell tests on powder give a family of yield loci.

(ii) Mohr's circle stress analysis gives pairs of values of unconfined yield stress σ_y, and compacting stress σ_C, and the value of the effective angle of internal friction, δ.

(iii) Pairs of values of σ_y and σ_C give the powder flow function.

(iv) Shear cell tests on the powder and the material of the hopper wall give the kinematic angle of wall friction Φ_w.

(v) Φ_w and δ are used to obtain hopper flow factor ff, and semi-included angle of conical hopper wall slope θ.

(vi) Powder flow function and hopper flow factor are combined to give the stress corresponding to the critical flow – no flow condition σ_{crit}.

(vii) σ_{crit}, $H(\theta)$ and bulk density ρ_B, are used to calculate the minimum diameter of the conical hopper outlet B.

9.6 DISCHARGE AIDS

Several devices designed to facilitate flow of powders from silos and hoppers are commercially available. These are known as *discharge aids* or *silo activators*. These should not, however, be employed as an alternative to good hopper design.

Discharge aids may be used where proper design recommends an unacceptably large hopper outlet which is incompatible with the device immediately downstream. In this case, the hopper should be designed to deliver uninterrupted mass flow to the inlet of the discharge aid, i.e. the slope of the hopper wall and inlet dimensions of the discharge aid are those calculated according to the procedure outlined in this chapter.

9.7 PRESSURE ON THE BASE OF A TALL CYLINDRICAL BIN

It is interesting to examine the variation of stress exerted on the base of a bin with increasing depth of powder. For simplicity we will assume that the powder is non-cohesive (i.e. does not gain strength on compaction). Referring to Figure 9.20, consider a slice of thickness ΔH at a depth H below the surface of the powder. The downward force is:

$$\frac{\pi D^2}{4}\sigma_v \tag{9.7}$$

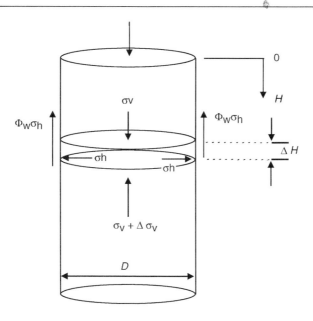

Figure 9.20   Forces acting on a horizontal slice of powder in a tall cylinder

where D is the bin diameter and σ_v is the stress acting on the top surface of the slice. Assuming stress increases with depth, the reaction of the powder below the slice acts upwards and is:

$$\frac{\pi D^2}{4}(\sigma_v + \Delta\sigma_V) \tag{9.8}$$

The net upward force on the slice is then

$$\frac{\pi D^2}{4}\Delta\sigma_v \tag{9.9}$$

If the stress exerted on the wall by the powder in the slice is σ_h and the wall friction is $\tan\Phi_w$, then the friction force (upwards) on the slice is:

$$\pi D\Delta H \tan\Phi_w\sigma_h \tag{9.10}$$

The gravitational force on the slice is:

$$\frac{\pi D^2}{4}\rho_B g\Delta H \text{ acting downwards} \tag{9.11}$$

where ρ_B is the bulk density of the powder, assumed to be constant throughout the powder (independent of depth).

   If the slice is in equilibrium the upward and downward forces are equal, giving:

$$D\Delta\sigma_v + 4\tan\Phi_w\sigma_h\Delta H = D\rho_B g\Delta H \tag{9.12}$$

If we assume that the horizontal stress is proportional to the vertical stress and that the relationship does not vary with depth,

$$\sigma_h = k\sigma_v \tag{9.13}$$

it follows that as ΔH tends to zero,

$$\frac{d\sigma_v}{dH} + \left(\frac{4\tan\Phi_w k}{D}\right)\sigma_v = \rho_B g \tag{9.14}$$

Noting that this is the same as:

$$\frac{d}{dH}\left[\sigma_v e^{(4\tan\Phi_w k/D)H}\right] = \rho_B g e^{(4\tan\Phi_w k/D)H} \tag{9.15}$$

and integrating, we have:

$$\sigma_v e^{(4\tan\Phi_w k/D)H} = \frac{D\rho_B g}{4\tan\Phi_w k} e^{(4\tan\Phi_w k/D)H} + \text{constant} \tag{9.16}$$

If, in general, the stress acting on the surface of the powder is σ_{v0} (at $H = 0$) the result is

$$\sigma_v = \frac{D\rho_B g}{4\tan\Phi_w k}\left[1 - e^{-(4\tan\Phi_w k/D)H}\right] + \sigma_{v0} e^{-(4\tan\Phi_w k/D)H} \tag{9.17}$$

This result was first demonstrated by Janssen (1895).

If there is no force acting on the free surface of the powder $\sigma_{v0} = 0$ and so:

$$\sigma_v = \frac{D\rho_B g}{4\tan\Phi_w k}\left(1 - e^{-(4\tan\Phi_w k/D)H}\right) \tag{9.18}$$

When H is very small:

$$\sigma_v \cong \rho_B H g \tag{9.19}$$
$$\text{(since for very small } z, e^{-z} \cong 1 - z)$$

which is equivalent to the static pressure at a depth H in a fluid of density ρ_B.

When H is large, an inspection of Equation (9.18) gives:

$$\sigma_v \cong \frac{D\rho_B g}{4\tan\Phi_w k} \tag{9.20}$$

and so the vertical stress developed becomes independent of depth of powder above. The variation in stress with depth of powder for the case of no force acting on the free surface of the powder ($\sigma_{v0} = 0$) is shown in Figure 9.21. Thus, contrary to intuition (which is usually based on our experience with fluids), the force exerted by a bed of powder becomes independent of depth if the bed is deep enough. If that is the case, most of the weight of the powder is supported by the walls of the bin. In practice, the stress becomes

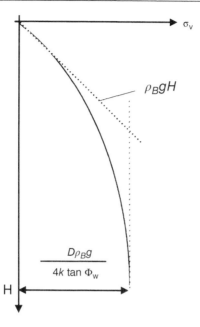

Figure 9.21   Variation in vertical pressure with depth of powder (for $\sigma_{v0} = 0$)

independent of depth (and also independent of any load applied to the powder surface) beyond a depth of about 4D.

9.8   MASS FLOW RATES

The rate of discharge of powder from an orifice at the base of a bin is found to be independent of the depth of powder unless the bin is nearly empty. This means that the observation for a static powder that the pressure exerted by the powder is independent of depth for large depths is also true for a dynamic system. It confirms that fluid flow theory cannot be applied to the flow of a powder.

For flow through an orifice in the flat-based cylinder, experiment shows that:

Mass flow rate, $M_p \propto (B - a)^{2.5}$ for a circular orifice of diameter B

where a is a correction factor dependent on particle size, and where the orifice size is sufficiently large that there is no permanent bridging, i.e. for orifice sizes larger than about 6 times the particle size.

For solids discharge from conical apertures in flat-based cylinders, Beverloo *et al.* (1961) give:

$$M_\mathrm{p} = 0.58\rho_\mathrm{B}g^{0.5}(B - kx)^{2.5} \tag{9.21}$$

This is widely known as the Beverloo equation; the proportionality with $g^{0.5}$ and near proportionality with $B^{2.5}$ have a good theoretical basis. Equation (9.21) is recommended for prediction of discharge rate. However, it should be noted that it only works well with

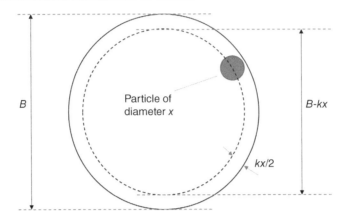

Figure 9.22 The empty annulus effect

that constant (0.58) if the density used is the *flowing* bulk density. The error in prediction is somewhat larger if the filled density is used. The term kx represents the so-called *empty annulus effect* which argues that no particle centre can be closer than a distance of $x/2$ from the edge of the orifice, or $kx/2$, allowing for packing effects, and so all particle centres must pass through a circle of diameter $(B - kx)$ – see Figure 9.22. k is ~1.5 for spherical or near-spherical particles.

For cohesionless coarse particles free falling over a distance h their velocity, neglecting drag and interaction, will be $u = \sqrt{2gh}$.

If these particles are flowing at a bulk density ρ_B through a circular orifice of diameter B, then the theoretical mass flow rate will be:

$$M_P = \frac{\pi}{4}\sqrt{2}\rho_B g^{0.5} h^{0.5} B^2 \tag{9.22}$$

The practical observation that flow rate is proportional to $B^{2.5}$ suggests that, in practice, particles only approach the free fall model when h is the same order as the orifice diameter.

This approach and the Beverloo equation only apply for bulk solids with negligible cohesion and in the absence of any pressure gradient due to fluid flow, which in practice means that it is restricted to use with particles above about 500 μm. In the case of smaller particles, the flow of air (or other gas) must be considered. For example, if the hopper top is sealed, gas must flow in the counter-current direction and will impede the solids discharge. On the other hand, an imposed gas flow in the discharge direction can aid flow. These issues are considered further by Seville *et al.* (1997).

9.9 CONCLUSIONS

Within the confines of a single chapter, it has been possible only to outline the principles involved in the analysis of the flow of unaerated powders. This has been done by reference to the specific example of the design of conical hoppers for mass flow. Other important considerations in the design of hoppers such as time consolidation effects and determination of the stresses acting on the hopper and bin wall have been omitted. These aspects together with the details of shear cell testing procedure are covered in the reference texts at the end of this book.

9.10   WORKED EXAMPLES

WORKED EXAMPLE 9.1

The results of shear cell tests on a powder are shown in Figure 9.W1.1. In addition, it is known that the angle of friction on stainless steel is 19° for this powder, and under flow conditions the bulk density of the powder is 1300 kg/m³. A conical stainless steel hopper is to be designed to hold the powder.

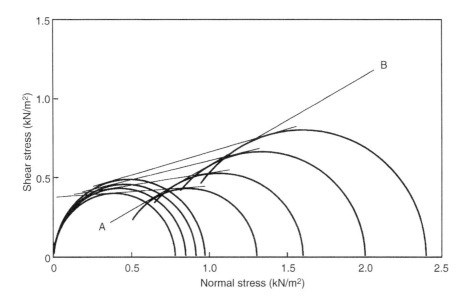

Figure 9.W1.1   Shear cell test data

Determine:
(a) The effective angle of internal friction.

(b) The maximum semi-included angle of the conical hopper which will confidently give mass flow.

(c) The minimum diameter of the circular hopper outlet necessary to ensure flow when the outlet slide valve is opened.

Solution

(a) From Figure 9.W1.1, determine the slope of the effective yield locus (line AB). Slope = 0.578.

Hence, the effective angle of internal friction, $\delta = \tan^{-1}(0.578) = 30°$.

(b) From Figure 9.W1.1, determine the pairs of values of σ_C and σ_y necessary to plot the powder flow function (Figure 9.W1.2).

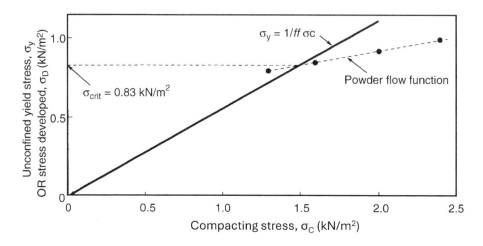

Figure 9.W1.2   Determination of critical stress

σ_C	2.4	2.0	1.6	1.3
σ_y	0.97	0.91	0.85	0.78

Using the flow factor chart for $\delta = 30°$ [Figure 9.18(a)] with $\Phi_v = 19°$ and a 3° margin of safety gives a hopper flow factor $ff = 1.8$, and the semi-included angle of hopper wall $\theta = 27.5°$ (see Figure 9.W1.3).

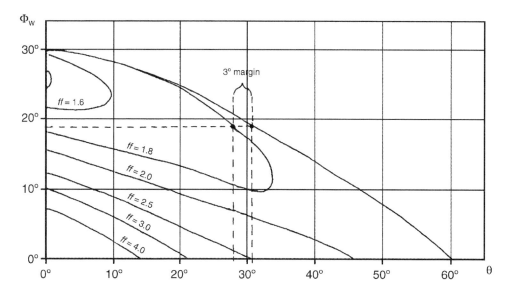

Figure 9.W1.3   Determination of θ and ff

(c) The relationship $\sigma_y = \sigma_c/ff$ is plotted on the same axes as the powder flow function (Figure 9.W1.2) and where this line intersects the powder flow function we find a value of critical unconfined yield stress $\sigma_{crit} = 0.83\,\text{kN/m}^2$. From Equation (9.5),

$$H(\theta) = 2.46 \text{ when } \theta = 27.5°$$

and from Equation (9.4), the minimum outlet diameter for mass flow B, is:

$$B = \frac{2.46 \times 0.83 \times 10^3}{1300 \times 9.81} = 0.160\,\text{m}$$

Summarizing, then, to achieve mass flow without risk of blockage using the powder in question we require a stainless steel conical hopper with a maximum semi-included angle of cone, 27.5° and a circular outlet with a diameter of at least 16.0 cm.

By the alternative method using the correlations for θ and ff in terms of δ and Φ_w (Tables 9.1 and 9.2) instead of the hopper flow factor charts:

Using the correlations in Table 9.1 with $\delta = 30°$ and $\Phi_w = 19°$, $\theta = 27.8°$

Using the correlations in Table 9.2 with $\delta = 30°$ and $\Phi_w = 19°$, $ff = 1.83$

From Equation (9.5): $H(\theta) = 2.0 + \dfrac{\theta}{60} = 2.463$

Fitting a linear equation to the powder flow function data:

$$\sigma_y = 0.169\sigma_c + 0.569$$

Solving this simultaneously with $\sigma_y = \dfrac{1}{ff}\rho_c$ gives:

$$\sigma_{crit} = 0.824\,\text{kN/m}^2$$

and from Equation (9.4), the minimum outlet diameter for mass flow B, is:

$$B = \frac{H(\theta)\sigma_{crit}}{\rho_B g} = \frac{2.46 \times 0.824 \times 10^3}{1300 \times 9.81} = 0.159\,\text{m}$$

WORKED EXAMPLE 9.2

Shear cell tests on a powder give the following information:
Effective angle of internal friction $\delta = 40°$.
Kinematic angle of wall friction on mild steel $\Phi_W = 16°$.
Bulk density under flow condition $\rho_B = 2000\,\text{kg/m}^3$.
The powder flow function can be represented by the relationship $\sigma_y = \sigma_C^{0.6}$, where σ_y is unconfined yield stress (kN/m^2) and σ_C is consolidating stress (kN/m^2).
Determine (a) the maximum semi-included angle of a conical mild steel hopper that will confidently ensure mass flow, and (b) the minimum diameter of a circular outlet to ensure flow when the outlet is opened.

Solution

(a) With an effective angle of internal friction $\delta = 40°$ we refer to the flow factor chart in Figure 9.18(b), from which at $\Phi_W = 16°$ and with a safety margin of 3° we obtain the hopper flow factor $ff = 1.5$ and hopper semi-included angle for mass flow $\theta = 30°$ (Figure 9.W2.1).

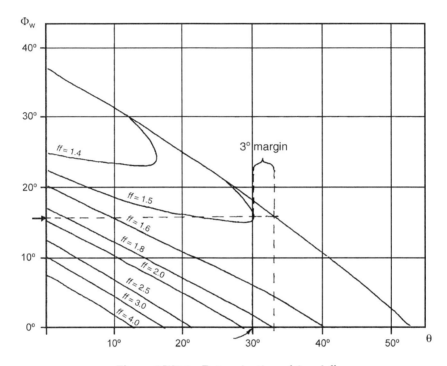

Figure 9.W2.1   Determination of θ and ff

(b) For flow: $\dfrac{\sigma_C}{ff} > \sigma_y$ [Equation (9.3)]

but for the powder in question σ_y and σ_C are related by the material flow function:

$$\sigma_y = \sigma_C^{0.6}.$$

Thus, the criterion for flow becomes

$$\left(\frac{\sigma_y^{1/0.6}}{ff}\right) > \sigma_y$$

and so, the critical value of unconfined yield stress σ_{crit} is found when $\left(\dfrac{\sigma_y^{1/0.6}}{ff}\right) = \sigma_y$;

hence,

$$\sigma_{crit} = 1.837 \text{ kN/m}^2.$$

From Equation (9.5), $H(\theta) = 2.5$ when $\theta = 30°$ and hence, from Equation (9.4), minimum diameter of circular outlet,

$$B = \frac{2.5 \times 1.837 \times 10^3}{2000 \times 9.81} = 0.234 \text{ m}$$

Summarizing, mass flow without blockages is ensured by using a mild steel hopper with a maximum semi-included cone angle 30° and a circular outlet diameter of at least 23.4 cm.

WORKED EXAMPLE 9.3

A powder has an effective angle of internal friction of 43° and has a powder flow function represented by the expression:

$$\sigma_y = \sigma_c^{0.60}$$

where the units of stress are kN/m^2.

If the bulk density of the powder is 1320 kg/m^3 and its angle of friction on a sample of hopper wall is 16°, determine, for a mild steel hopper, the maximum semi-included angle of cone required to safely ensure mass flow, and the minimum size of circular outlet to ensure flow when the outlet is opened.

Since the angle of internal friction is between the charts for 40° and 50°, we will use the method based on the correlations shown in Tables 9.1 and 9.2.

With $\delta = 43°$ and angle of wall friction $\Phi_w = 16°$, we need to interpolate. Based on the values of theta for delta values of 30°, 40°, 50° and 60°, we develop a third-order polynomial expression from which the value of θ at $\delta = 43°$ may be found:

$$\theta = 0.0001325\Phi_w^3 - 0.01237\Phi_w^2 + 0.06724\Phi_w + 38.882$$

Hence, for $\delta = 43°$ and $\Phi_w = 16°$, $\theta = 29.4°$.

Similarly, for hopper flow factor ff, the third-order polynomial for interpolation is:

$$ff = -1.7727 \times 10^{-5}\Phi_w^3 + 0.003076\Phi_w^2 - 0.18640\Phi_w + 5.1875$$

Hence, for $\delta = 43°$, and $\Phi_w = 16°$, $ff = 1.451$.

The powder flow function is described by the given expression: $\sigma_y = \sigma_c^{0.60}$.

Solving this simultaneously with $\sigma_y = \dfrac{1}{ff}\rho_c$ gives σ_{crit}

$$\sigma_{crit} = ff\sigma_{crit}^{0.60}$$

Hence,

$$\sigma_{crit} = (ff)^{0.6/(1-0.60)}$$

and so,

$$\sigma_{\text{crit}} = 1.748 \text{ kN/m}^2$$

From Equation (9.4), the minimum outlet diameter for mass flow, B, is:

$$B = \frac{H(\theta)\sigma_{\text{crit}}}{\rho_B g} = \frac{\left(2 + \dfrac{29.4}{60}\right) \times 1.748 \times 10^3}{1320 \times 9.81} = 0.336 \text{ m}$$

Summarizing, the maximum semi-included angle of cone = 29.4°, minimum outlet diameter = 0.336 m, and hopper flow factor ff = 1.451.

TEST YOURSELF

9.1 Explain with the aid of sketches what is meant by the terms *mass flow* and *core flow* with respect to solids flow in storage hoppers.

9.2 The starting point for the design philosophy presented in this chapter is the *flow – no flow criterion*. What is the flow – no flow criterion?

9.3 Which quantity describes the strength developed by a powder in an arch preventing flow from the base of a hopper? How is this quantity related to the hopper flow factor?

9.4 What is the *powder flow function?* Is the powder flow function dependent on (a) the powder properties, (b) the hopper geometry, and (c) both the powder properties and the hopper geometry?

9.5 Show how the critical value of stress is determined from a knowledge of the *hopper flow factor* and the *powder flow function*.

9.6 What is meant by critical failure (yield) of a powder? What is its significance?

9.7 With the aid of a sketch plot of shear stress versus normal stress, show how the effective angle of internal friction of a powder is determined from a family of yield loci.

9.8 What is the kinematic angle of wall friction and how is it determined?

9.9 A powder is poured gradually into a measuring cylinder of diameter 3 cm. At the base of the cylinder is a load cell which measures the normal force exerted by the powder on the base. Produce a sketch plot showing how the normal force on the cylinder base would be expected to vary with powder depth, up to a depth of 18 cm.

9.10 How would you expect the mass flow rate of particulate solids from a hole in the base of a flat-bottomed container to vary with (a) the hole diameter and (b) the depth of solids?

EXERCISES

9.1 Shear cell tests on a powder show that its effective angle of internal friction is 40° and its powder flow function can be represented by the equation: $\sigma_y = \sigma_C^{0.45}$, where σ_y is the unconfined yield stress and σ_c is the compacting stress, both in kN/m². The bulk density of the powder is 1000 kg/m^3 and the angle of friction on a mild steel plate is 16°. It is proposed to store the powder in a mild steel conical hopper of semi-included angle 30° and having a circular discharge opening of 0.30 m diameter. What is the critical outlet diameter to give mass flow? Will mass flow occur?

[Answer: 0.355 m; no flow]

9.2 Describe how you would use shear cell tests to determine the effective angle of internal friction of a powder.

A powder has an effective angle of internal friction of 60° and has a powder flow function represented in the graph shown in Figure 9.E2.1. If the bulk density of the powder is 1500 kg/m^3 and its angle of friction on mild steel plate is 24.5°, determine, for a mild steel hopper, the maximum semi-included angle of cone required to safely ensure mass flow, and the minimum size of circular outlet to ensure flow when the outlet is opened.

[Answer: 17.5°; 18.92 cm]

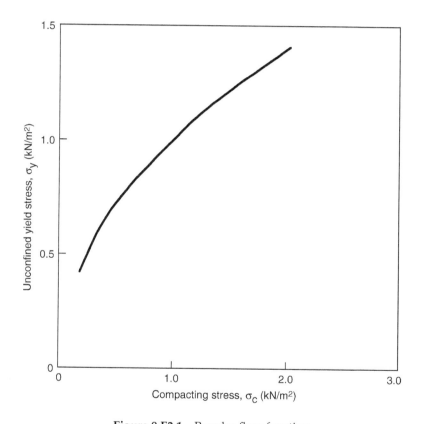

Figure 9.E2.1 Powder flow function

9.3

(a) Summarize the philosophy used in the design of conical hoppers to ensure flow from the outlet when the outlet valve is opened.

(b) Explain how the powder flow function and the effective angle of internal friction are extracted from the results of shear cell tests on a powder.

(c) A factory having serious hopper problems takes on a chemical engineering graduate. The hopper in question feeds a conveyor belt and periodically blocks at the outlet and needs to be 'encouraged' to restart. The graduate makes an investigation on the hopper, commissions shear cell tests on the powder and recommends a minor modification to the hopper. After the modification the hopper gives no further trouble and the graduate's reputation is established. Given the information below, what was the graduate's recommendation?

Existing design: Material of wall – mild steel.

Semi-included angle of conical hopper – 33°.

Outlet – circular, fitted with a 25 cm diameter slide valve.

Shear cell test data: Effective angle of internal friction $\delta = 60°$.

Angle of wall friction on mild steel $\Phi_W = 8°$.

Bulk density $\rho_B = 1250 \, \text{kg/m}^{3.}$

Powder flow function: $\sigma_y = \sigma_C^{0.55}$ (σ_y and σ_C in kN/m^2).

9.4 Shear cell tests are carried out on a powder for which a stainless steel conical hopper is to be designed. The results of the tests are shown graphically in Figure 9.E4.1. In addition, it is found that the friction between the powder and a stainless steel surface can be described by an angle of wall friction of 11°, and that the relevant bulk density of the powder is 900 kg/m^3.

(a) From the shear cell results of Figure 9.E4.1, deduce the effective angle of internal friction δ of the powder.

(b) Determine:

   (i) The semi-included hopper angle safely ensuring mass flow.

   (ii) The hopper flow factor *ff*.

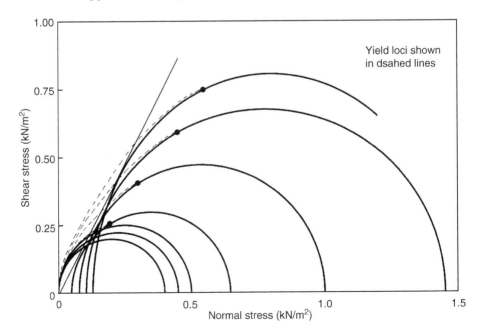

Figure 9.E4.1   Shear cell test data

(c) Combine this information with further information gathered from Figure 9.E4.1 in order to determine the minimum diameter of circular outlet to ensure flow when required. (*Note:* Extrapolation is necessary here.)

(d) What do you understand by 'angle of wall friction' and 'effective angle of internal friction'?

[Answer: (a) 60°; (b) (i) 32.5°, (ii) 1.29; (c) 0.110 m.]

9.5 The results of shear cell tests on a powder are given in Figure 9.E5.1. An aluminium conical hopper is to be designed to suit this powder. It is known that the angle of wall friction between the powder and aluminium is 16° and that the relevant bulk density is 900 kg/m³.

(a) From Figure 9.E5.1 determine the effective angle of internal friction of the powder.

(b) Determine:

   (i) The semi-included hopper angle safely ensuring mass flow.

   (ii) The hopper flow factor *ff*.

(c) Combine the information with further data gathered from Figure 9.E5.1 in order to determine the minimum diameter of circular outlet to ensure flow when required. (*Note:* Extrapolation of these experimental results may be necessary.)

[Answer: (a) 40°; (b) (i) 29.5°; (ii) 1.5; (c) 0.5 m ± approximately 7% depending on the extrapolation.]

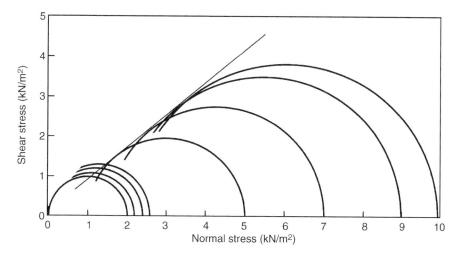

Figure 9.E5.1   Shear cell test data

9.6 A powder has an effective angle of internal friction of 38° and has a powder flow function represented by the expression:

$$\sigma_y = \sigma_c^{0.54}$$

where the units of stress are kN/m²

If the bulk density of the powder is 1200 kg/m^3 and its angle of friction on steel plate is 24°, determine, for a mild steel hopper, the maximum semi-included angle of cone required to safely ensure mass flow, and the minimum size of circular outlet to ensure flow when the outlet is opened.

[Answer: maximum semi-included angle of cone = 19.1°, minimum outlet diameter = 0.316 m, hopper flow factor ff = 1.49].

10

Mixing and Segregation

Achieving good mixing of particulate solids of different size and density is important in many of the process industries, and yet it is not a trivial exercise. For free-flowing powders, the preferred state for particles of different size and density is to remain segregated. That is why in a packet of muesli breakfast cereal the large particles come to the top as a result of the vibration caused by handling of the packet. An extreme example of this *segregation* is that a large steel ball can be made to rise to the top of a beaker of sand by simply shaking the beaker up and down – this must be seen to be believed! Since the preferred state for free-flowing powders is to segregate by size and density, it is not surprising that many processing steps give rise to segregation. Care needs to be taken to avoid processing steps which inadvertently promote segregation following steps in which mixing is promoted. In this chapter, we will examine mechanisms of segregation and mixing in particulate solids, and briefly look at how mixing is carried out in practice and how the quality of a mixture is assessed.

10.1 TYPES OF MIXTURE

A perfect mixture of two types of particles is one in which a group of particles taken from any position in the mixture will contain the same proportions of each particle as the proportions present in the whole mixture. In practice, a perfect mixture cannot be obtained. Generally, the aim is to produce a *random mixture*, i.e. a mixture in which the probability of finding a particle of any component is the same at all locations and equal to the proportion of that component in the mixture as a whole. When attempting to mix particles which are not subject to segregation, this is generally the best quality of mixture that can be achieved. If the particles to be mixed differ in physical properties, then segregation may occur. In this case particles of one component have a greater probability of being

Introduction to Particle Technology, Third Edition. Martin Rhodes and Jonathan Seville.
© 2024 John Wiley & Sons Ltd. Published 2024 by John Wiley & Sons Ltd.
Website: www.wiley.com/go/rhodes/particle3e

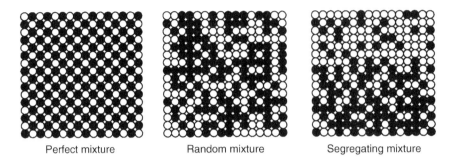

| Perfect mixture | Random mixture | Segregating mixture |

Figure 10.1 Types of mixture

found in one part of the mixture and so a random mixture cannot be achieved. In Figure 10.1 examples are given of what is meant by perfect, random and segregating mixtures of two components. The random mixture was obtained by tossing a coin – heads give a black particle at a given location and tails give a white particle. For the segregating mixture the coin is replaced by a die. In this case the black particles differ in some property which causes them to have a greater probability of appearing in the lower half of the box. In this case, in the lower half of the mixture there is a chance of two in three that a particle will be black (i.e. a throw of 1, 2, 3 or 4) whereas in the upper half the probability is one in three (a throw of 5 or 6). It is possible to produce mixtures with better than random quality by taking advantage of the natural attractive forces between particles; such mixtures are achieved through *ordered* or *interactive* mixing (see below).

10.2   SEGREGATION

10.2.1   Causes and Consequences of Segregation

When particles to be mixed have the same important physical properties (size distribution, shape, density) then, provided the mixing process goes on for long enough, a random mixture will be obtained. However, in many common systems, the particles to be mixed have different properties and tend to exhibit segregation. Particles with the same physical property then collect together in one part of the mixture and the random mixture is not a natural state for such a system of particles. Even if particles are originally mixed by some means, they will tend to unmix on handling (moving, pouring, conveying, processing).

Although differences in size, density and shape of the constituent particles of a mixture may give rise to segregation, difference in particle size is by far the most important of these. Density difference is comparatively unimportant (see the steel ball in sand example below) except in gas fluidization where density difference is more important than size difference. Many industrial problems arise from segregation. Even if satisfactory mixing of constituents is achieved in a powder mixing device, unless great care is taken, subsequent processing and handling of the mixture will result in demixing or segregation.

This can give rise to variations in bulk density of the powder going to packaging (e.g. it becomes impossible to fit 25 kg of mixture into a bag designed to hold 25 kg) or, more seriously, the chemical composition of the product may be off specification (e.g. in blending of constituents for detergents or drugs).

10.2.2 Mechanisms of Segregation

Four mechanisms of segregation according to size may be identified (Williams, 1990):

(1) *Trajectory segregation*: If particles are thrown horizontally into a gas, as shown in Figure 10.2, in such a way that they can move independently, it is intuitively obvious that if they all have the same initial velocity, the larger ones will travel further than the smaller ones. This is a significant cause of segregation in operations such as mechanical belt conveyors, where particles are ejected from the belt horizontally. The effect can be quantified as follows. Consider a small particle of diameter x and density ρ_p, whose drag is governed by Stokes' law [Equation (3.7)], which is projected horizontally with a velocity U into a fluid of viscosity μ and density ρ_f,

---

From Chapter 3, the retarding force on the particle $= C_D \frac{1}{2} \rho_f U^2 \left(\dfrac{\pi x^2}{4} \right)$

Deceleration of the particle $= \dfrac{\text{retarding force}}{\text{mass of particle}}$

In Stokes' law region, $C_D = 24/Re_p$

Hence, deceleration $= \dfrac{18\mu}{\rho_p x^2}$

From the equation of motion, a particle with an initial velocity U and constant deceleration $18U\mu/\rho_p x^2$ will travel a distance $U\rho_p x^2/18\mu$ before coming to rest.

A particle of diameter $2x$ would therefore travel four times as far as one of diameter x before coming to rest.

(2) *Percolation of fine particles*: If a mass of particles is disturbed in such a way that individual particles move, a rearrangement in the packing of the particles occurs. The spaces created allow particles from above to fall into gaps, and particles in some other place to move upwards. If the powder is composed of particles of different size, it will be easier for small particles to fall down and so there will be a tendency for small particles to move downwards leading to segregation. Even a very small difference in particle size can give rise to significant segregation due to repeated small disturbances.

Segregation by percolation of fine particles can occur whenever the mixture is disturbed, causing rearrangement of particles. This can happen during stirring, shaking, vibration or when pouring particles into a heap. Note that stirring, shaking and vibration would all be expected to promote mixing in liquids or gases, but cause segregation in free-flowing particle mixtures. Figure 10.3(a) shows segregation in the heap formed by pouring a mixture of two sizes of particles. The shearing caused when a particle mixture is rotated in a drum can also give rise to segregation by percolation [Figure 10.3(b)].

Segregation by percolation occurs in charging and discharging storage hoppers (see Chapter 9). As particles are fed into a hopper, normally in free flow from above, they generally pour onto a conical heap, resulting in segregation if there is a size distribution and the powder is free-flowing, because larger particles can roll further

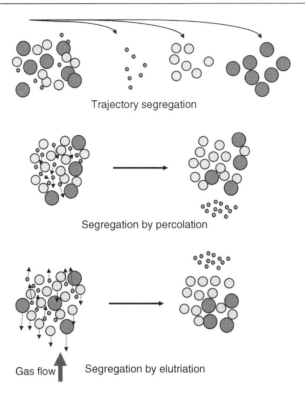

Figure 10.2   Mechanisms of segregation

down the slope than smaller ones. There are some devices and procedures available to minimize this effect if segregation is a particular concern. However, angled surfaces also form during discharge of a core flow hopper (see Chapter 9), along which

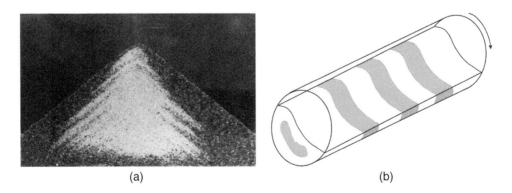

Figure 10.3   (a) Segregation pattern formed by pouring a free-flowing mixture of two sizes of particles into a heap (smaller particles are light coloured). (b) Schematic representation of a typical segregation pattern formed by rotating a free-flowing mixture of two sizes in a drum (smaller particles are blue, coarser particles are white)

particles roll, and this gives rise to segregation in free-flowing powders. Therefore if segregation is a cause for concern, core flow hoppers should be avoided.

(3) *Rise of coarse particles on vibration*: If a mixture of particles of different size is vibrated, the larger particles move upwards. This can be demonstrated by placing a single large ball at the bottom of a bed of sand (for example a 20 mm steel ball or similarly sized pebble in a beaker of sand from the beach). On shaking the beaker up and down, the steel ball rises to the surface. Figure 10.4 shows a series of photographs taken from a 'two-dimensional' version of the steel ball experiment. This so-called 'Brazil nut

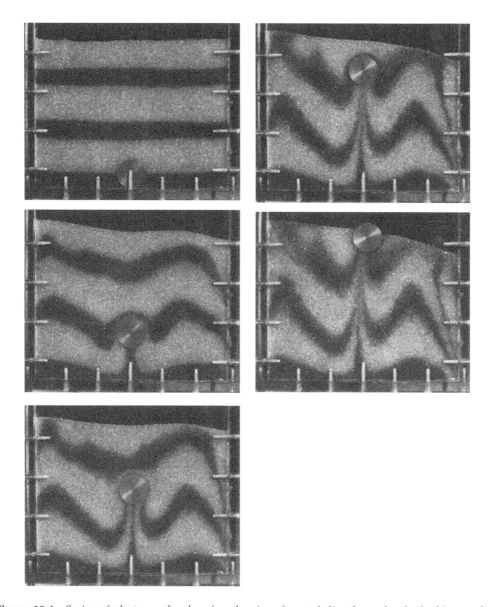

Figure 10.4 Series of photographs showing the rise of a steel disc through a bed of 2 mm glass spheres due to vibration (a 'two-dimensional' version of the rising steel ball experiment)

effect' has received much attention in the literature over recent years, but research and comments date back much further. The mechanisms causing it have long been a matter for debate, though there is a growing consensus that there are in fact multiple distinct mechanisms in play during the vibration-induced segregation of particles, the dominant mechanism depending on the relevant system parameters and material properties. The rise of one or more larger particles or 'intruders' within a bed of smaller particles has been explained by various possible mechanisms. These include creation and filling of voids beneath the intruder, the establishment of convection cells within the bed of smaller particles, granular equivalents to buoyancy and thermal diffusion, a *condensation* effect where the kinetic energy of the system is sufficient to mobilize smaller particles but not larger particles, and other effects due to the non-equipartition of energy between large and small grains. It is also important to note that the larger components of a vibrated granular system do not always rise to the top – much work has also been performed exploring the so-called *reverse Brazil nut effect*, in which the opposite is found to occur. Nor is all segregation driven by differences in particle size – vibration-induced segregation may also be driven by differences in a number of other particle properties including (but not limited to) density, elasticity, friction, and shape. External factors such as interstitial fluids can also play a strong role in determining the type, speed, and extent of segregation. Further details regarding all of the above mechanisms, and others besides, may be found in Rosato and Windows-Yule (2020).

(4) *Elutriation segregation*: When a powder containing an appreciable proportion of particles under 50 µm is charged into a storage vessel or hopper, air (or some other gas) is displaced upwards. The upward velocity of this air may exceed the terminal velocity (see Chapter 3) of some of the finer particles, which may then remain in suspension after the larger particles have settled to the surface of the hopper contents (Figure 10.2). For particles in this size range in air, the terminal velocity will be typically of the order of a few centimetres per second and will increase as the square of particle diameter (e.g. for 30 µm sand particles in air the terminal velocity is 7 cm/s). Thus, a packet of fine particles is generated in the hopper each time solids are charged.

10.3   REDUCTION OF SEGREGATION

As discussed above, segregation occurs primarily as a result of size differences. The difficulty of mixing two components can therefore be reduced by making the size of the components as similar as possible and by reducing the absolute size of both components. Segregation is generally not a serious problem when all particles are less than 30 µm (for particle densities in the range 2000–3000 kg/m^3). In such fine powders, the interparticle forces generated by electrostatic charging, van der Waals forces and forces due to moisture are large compared with the gravitational and inertial forces on the particles (see Chapter 2). This causes the particles to stick together, preventing segregation as the particles are not free to move relative to one another. These powders are referred to as cohesive powders (Geldart's classification of powders for fluidization is relevant here – see Chapter 6). The lack of mobility of individual particles in cohesive powders is one reason why they give a better quality of mixing. The other reason is that if a random mixture is approached, the standard deviation of the composition of samples taken from the mixture will decrease in inverse proportion to the number of particles in the sample.

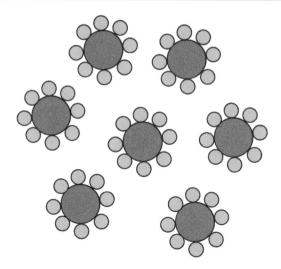

Figure 10.5 An ordered mixture of small particles on carrier particles

Therefore, for a given mass of sample the standard deviation decreases and mixture quality increases with decreasing particle size. The mobility of particles in free-flowing powders can be reduced by the addition of small quantities of liquid. The reduction in mobility reduces segregation and permits better mixing.

It is possible to take advantage of this natural tendency for particles to adhere in order to produce mixtures of quality better than random mixtures. As mentioned earlier, such mixtures are known as ordered or interactive mixtures; they are made up of small particles (e.g. <5 μm) adhered to the surface of a carrier particle in a controlled manner (Figure 10.5). By careful selection of particle size and engineering of interparticle forces, high-quality mixtures with very small variance can be achieved. This technique is used in the pharmaceutical industry where quality control standards are exacting. For further details on ordered mixing and on the mixing of cohesive powders the reader is referred to Harnby *et al.* (1992).

If it is not possible to alter the size of the components of the mixture or to add liquid, then in order to avoid serious segregation, care should be taken to avoid situations which are likely to promote it. In particular, pouring operations and the formation of a moving sloping powder surface should be avoided.

10.4   EQUIPMENT FOR PARTICULATE MIXING

10.4.1   Mechanisms of Mixing

Lacey (1954) identified three mechanisms of powder mixing:

(1) Shear mixing.

(2) Diffusive mixing.

(3) Convective mixing.

In shear mixing, shear stresses give rise to slip zones and mixing takes place by interchange of particles between layers within the zone. Diffusive mixing occurs when particles roll down a sloping surface. Convective mixing occurs by deliberate bulk movement of packets of powder around the powder mass.

In free-flowing powders both diffusive mixing and shear mixing give rise to size segregation and so for such powders convective mixing is the major mechanism which is adopted in mixers.

10.4.2   Types of Mixer

(1) *Tumbling mixers*: A tumbling mixer comprises a closed vessel rotating about one axis. Common shapes for the vessel are cube, double cone and V (Figure 10.6). The primary mechanism in these mixers is diffusive mixing, although significant convective mixing may be involved as strong internal circulation patterns are established, especially in the V-mixer. Since this can give rise to segregation in free-flowing powders the quality of mixture achievable with such powders in tumbling mixers is limited. Baffles may be installed in an attempt to reduce segregation but generally have little effect.

(2) *Mixers with slow-moving internals*: In these mixers circulation patterns are set up within a static shell by rotating blades or paddles. The main mechanism is convective mixing, although this is accompanied by some diffusive and shear mixing. One of the most common convective mixers is the ribbon blender in which helical blades or ribbons rotate on a horizontal axis in a static cylinder or trough (Figure 10.7). Rotational speeds are typically less than one revolution per second. A somewhat different type of convective mixer is the Nauta mixer (Figure 10.8), in which an Archimedean screw lifts material from the base of a conical hopper and progresses around the hopper wall.

(3) *Fluidized bed mixers*: These rely on the natural mobility of particles in the fluidized bed. The mixing is largely convective with the circulation patterns set up by the bubble motion within the bed. An important feature of the fluidized bed mixer is that

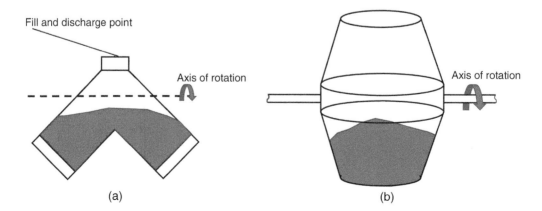

(a)   (b)

Figure 10.6   Tumbling mixers (a) V-mixer, (b) double cone mixer

Figure 10.7 Ribbon blender – With acknowledgments to Winkworth Machinery

Figure 10.8 Nauta mixer – (image reproduced with permission from Hosokawa Micron B.V. (Nauta® is a registered trademark of Hosokawa Micron B.V.)

several processing steps (e.g. mixing, reaction, and coating drying) may be carried out in the same vessel.

(4) *High shear mixers*: Local high shear stresses are created by devices similar to those used in comminution; for example, high velocity rotating blades, low velocity – high compression rollers. In the high shear mixer the emphasis is on breaking down agglomerates of cohesive powders rather than breaking individual particles. The dominant mechanism is shear mixing.

10.5   ANALYSIS OF MIXER PERFORMANCE

10.5.1   Simple V-Mixer

One of the simplest rotating tumbling types of mixer is the V-mixer, as used in the pharmaceutical industry, which simply consists of two cylindrical tubes joined at an angle. The vessel rotates about an axis in the plane of the paper such that material is exchanged between the two 'legs' during each rotation (Figure 10.9).

Figure 10.10 shows a DEM simulation (see Chapter 4) of the progress of mixing over several rotations of the device.

The V-mixer is an extremely simple device and is not 'state-of-the-art' for solids mixing. Nevertheless, it represents a generic type of equipment in which mixing occurs by repeated splitting and recombination of material. In fact, this mixing mechanism is important in most types of bladed mixers too. Closer study of the motion of powders in the V-mixer reveals a flow pattern in the plane of the V as shown in Figure 10.11.

While mixing within each of the two arms is quite rapid, exchange of material between them is slower. This suggests a simple model for the rate of mixing, based on a constant mass exchanged between the arms per revolution (Brone *et al.*, 1998). A mass balance on the left-hand arm from N to $N + dN$ revolutions gives:

$$M_1 \times \frac{dm_1}{dN} = q(m_r - m_1) \tag{10.1}$$

where M_1 is the mass of particles in the left arm; m_r is the mass fraction of the constituent of interest in the right arm; m_1 is the mass fraction of the constituent of interest in the left arm; q is the mass of material that crosses the plane of symmetry per revolution, which is assumed to be independent of the flow direction across the plane; and N is the number of

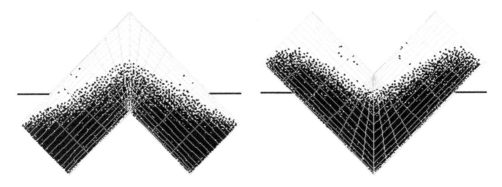

Figure 10.9   Splitting and recombination of material in the V-mixer. Adapted from Kuo *et al.* (2002)

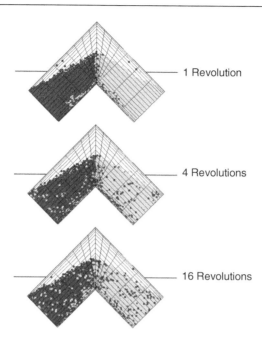

Figure 10.10 Progress of mixing with increasing revolutions of the V-mixer (Kuo *et al.*, 2002)

revolutions. If the two arms are internally well-mixed, the concentration within the mass transferred q is the same as that of the bulk in each arm.

Considering that the mass of material in each arm is half of the total and defining m_0 as the mass fraction of the constituent of interest in the whole vessel, we can show that:

$$\frac{M}{4}\frac{dm_1}{dN} = q(m_0 - m_1) \qquad (10.2)$$

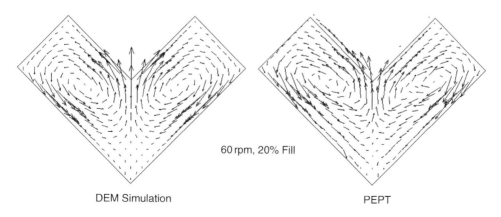

Figure 10.11 Flow patterns within the V-mixer as revealed by DEM simulation and PEPT (positron emission particle tracking) analysis (Kuo *et al.*, 2002)

where M is the total mass of particles in the vessel. Integrating with the initial conditions $m_1 = 0$ at $N = 0$ gives:

$$m_1 = m_0[1 - exp\,(-4qN/M)] \tag{10.3}$$

Typical trends predicted by this equation are shown in Figure 10.12 for $m_0 = 0.05$ and q/M values of 0.01, 0.015, 0.02 and 0.025 (the fraction of the total mass that crosses the plane of symmetry per revolution). The figure shows that (a) the mass fraction of the component of interest tends asymptotically to m_0 with increasing number of revolutions and (b) increasing the value of q/M reduces the number of revolutions required to reach m_0.

This equation has been found to be in reasonable agreement with experiment for the V-mixer; an exponential approach to a final mixed state is typical in such tumbling devices.

It might be thought that increasing the amount of material in the mixer might increase the rate of production of mixed products. However, experiments show that the exchange rate between the two arms of the mixer decreases strongly as the fill level is increased – typically by about 50% as the fill percentage is increased from 40% to 60% (Brone *et al.*, 1998). This is because the empty space above the fill is necessary to allow the material to circulate. The other variable of interest is rotation speed. At low speeds the extent of mixing depends on the number of rotations, as Equation (10.3) indicates, and is independent of speed. At higher speeds, however, centrifugal effects start to become important, and the rate of mixing is reduced (Brone *et al.*, 1998).

There are four implications for the use of V-mixers (Brone *et al.*, 1998):

(1) For the most rapid approach to a desired mixture, the V-mixer should be loaded with equal amounts of each component on each side of the plane of symmetry, because the mixing within each arm is more rapid than the mixing between them.

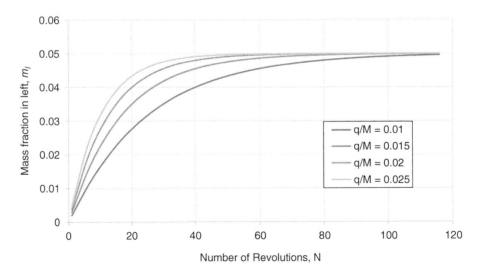

Figure 10.12 Mass fraction of component of interest as a function of number of passes, N and fraction of total mass transferred per pass, q/M, according to Equation (10.3)

(2) For the same reason, sampling of the mixture should be done equally on both sides of the plane of symmetry.

(3) Scale-up procedures should either keep the fractional volumetric fill constant or consider the effect of fill level on mixing rate, which is considerable.

(4) Mixers should not be driven too fast, because centrifugal effects are not usually helpful.

10.5.2 Analysing More Complex Mixers

The mixing analysis presented for the V-mixer considers only one dimension. More complex machines such as the bladed mixer shown in Figure 10.13(a) need a fully three-dimensional approach.

Martin *et al.* (2007) studied mixing in a bladed mixer using PEPT (positron emission particle tracking). In their analysis, these authors considered the volume element shaded blue in Figure 10.13(b). Particles within this volume element at $t = 0$ move under the influence of the blades to new positions, each marked by an X, at $t = t_1$. The dispersion in the end points X is an indication of the extent of mixing local to that volume element. Given that each end point X has a different set of coordinates the easiest way to describe the extent of mixing is to calculate the variance on the mean location:

$$\sigma^2 = \frac{1}{n} \sum_{i=1}^{n} (x_i - \bar{x})^2 + (y_i - \bar{y})^2 + (z_i - \bar{z})^2 \tag{10.4}$$

where σ is the standard deviation, $\bar{x}, \bar{y}, \bar{z}$ represent the mean location and n is the number of passes through the starting region. Quantifying this dispersion for every volume element in the mixer results in a map of the variance associated with each volume element in

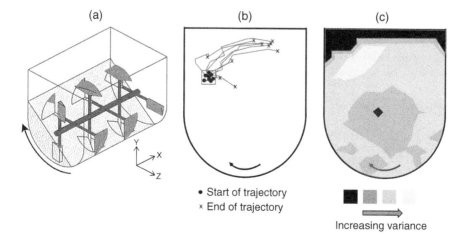

• Start of trajectory
× End of trajectory

Increasing variance

Figure 10.13 Analysis of bladed mixer performance (a) schematic representation of a bladed mixer, (b) identification of volume element for study and (c) map of the variance associated with each volume element in the cross-section of the mixer Martin *et al.* (2007) / with permission of Elsevier

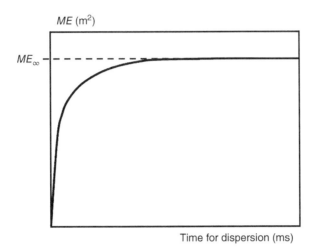

Figure 10.14 Effectiveness of a bladed mixer over time. Adapted from Martin *et al.* (2007) / with permission of Elsevier

the cross-section of a bladed mixer as shown in Figure 10.13(c). The light area top left is associated with the largest contribution to mixing. This is the point at which the rotating blades break through the bed surface and scatter the particles across the free surface.

Summing all the local contributions to mixing, as in Equation (10.5), enables an overall *mixer effectiveness* to be calculated for the machine, which shows an exponential approach to a steady state when assessed against the time allowed for dispersion (Figure 10.14).

$$ME = \frac{1}{N_P} \sum_{t=1}^{n} \sigma_i^2 n_{pi} \qquad (10.5)$$

where, σ_1^2 is the variance for element i, n_{pi} is the number of passes through element i, n is the number of elements in the system, N_p is the total number of passes.

10.6 ASSESSING THE MIXTURE

10.6.1 Quality of a Mixture

The end use of a particle mixture will determine the quality of mixture required. The end use imposes a scale of scrutiny on the mixture. *Scale of scrutiny* was a term used by Danckwerts (1952) meaning 'the maximum size of the regions of segregation in the mixture which would cause it to be regarded as imperfectly mixed'. For example, the appropriate scale of scrutiny for a detergent powder composed of active ingredients in particulate form is the quantity of detergent in the scoop used to dispense it into the washing machine. The composition should not vary significantly between the first and last scoops taken from the box. At another extreme, the scale of scrutiny for a pharmaceutical drug is the quantity of material making up the tablet or capsule. The quality of a mixture decreases with decreasing scale of scrutiny until in the extreme we are scrutinizing only individual particles. An example of this is the image on a smartphone or laptop screen, which at normal viewing distance appears as a lifelike image or legible text, but which under close 'scrutiny' is made up of tiny dots or pixels of colour.

10.6.2 Sampling

To determine the quality of a mixture it is generally necessary to take samples. In order to avoid bias in taking samples from a particulate mixture, the guidelines for sampling powders set out in Chapter 1 must be followed. The size of the sample required to determine the quality of the mixture is governed by the scale of scrutiny imposed by the intended use of the mixture.

10.6.3 Statistics Relevant to Mixing

It is evident that the sampling of mixtures and the analysis of mixture quality require the application of statistical methods. The statistics relevant to random binary mixtures are summarized below:

- *Mean composition*: The true composition of a mixture μ is often not known but an estimate $\bar{y}$ may be found by sampling. If we have N samples of composition y_1 to y_N in one component, the estimate of the mixture composition $\bar{y}$ is given by:

$$\bar{y} = \frac{1}{N} \sum_{i=1}^{N} y_i \tag{10.6}$$

- *Standard deviation and variance*: The true standard deviation, σ, and the true variance, σ^2, of the composition of the mixture are quantitative measures of the quality of the mixture. The true variance is usually not known but an estimate S^2 is defined as:

$$S^2 = \frac{\sum_{i=1}^{N} (y_i - \mu)^2}{N} \quad \text{if the true composition } \mu \text{ is known} \tag{10.7}$$

$$S^2 = \frac{\sum_{i=1}^{N} (y_i - \bar{y})^2}{N-1} \quad \text{if the true composition } \mu \text{ is unknown} \tag{10.8}$$

The standard deviation is equal to the square root of variance.

- *Theoretical limits of variance*: For a two-component system the theoretical upper and lower limits of mixture variance are as follows:

$$\text{(a) upper limit (completely segregated)} \quad \sigma_0^2 = p(1-p) \tag{10.9}$$

$$\text{(b) lower limit (randomly mixed)} \quad \sigma_R^2 = \frac{p(1-p)}{n} \tag{10.10}$$

where p and $(1 - p)$ are the proportions of the two components determined from samples and n is the number of particles in each sample.

Actual values of mixture variance lie between these two extreme values.

- *Mixing indices*: A measure of the degree of mixing is the Lacey mixing index (Lacey, 1954):

$$\text{Lacey mixing index} = \frac{\sigma_0^2 - \sigma^2}{\sigma_0^2 - \sigma_R^2} \qquad (10.11)$$

In practical terms the Lacey mixing index is the ratio of 'mixing achieved' to 'mixing possible'. A Lacey mixing index of zero would represent complete segregation and a value of unity would represent a completely random mixture. Practical values of this mixing index, however, are found to lie in the range 0.75–1.0 and so the Lacey mixing index does not provide sufficient discrimination between mixtures.

A further mixing index suggested by Poole *et al.* (1964) is defined as:

$$\text{Poole } et \ al. \text{mixing index} = \frac{\sigma}{\sigma_R} \qquad (10.12)$$

This index gives better discrimination for practical mixtures and approaches unity for completely random mixtures.

- *Standard error*: When the sample compositions have a normal distribution the sampled variance values will also have a normal distribution. The standard deviation of the variance of the sample compositions is known as the 'standard error' of the variance $E(S^2)$.

- *Tests for precision of mixture composition and variance*: The mean mixture composition and variance which we measure from sampling are only samples from the normal distribution of mixture compositions and variance values for that mixture. We need to be able to assign a certain confidence to this estimate and to determine its precision.

Assuming that the sample compositions are normally distributed:

(1) *Sample composition*

Based on N samples of mixture composition with mean $\bar{y}$ and estimated standard deviation S, the true mixture composition μ may be stated with precision:

$$\mu = \bar{y} \pm \frac{tS}{\sqrt{N}} \qquad (10.13)$$

where t is from the *Student's t-test* for statistical significance. The value of t depends on the confidence level required. For example, at a 95% confidence level, $t = 2.0$ for $N = 60$, and so there is a 95% probability that the true mean mixture composition lies in the range: $\bar{y} \pm 0.258\,S$. In other words, 1 in 20 estimates of mixture variance estimates would lie outside this range.

(2) *Variance*

(a) When more than 50 samples are taken (i.e. $n > 50$), the distribution of variance values can also be assumed to be normal and the Student's t-test may be used. The best estimate of the true variance σ^2 is then given by:

$$\sigma^2 = S^2 \pm \left[t \times E(S^2) \right] \tag{10.14}$$

The standard error of the mixture variance required in this test is usually not known but is estimated from:

$$E(S^2) = S^2 \sqrt{\frac{2}{N}} \tag{10.15}$$

The standard error decreases as $1/\sqrt{N}$ and so the precision increases as $\sqrt{N}$.

(b) When less than 50 samples are taken (i.e. $n < 50$), the variance distribution curve may not be normal and is likely to be a χ^2 (*chi-squared*) distribution. In this case the limits of precision are not symmetrical. The range of values of mixture variance is defined by lower and upper limits:

$$\text{Lower limit}: \sigma_L^2 = \frac{S^2(N-1)}{\chi_\alpha^2} \tag{10.16}$$

$$\text{Upper limit}: \sigma_U^2 = \frac{S^2(N-1)}{\chi_{1-\alpha}^2} \tag{10.17}$$

where α is the significance level [for a 90% confidence range, $\alpha = 0.5(1 - 90/100) = 0.05$; for a 95% confidence range, $\alpha = 0.5(1 - 95/100) = 0.025$]. The lower and upper χ^2 values, χ_α^2 and $\chi_{1-\alpha}^2$, for a given confidence level are found in χ^2 distribution tables.

Note: Many spreadsheet programs and statistics packages include implementations of Student's t-test and the χ^2 test.

10.7   WORKED EXAMPLES

WORKED EXAMPLE 10.1 (AFTER WILLIAMS, 1990)

A random mixture consists of two components A and B in proportions 60% and 40% by mass, respectively. The particles are spherical and A and B have particle densities 500 and 700 kg/m³, respectively. The cumulative undersize mass distributions of the two components are shown in Table 10.W1.1.

Table 10.W1.1   Size distributions of particles A and B

Size x (μm)	2057	1676	1405	1204	1003	853	699	599	500	422
$F_A(x)$	1.00	0.80	0.50	0.32	0.19	0.12	0.07	0.04	0.02	0
$F_B(x)$			1.00	0.88	0.68	0.44	0.21	0.08	0	

If samples of 1 g are withdrawn from the mixture, what is the expected value for the standard deviation of the composition of the samples?

Solution

The first step is to estimate the number of particles per unit mass of A and B. This is done by converting the size distributions into differential frequency number distributions and using:

$$\left(\begin{array}{c} \text{mass of particles} \\ \text{in each size range} \end{array}\right) = \left(\begin{array}{c} \text{number of particles} \\ \text{in size range} \end{array}\right) \times \left(\begin{array}{c} \text{mass of one} \\ \text{particle} \end{array}\right)$$

$$\mathrm{d}m \quad = \quad \mathrm{d}n \quad \quad \frac{\rho_p \pi x^3}{6}$$

where ρ_p is the particle density and x is the arithmetic mean of adjacent sieve sizes.

These calculations are summarized in Tables 10.W1.2 and 10.W1.3.

Table 10.W1.2   A particles

Mean size of range x (µm)	dm	dn
1866.5	0.20	117 468
1540.5	0.30	334 081
1304.5	0.18	309 681
1103.5	0.13	369 489
928	0.07	334 525
776	0.05	408 658
649	0.03	419 143
54.5	0.02	460 365
461	0.02	779 655
Totals	1.00	3.51×10^6

Table 10.W1.3   B particles

Mean size of range x (µm)	dm	dn
1866.5	0	0
1540.5	0	0
1304.5	0.12	0.147×10^6
1103.5	0.20	0.406×10^6
928	0.24	0.819×10^6
776	0.23	1.343×10^6
649	0.13	1.297×10^6
54.5	0.08	1.315×10^6
461	0	0
Totals	1.00	5.33×10^6

Thus $n_\mathrm{A} = 3.51 \times 10^6$ particles per kg and

$n_\mathrm{B} = 5.33 \times 10^6$ particles per kg

And in samples of 1 g (0.001 kg) we would expect a total number of particles:

$$n = 0.001 \times \left(3.51 \times 10^6 \times 0.6 + 5.33 \times 10^6 \times 0.4\right)$$
$$= 4238 \text{ particles}$$

And so, from Equation (10.5) for a random mixture,

$$\text{Standard deviation } \sigma = \sqrt{\frac{0.6 \times 0.4}{4238}} = 0.0075$$

WORKED EXAMPLE 10.2 (AFTER WILLIAMS, 1990)

Sixteen samples are removed from a binary mixture and the percentage proportions of one component by mass are as follows:

$$41, 37, 41, 39, 45, 37, 39, 40$$
$$41, 43, 40, 38, 39, 37, 43, 40$$

Determine the upper and lower 95% and 90% confidence limits for the standard deviation of the mixture.

Solution

From Equation (10.1), the mean value of the sample composition is:

$$\bar{y} = \frac{1}{16} \sum_{i=1}^{16} y_i = 40\%$$

Since the true mixture composition is not known, an estimate of the standard deviation is found from Equation (10.8):

$$S = \sqrt{\left[\frac{1}{16-1} \sum_{i=1}^{16} (y_i - 40)^2\right]} = 2.31$$

Since there are fewer than 50 samples, the variance distribution curve is more likely to be a χ^2 distribution. Therefore, from Equations (10.16) and (10.17):

$$\text{Lower limit}: \sigma_L^2 = \frac{2.31^2(16-1)}{\chi_\alpha^2}$$
$$\text{Upper limit}: \sigma_U^2 = \frac{2.31^2(16-1)}{\chi_{1-\alpha}^2}$$

At the 90% confidence level $\alpha = 0.05$ and so referring to the χ^2 distribution tables with 15 degrees of freedom $\chi_\alpha^2 = 24.996$ and $\chi_{1-\alpha}^2 = 7.261$.

Hence, $\sigma_L^2 = 3.2$ and $\sigma_U^2 = 11.02$.

At the 95% confidence level $\alpha = 0.025$ and so referring to the χ^2 distribution tables with 15 degrees of freedom $\chi_\alpha^2 = 27.49$ and $\chi_{1-\alpha}^2 = 6.26$.

Hence, $\sigma_L^2 = 2.91$ and $\sigma_U^2 = 12.78$.

WORKED EXAMPLE 10.3

During the mixing of a drug with an excipient the standard deviation of the compositions of 100 mg samples tends to a constant value of ±0.005. The size distributions of drug (D) and excipient (E) are given in Table 10.W3.1.

Table 10.W3.1   Size distributions of drug and excipient

Size x (μm)	420	355	250	190	150	75	53	0
$F_D(x)$	1.00	0.991	0.982	0.973	0.964	0.746	0.047	0
$F_E(x)$	1.00	1.00	0.977	0.967	0.946	0.654	0.284	0

The mean proportion by mass of the drug is known to be 0.2. The densities of drug and excipient are 1100 and 900 kg/m^3, respectively.

Determine whether the mixing is satisfactory (a) if the criterion is a random mixture and (b) if the criterion is an in-house specification that the composition of 95% of the samples should lie within ±15% of the mean.

Solution

The number of particles of drug (Table 10.W3.2) and excipient (Table 10.W3.3) in each sample is first calculated as shown in Worked Example 10.1.

$$\text{Thus } n_D = 8.96 \times 10^9 \text{ particles per kg}$$

and

$$n_E = 3.37 \times 10^{10} \text{ particles per kg}$$

Table 10.W3.2   Number of drug particles in each kg of sample

Mean size of range x (μm)	dm	dn
388	0.009	2.67×10^5
303	0.009	5.62×10^5
220	0.009	1.47×10^6
170	0.009	3.18×10^6
113	0.218	2.62×10^8
64	0.700	4.64×10^9
27	0.046	4.06×10^9
20	0.00	0
0	0.00	0
Totals	1.00	8.96×10^9

Table 10.W3.3 Number of excipient particles in each kg of sample

Mean size of range x (µm)	dm	dn
388	0	0
303	0.023	1.75×10^6
220	0.010	1.99×10^6
170	0.021	9.07×10^6
113	0.292	4.29×10^8
64	0.374	3.03×10^9
27	0.28	3.02×10^{10}
20	0.00	0
0	0	0
Totals	1.00	3.37×10^{10}

And in samples of 1 g (0.001 kg) we would expect a total number of particles:

$$n = 100 \times 10^{-6} \times \left(8.96 \times 10^9 \times 0.2 + 3.37 \times 10^{10} \times 0.8\right)$$
$$= 2.88 \times 10^6 \text{ particles}$$

And so, from Equation (10.10) for a random mixture,

$$\text{Standard deviation } \sigma_R = \sqrt{\frac{0.2 \times 0.8}{2.88 \times 10^6}} = 0.000235$$

Conclusion: The actual standard deviation of the mixture is greater than that for a random mixture and so the criterion for random mixing is not achieved.

For a normal distribution the in-house criterion that 95% of samples should lie within ±15% of the mean suggests that:

$$1.96\sigma = 0.15 \times 0.2$$

(Since, for a normal distribution, 95% of the values lie within ±1.96 standard deviations of the mean.)

Hence, $\sigma = 0.0153$. So, the in-house criterion is achieved.

WORKED EXAMPLE 10.4

A V-mixer is required to mix a small quantity of component A with component B. The mixer is known to transfer 4% of the total mixer inventory between arms per revolution. Determine the minimum number of revolutions of the mixer that would be required for the composition of component A in the left arm to reach 95% of the composition of that component in the whole vessel.

Solution

Equation (10.3) models the number of revolutions of the V-mixer required to achieve a certain degree of mixing i.e. the number of revolutions of the mixer required for the composition of a component in the left arm to reach a certain fraction of the composition of that component in the whole vessel:

$$m_1 = m_0[1 - exp\,(-4qN/M)] \qquad\qquad (10.3)$$

In this case, $m_1/m_0 = 0.95$ and $q/M = 0.04$

Hence,

$$0.95 = [1 - exp(-0.16N)]$$

Solving, $N = 18.7$

Therefore, minimum number of revolutions required = 19

TEST YOURSELF

10.1 Explain the difference between a *random mixture* and a *perfect mixture*. Which of these two types of mixture is more likely to occur in an industrial process?

10.2 Explain how *trajectory segregation* occurs. Give examples of two practical situations that might give rise to trajectory segregation of powders in the process industries.

10.3 What type of segregation is produced when a free-flowing mixture of particles is poured into a heap? Describe the typical segregation pattern produced.

10.4 Explain how core flow of free-flowing particulate mixture from a hopper gives rise to a size-segregated discharge.

10.5 What is the *Brazil nut effect?* Under what conditions, relevant to the process industries, might it occur?

10.6 Explain why size segregation is generally not a problem if all components of the particulate mixture are smaller than around 30 μm.

10.7 Describe two types of industrially relevant mixers. Which mixing mechanism dominates in each type of mixer?

10.8 Explain what is meant by *scale of scrutiny* of a particulate mixture. What scale of scrutiny would be appropriate for (a) the active drug in a powder mixture fed to the tableting machine, (b) muesli breakfast cereal, (c) a health supplement fed to chickens?

10.9 For a two-component mixture, write down expressions for (a) *mean* composition, (b) *estimated variance* when the true mean is unknown, (c) upper and lower theoretical limits of mixture variance. Define all symbols used.

10.10 Explain how one would go about determining whether the mixture produced by an industrial process is satisfactory.

EXERCISES

10.1 Thirty-one samples are removed from a binary mixture and the percentage proportions of one component by mass are as follows:

$$19, 22, 20, 24, 23, 25, 22, 18, 24, 21, 27, 22, 18, 20, 23, 19,$$

$$20, 22, 25, 21, 17, 26, 21, 24, 25, 22, 19, 20, 24, 21, 23$$

Determine the upper and lower 95% confidence limits for the standard deviation of the mixture.

[Answer: 0.355 to 0.595]

10.2 A random mixture consists of two components A and B in proportions 30% and 70% by mass, respectively. The particles are spherical and components A and B have particle densities 500 and 700 kg/m^3, respectively. The cumulative undersize mass distributions of the two components are shown in Table 10.E2.1.

Table 10.E2.1 Size distributions of particles A and B

Size x (μm)	2057	1676	1405	1204	1003	853	699	599	500	422	357
F_A (x)	1.00	1.00	0.85	0.55	0.38	0.25	0.15	0.10	0.07	0.02	0.00
F_B (x)		1.00	0.80	0.68	0.45	0.25	0.12	0.06	0.00	0.00	

If samples of 5 g are withdrawn from the mixture, what is the expected value for the standard deviation of the composition of the samples?

[Answer: 0.0025]

10.3 During the mixing of a drug with an excipient the standard deviation of the compositions of 10 mg samples tends to a constant value of ±0.005. The size distributions by mass of drug (D) and excipient (E) are given in Table 10.E3.1.

Table 10.E3.1 Size distributions of drug and excipient

Size x (μm)	499	420	355	250	190	150	75	53	0
F_D (x)	1.00	0.98	0.96	0.94	0.90	0.75	0.05	0.00	0.00
F_E (x)	1.00	1.00	0.97	0.96	0.93	0.65	0.25	0.05	0.00

The mean proportion by mass of the drug is known to be 0.1. The densities of the drug and the excipient are 800 and 1000 kg/m^3, respectively.

Determine whether the mixing is satisfactory if:

(a) The criterion is a random mixture.

(b) The criterion is an in-house specification that the composition of 99% of the samples should lie within ±20% of the mean.

[Answer: (a) 0.00118, criterion not achieved; (b) 0.00775, criterion achieved.]

10.4 A V-mixer is required to mix a small quantity of component A with component B. The mixer is known to transfer 2% of the total mixer inventory between arms per revolution. Determine the minimum number of revolutions of the mixer that would be required for the composition of component A in the left arm to reach 99% of the mass fraction of component A in the whole vessel.

[Answer: 58 revolutions]

11

Particle Size Reduction

Comminution is the deliberate reduction of particle size to achieve desired powder and process outcomes. *Attrition* is taken here to refer to the accidental and unwanted reduction in particle size.

Comminution is an important step in the processing of many solid materials. It may be used to create particles of a certain size and shape, to increase the surface area available for chemical reaction, or to liberate valuable minerals held within particles. Comminution is also of increasing importance in recycling processes, where useful materials must be liberated and separated for further processing. The size reduction of solids is an energy-intensive and highly inefficient process: it is claimed that around 5% of all electricity generated is used in size reduction and that based on the energy required for the creation of new surfaces (see Section 2.3), the industrial scale process is generally less than 1% efficient. The two statements would indicate that there is a great incentive to improve the efficiency of comminution processes. However, despite a considerable research effort over many years, comminution processes have remained stubbornly inefficient. Also, despite the existence of a well-developed theory for a strength and breakage mechanism of solids, the design and scale-up of comminution processes is usually based on experience and testing and is very much in the hands of the manufacturer of comminution equipment.

Attrition is widespread in industrial processes wherever powders are handled and transported. Problems arising from the creation of fines as a result of attrition include loss of product, requirement for additional separation and recycling, off-specification product and loss of catalyst. All of these have a significant impact on the economics of processes involving powders as raw materials, intermediates, or products.

This chapter is intended as an introduction to the topic of size reduction. The fundamental science of the reduction of particle size by breakage and abrasion is generally applicable to both comminution and attrition and so this is covered first. Next is a section on attrition, covering sources in processes, problems arising, measurement and testing. Finally, there is a section on practical equipment and systems for

Introduction to Particle Technology, Third Edition. Martin Rhodes and Jonathan Seville.
© 2024 John Wiley & Sons Ltd. Published 2024 by John Wiley & Sons Ltd.
Website: www.wiley.com/go/rhodes/particle3e

comminution and how the choice of machine might be matched to the material and the required duty.

11.1   MECHANISMS OF PARTICLE SIZE REDUCTION

11.1.1   Breakage

Consider a crystal of sodium chloride (common salt) as a simple and convenient model of a brittle material. Such a crystal is composed of a lattice of positively charged sodium ions and negatively charged chloride ions arranged such that each ion is surrounded by six ions of the opposite sign, as shown in Figure 13.1. Between the oppositely charged ions there is an attractive force whose magnitude is inversely proportional to the square of the separation of the ions. There is also a repulsive force between the negatively charged electron clouds of these ions which becomes important at very small interatomic distances. Therefore, two oppositely charged ions have an equilibrium separation such that the attractive and repulsive forces between them are equal and opposite. Figure 11.1 shows how the sum of the attractive and repulsive forces varies with changes in the separation of the ions. It can be appreciated that if the separation of the ions is increased or decreased by a small amount from the equilibrium separation there will be a resultant net force restoring the ions to the equilibrium position. Over a small range of interatomic distance, the relationship between applied tensile or compressive force and resulting change in ion separation is linear. That is, in this region (*AB* in Figure 11.1) Hooke's law applies: strain is directly proportional to applied stress (see Chapter 2). The Young's modulus of the material (stress/strain) describes this proportionality. In this Hooke's law range the deformation of the crystal is elastic, i.e. the original shape of the crystal is recovered upon removal of the stress.

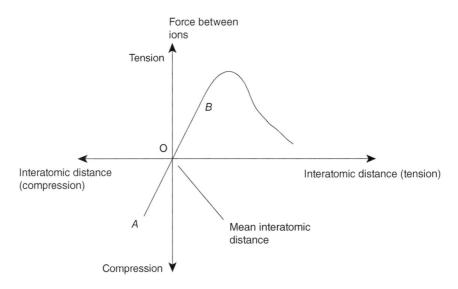

Figure 11.1   Force versus distance on an atomic scale

In order to break the crystal, it is necessary to separate adjacent layers of ions in the crystal and this involves increasing the separation of the ions beyond the region where Hooke's law applies, i.e. beyond point B in Figure 11.1. One way of viewing crystal breakage would be to assume that under tensile stress all bonds in the crystal planes perpendicular to the applied stress are stretched until they simultaneously yield and the material splits into many planes one atom thick – but this gives a theoretical yield strength much greater than that measured in reality. In practice the true failure mechanism for these materials turns out to be more involved and more interesting.

A body under tension stores elastic strain energy. The amount of strain energy stored by a brittle material under tension is given by the area under the appropriate stress–strain curve. This strain energy is not uniformly distributed throughout the body but is concentrated around holes, corners and cracks. Inglis (1913) proposed that the stress concentration factor K, around a hole, crack or corner could be calculated according to the formula:

$$K = \left(1 + 2\sqrt{\frac{L}{R}}\right) \tag{11.1}$$

where L is half the length of the crack, R is the radius of the crack tip or hole and K is the stress concentration factor (local stress/mean stress in body).

Thus, for a round hole $K = 3$.

For a 2 µm long crack with tip radius equal to half the interatomic distance ($R = 10^{-10}$ m), $K \approx 200$.

Real materials fail in tension at loads well below those theoretically necessary to cause simultaneous failure of all the intermolecular bonds across the failure surface. The science of *fracture mechanics* starts with a recognition that flaws or imperfections in the structure act to concentrate the applied stress so that failure is initiated at those points. As a simple example, consider a bar under tension as in Figure 11.2, containing a sharp crack of length A.

As the crack lengthens, there are two changes in energy:

(a) Strain energy is released around the crack, in the area shown, and

(b) Additional new surface of the material is created, and this is accompanied by a change in the total surface energy, as discussed in Section 2.3

The crack will extend if the energy released around the crack as it extends is sufficient to balance the energy required to create the new surface:

$$\frac{1}{B}\frac{\partial U}{\partial A} = 2\gamma^* = G_C \tag{11.2}$$

where $\partial U/\partial A$ is the rate of release of strain energy with respect to crack extension, $\gamma*$ is the free energy per unit area of fresh surface and G_C is known as the *critical strain energy release rate*. For materials, which can be formed into bars in this way, G_C can be measured in, for example, a three-point bend test (Figure 11.3a). For single particles, fracture testing is more difficult and usually involves some form of crushing test in which the tensile stress is developed indirectly (Figure 11.3b). In both cases, it is important to arrange

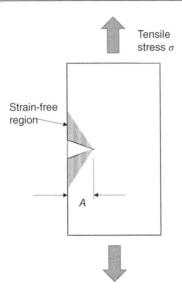

Figure 11.2 Elastic body under tensile stress, showing strain release around a crack. (Depth B into the paper.)

the points at which the force is applied so as not to develop undesired local effects; standard test geometries are available.

This analysis was developed by Griffith[1], who showed that for a given mean stress applied to a body there should be a critical minimum crack length for which the stress concentration at the tip will just be sufficient to cause the crack to propagate. Under the action of this mean stress, a crack initially longer than the critical crack length for that stress will rapidly extend until the body is broken. As the crack grows, provided the mean stress remains constant, there is strain energy excess over that required to propagate that

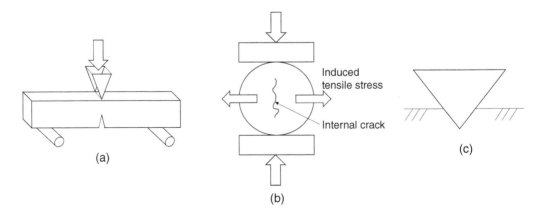

Figure 11.3 Schematic arrangement of common fracture test geometries: (a) three-point bend, (b) diametral-compression fracture test and (c) indentation test

---

[1] Alan Arnold Griffith (1893–1963), English engineer.

crack. This excess strain energy is dissipated at the velocity of sound in the material to concentrate at the tips of other cracks, causing them to propagate. The rate of crack propagation is lower than the velocity of sound in the material and so other cracks begin to propagate before the first crack brings about failure. Thus, in brittle materials multiple fracture is common. If cracks in the surface of brittle materials can be avoided, then the material strength increases to become nearer to the theoretical value. This can be demonstrated by heating a glass rod until it softens and then drawing it out to create a new surface. As soon as the rod is cooled it can withstand surprisingly high tensile stress, as demonstrated by bending. Once the new surface is handled or even exposed to the normal environment for a short period, its tensile strength diminishes due to the formation of microscopic cracks in the surface. In practice, all real materials have surface flaws or cracks of some kind.

Gilvary (1961) proposed the concept of volume, facial and edge flaws (cracks) in order to calculate the size distribution of breakage products. Assuming that all flaws were randomly distributed and independent of each other and that the initial stress system is removed once the first flaws begin to propagate, Gilvary showed that the product size distributions common to comminuted materials could be predicted.

Evans *et al.* (1961) showed that for a disc acted upon by opposing diametrical loads, as in Figure 11.3b, there is a uniform tensile stress acting at 90° to the diameter. Under sufficiently high compressive loads, therefore, the resulting tensile stress could exceed the cohesive strength of the material and the disc would split across the diameter. This approach was taken up and applied particularly to strength measurement of pharmaceutical tablets by Fell and Newton (1970). Under ideal conditions, the tensile stress is constant over the whole of the diameter and the tensile strength σ_T is given by:

$$\sigma_T = \frac{2F}{\pi Dh} \tag{11.3}$$

where F is the force at fracture and D and h are the diameter and height of the tablet.

Evans extended the analysis to three-dimensional particles to show that even when particles are stressed compressively, the stress pattern set up by virtue of the shape of the particle may cause it to fail in tension, whether cracks exist or not.

Cracks are less important for 'tough' materials (e.g. rubber, plastics and metals) since excess strain energy is used in plastic deformation of the material in the 'process zone' around the crack tip rather than in crack propagation. Thus, in ductile metals, for example, the stress concentration at the tip of a crack will cause deformation of the material around the tip, resulting in a larger tip radius (blunting) and lower stress concentration. Brittle materials often exhibit unstable fracture that is characterized by an instantaneous reduction in the loading force corresponding to the initiation of crack propagation, whereas plastically deforming materials often show stable cracks such that the forces decrease gradually after fracture and the crack will cease to propagate if the load is removed.

The observation that small particles are more difficult to break than large particles can be explained using the concept of failure by crack propagation. First, the length of a crack is limited by the size of the particle and so one would expect lower maximum stress concentration factors to be achieved in small particles. Lower stress concentrations mean that higher mean stresses must be applied to the particles to cause failure. Secondly, the Inglis equation [Equation (11.1)] overpredicts K in the case of small particles since in these particles there is less room for the stress distribution patterns to develop. This effectively limits the maximum stress concentration possible and means that a higher mean stress is necessary to cause crack propagation. Kendall (1978) showed that as particle size decreases, the fracture strength increases until a critical size is reached when crack propagation becomes impossible, and he offered a way of predicting this critical particle size. The probability of a critical crack existing becomes less with reducing size.

11.1.2 Abrasion

Abrasion is the removal of material from the exterior of a body by frictional wear processes such that the material removed is very much smaller than the parent. As a mechanism for size reduction, abrasion is distinct from particle breakage. Abrasion can take place at low stresses and gives a product composed of the original particles (somewhat smoother and with a slight size reduction) plus lots of very fine material. This can be distinguished from gross particle breakage, which generally requires higher stresses and results in fragmentation and a product with a wide range of particle sizes. Angular particles and particles with surface asperities are more susceptible to attrition than smoother, rounder particles. Depending on the situation, either or both of these mechanisms may be important.

11.1.3 Stressing Mechanisms for Size Reduction

It is possible to identify three stressing mechanisms responsible for particle size reduction:

(1) Stress applied between two surfaces (either surface–particle or particle–particle) at low velocity, 0.01–10 m/s, resulting in crushing plus abrasion, as in Figure 11.4a.

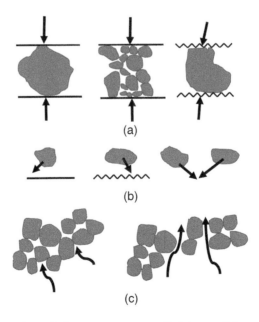

Figure 11.4 Mechanisms by which stresses are applied in particle comminution: (a) stresses applied between two surfaces, (b) stresses applied at a single surface and (c) stresses applied by the medium carrying the particles

(2) Stress applied at a single solid surface (surface–particle or particle–particle) at high velocity, 10–200 m/s, resulting in impact fracture plus abrasion, as in Figure 11.4b.

(3) Stress applied directly by the medium carrying the particles (applicable for liquids – for example in wet grinding to bring about disaggregation – Figure 11.4c).

It is important to note that repeated stressing of particles has an important role to play in achieving breakage, since cracks may extend to a point where they become critical and can cause fracture. Machines are therefore generally designed to recycle material through high-stress regions.

11.2 MODEL PREDICTING ENERGY REQUIREMENT AND PRODUCT SIZE DISTRIBUTION

11.2.1 Energy Requirement

There are three well-known postulates which are used to predict energy requirements for particle size reduction. They are covered here in the chronological order in which they were proposed (Wills and Finch, 2015). Rittinger proposed that the energy required for particle size reduction should be directly proportional to the area of new surface created, which follows the definition of surface energy discussed in Section 2.3. Thus, if the initial and final particle sizes are x_1 and x_2, respectively, then assuming a volume shape factor k_v independent of size,

$$\text{volume of initial particle} = k_v x_1^3$$
$$\text{volume of final particle} = k_v x_2^3$$

each particle of size x_1 will give rise to x_1^3/x_2^3 particles of size x_2.

If the surface shape factor k_s is also independent of size, then for each original particle, the new surface created upon reduction is given by the expression:

$$\left(\frac{x_1^3}{x_2^3}\right) k_s x_2^2 - k_s x_1^2 \tag{11.4}$$

which simplifies to:

$$k_s x_1^3 \left(\frac{1}{x_2} - \frac{1}{x_1}\right) \tag{11.5}$$

Therefore, new surface created per unit mass of original particles:

$$= k_s x_1^3 \left(\frac{1}{x_2} - \frac{1}{x_1}\right) \times (\text{number of original particles per unit mass})$$

$$= k_s x_1^3 \left(\frac{1}{x_2} - \frac{1}{x_1}\right) \times \left(\frac{1}{k_v x_1^3 \rho_p}\right)$$

$$= \frac{k_s}{k_v} \frac{1}{\rho_p} \left(\frac{1}{x_2} - \frac{1}{x_1}\right)$$

where ρ_p is the particle density. Hence assuming shape factors and density are independent of size, Rittinger's postulate may be expressed as:

$$\text{breakage energy per unit mass of feed}, E = C_R\left(\frac{1}{x_2} - \frac{1}{x_1}\right) \tag{11.6}$$

where C_R is a constant. If this is the integral form, then in differential form, Rittinger's postulate becomes:

$$\frac{dE}{dx} = -C_R\frac{1}{x^2} \tag{11.7}$$

An obvious objection to Rittinger's approach is that it neglects many other sources of energy dissipation in addition to creating new surfaces, which are in practice many times larger than the new surface requirement.

A second theory is that proposed by Kick. It states that the energy required depends on the ratio of size reduction and not on the original particle size.

If Δx_1 is the change in particle size,

$$\frac{x_2}{x_1} = \frac{x_1 - \Delta x_1}{x_1} = 1 - \frac{\Delta x_1}{x_1}$$

and so, $\Delta x_1/x_1$ determines the energy requirement for particle size reduction from x_1 to $x_1 - \Delta x_1$. Or

$$\Delta E = C_K\left(-\frac{\Delta x}{x}\right) \tag{11.8}$$

As $\Delta x_1 \to 0$, we have:

$$\frac{dE}{dx} = -C_K\frac{1}{x} \tag{11.9}$$

This is Kick's theory in differential form (where C_k is a constant for a given material). Integrating, we have:

$$E = C_K \ln\left(\frac{x_1}{x_2}\right) \tag{11.10}$$

This proposal is unrealistic in most cases since it predicts that the same energy is required to reduce 10 μm particles to 1 μm particles as is required to reduce 1 m boulders to 10 cm blocks. This is clearly not true and Kick's theory gives unrealistically low values if data gathered for large product sizes are extrapolated to predict energy requirements for small product sizes.

Bond suggested a more useful formula, based on data obtained from industrial and laboratory scale processes involving many materials and presented in its basic form in Equation (11.11a):

$$E = C_B\left(\frac{1}{\sqrt{x_2}} - \frac{1}{\sqrt{x_1}}\right) \tag{11.11a}$$

Or more commonly as:

$$E_B = W_1 \left(\frac{10}{\sqrt{X_2}} - \frac{10}{\sqrt{X_1}} \right) \qquad (11.11b)$$

where E_B is the energy required to reduce the top particle size of the material from X_1 to X_2 and W_1 is the Bond work index.

Since the top size is difficult to define, in practice X_1 to X_2 are taken to be the sieve size in micrometres through which 80% of the material, in the feed and product, respectively, will pass. Bond attached particular significance to the 80% passing size.

The work index is determined as follows.

Both E_B and W_1 have the dimensions of energy per unit mass. The Bond work index, W_1, must be determined empirically through laboratory-scale experiments and is assumed to be independent of the final product size. Some common examples are bauxite, 37.5 kJ/kg; coke from coal, 82.1 kJ/kg and gypsum rock, 32.4 kJ/kg.

Bond's formula gives a fairly reliable first approximation to the energy requirement provided the product's top size is not less than 100 μm. In differential form Bond's formula becomes:

$$\frac{dE}{dx} = -C_B \frac{1}{x^{1.5}} \qquad (11.12)$$

Attempts have been made (e.g. Hukki, 1961) to find the general formula for which the relationships proposed by Rittinger, Kick and Bond are special cases. It can be seen from the results of the above analysis that these three relationships can be considered as being the integrals of the same differential equation:

$$\frac{dE}{dx} = -C \frac{1}{x^N} \qquad (11.13)$$

with

$$\begin{array}{lll} N = 2 & C = C_R & \text{for Rittinger} \\ N = 1 & C = C_K & \text{for Kick} \\ N = 1.5 & C = C_B & \text{for Bond} \end{array}$$

It has been suggested that the three approaches to the prediction of energy requirements mentioned above are each more applicable in certain areas of product size. It is common practice to assume that Kick's proposal is applicable for large particle sizes (coarse crushing and grinding), Rittinger's for very small particle sizes (ultrafine grinding, where surface areas are very large) and the Bond formula being suitable for intermediate particle size–the most common range for many industrial grinding processes. This is shown in Figure 11.5, in which specific energy requirement is plotted against particle size on logarithmic scales. For Rittinger's postulate, $E \propto 1/x$ and so $\ln E \propto -1 \ln(x)$ and hence the slope is −1. For Bond's formula, $E \propto 1/x^{0.5}$ and so $\ln E \propto -0.5 \ln(x)$ and hence the slope is −0.5. For Kick's law, the specific energy requirement is dependent on the reduction ratio x_1/x_2 irrespective of the actual particle size; hence the slope is zero. Typical specific energy values (in kJ/kg) are: primary crushing (i.e. 1000–100 mm), 0.36–0.54; secondary

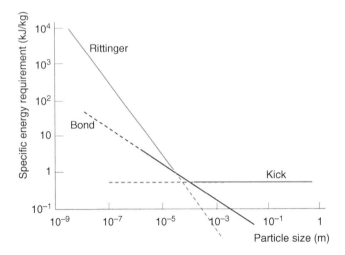

Figure 11.5 Specific energy requirement for breakage: relationship to laws of Rittinger, Bond and Kick

crushing (100–10 mm), 3.6–4.3; coarse grinding (10–1 mm), 10.8–12.6; and fine grinding (1–0.1 mm), 36 (Wills, 2015). For ultrafine grinding (100–10 μm and smaller) the specific energy requirement is typically in the range 100–360 kJ/kg.

In practice, however, it is generally advisable to rely on the experience of equipment manufacturers and on tests in order to predict energy requirements for the milling of a particular material.

11.2.2 Prediction of the Product Size Distribution

The result of particle size reduction operations is to alter the particle size distribution. A universal way of presenting this, which is applicable to all size-changing operations, is by means of the general *population balance equation*, which for a given size range in an element or vessel in which particle sizes are changing is:

$$
\begin{array}{l}
\text{Rate of increase in} \\
\text{mass/number of particles} \\
\text{in the range} \\
1
\end{array}
=
\begin{array}{l}
\text{Rate of \textbf{Flow} of this} \\
\text{size into the element} \\
-\text{Rate of \textbf{Flow} of this} \\
\text{size out of the element} \\
2
\end{array}
+
\begin{array}{l}
\text{Rate of \textbf{Growth}} \\
\text{into range} - \text{Rate of} \\
\textbf{Growth} \text{ out of range} \\
3
\end{array}
+
\begin{array}{l}
\text{Rate of \textbf{Breakage}} \\
\text{into range} - \text{Rate of} \\
\textbf{Breakage} \text{ out of range} \\
4
\end{array}
$$

(11.14)

In size reduction processes (attrition or comminution) growth rates are not relevant and so Term 3 in Equation (11.14) is zero:

$$
\begin{array}{l}
\text{Rate of increase in} \\
\text{mass/number of particles} \\
\text{in the range} \\
1
\end{array}
=
\begin{array}{l}
\text{Rate of \textbf{Flow} of this} \\
\text{size into the element} \\
-\text{Rate of \textbf{Flow} of this} \\
\text{size out of the element} \\
2
\end{array}
+
\begin{array}{l}
\text{Rate of \textbf{Breakage}} \\
\textbf{into} \text{ range} - \text{Rate of} \\
\textbf{Breakage out of} \text{ range} \\
4
\end{array}
$$

(11.15)

For a batch process, considering the whole vessel contents, Term 2 in Equation (11.14) is zero, and so:

$$
\begin{array}{ll}
\text{Rate of increase in} & \text{Rate of \textbf{Breakage}} \\
\text{mass/number of particles} = & \textbf{into} \text{ range-Rate of} \\
\text{in the range} & \textbf{Breakage out of} \text{ range} \\
\qquad\qquad 1 & \qquad\qquad 4
\end{array}
\qquad (11.16)
$$

In comminution and attrition processes, particle breakage in Term 4 may be modelled using two functions, the *specific rate of breakage* and the *breakage distribution function*. The specific rate of breakage S_j is the probability of a particle of size j being broken in unit time (in practice, 'unit time' may mean a certain number of mill revolutions, for example). The breakage distribution function $b(i, j)$ describes the size distribution of the product from the breakage of a given size of the particle. For example, $b(i, j)$ is the fraction of breakage product from size interval j which falls into size interval i. Figure 11.6 helps demonstrate the meaning of S_j and $b(i, j)$ when dealing with 10 kg of monosized particles in size interval i. If $S_1 = 0.6$ we would expect 4 kg of material to remain in size interval 1 after unit time. The size distribution of the breakage product would be described by the set of $b(i, j)$ values. Thus, for example, if $b(4, 1) = 0.25$ we would expect to find 25% by mass from size interval 1 to fall into size interval 4. The breakage distribution function may also be expressed in cumulative form as $B(i, j)$, the fraction of the breakage product from size interval j which falls into size intervals j to n, where n is the total number of size intervals. [$B(i, j)$ is thus a cumulative undersize distribution.]

Thus, remembering that S is a *rate* of breakage, we have Equation (11.17), which expresses the rate of change of the mass of particles in size interval i with time:

$$
\frac{\mathrm{d}m_i}{\mathrm{d}t} = \sum_{j=1}^{j=i-1} \left[b(i, j) S_j m_j \right] - S_i m_i
\qquad (11.17)
$$

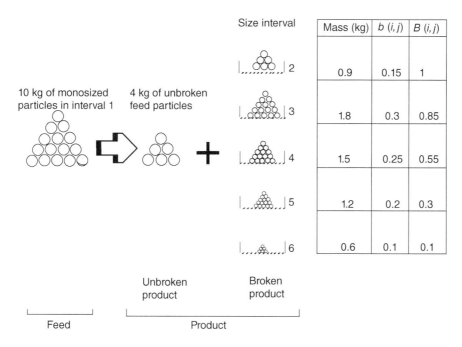

Figure 11.6 Meanings of specific rate of breakage and breakage distribution function

where

$\sum_{j=1}^{j=i-1} [b(i,j)S_j m_j]$ = mass broken per unit time **into** interval i from all intervals of $j > i$. $S_i m_i$ = mass broken per unit time **out of** interval i

Since $m_i = y_i M$ and $m_j = y_j M$, where M is the total mass of feed material and y_i is the mass fraction in size interval i, then we can write a similar expression for the rate of change of mass fraction of material in size interval i with time:

$$\frac{dy_i}{dt} = \sum_{j=1}^{j=i-1} [b(i,j)S_j y_j] - S_i y_i \qquad (11.18)$$

Thus, with a set of S and b values for a given feed material, the product size distribution after a given time in a mill may be determined. In practice, both S and b are dependent on particle size, material and machine. From the earlier discussion on mechanisms of size reduction it would be expected that the specific rate of breakage should decrease with decreasing particle size, and this is found to be the case. The aim of this approach is to be able to use values of S and b determined from small-scale tests to predict product size distributions on a large scale. This method is found to give quite reliable predictions.

11.3 ATTRITION

Attrition, taken here to refer to the accidental and unwelcome reduction in particle size, is widespread in industrial processes wherever powders are handled and transported. Problems arising from the creation of fines as a result of attrition include loss of product, requirement for additional separation and recycling, off-specification product and loss of catalyst. All of these have a significant impact on the economics of processes involving powders as raw materials, intermediates or products. The mechanisms of attrition are complex and are influenced by many variables of the particles and their environment (Table 11.1).

Attrition results from the breakage and abrasion of particles arising from particle–surface contact, particle–particle contact and from thermal shock. Relative motion may be great, for example in pneumatic transport of particulates, where the kinetic energy is absorbed in the collision. Alternatively, the relative motion may be small as in storage and discharge from storage, but the contact forces may be significant.

Table 11.1 Some variables affecting attrition

Properties of particles	Properties of the environment
Size	Residence time
Shape	Velocity
Surface area	Pressure
Porosity	Shear
Hardness	Temperature
Cracks	

Adapted from Bemrose and Bridgwater (1987)

11.3.1 Sources of Attrition in Processes

Storage – Stresses within the stored bulk solids can be significant during storage and discharge. Particle–particle contacts and particle–surface contacts under these conditions can result in attrition.

Fluidized beds – Fluidized bed equipment contains many sites where attrition might take place. These include high-velocity regions around the point of gas entry at the distributor, motion of particles against tubes and surfaces, particle–particle impacts within the bed and in the freeboard, and high-velocity ejection of particles from bursting bubbles. The attrition of catalysts in fluidized bed processes has been a cause for concern and a subject of much research since the early days of the application of fluidized beds. Catalysts are usually expensive and loss from the process through fines production is both an economic and environmental issue, and especially challenging in processes where the catalyst is expected to have a lifetime of many months. In refining, much attention has been devoted to the fluid catalytic cracking (FCC) process, where many tonnes/hr. of catalyst particles circulate between the riser reactor and the fluidized bed regenerator via transport lines, standpipes and multi-stage cyclones. In general, strategies for minimizing attrition in fluidized beds include redesign of the gas distributor to reduce gas entry velocities and/or the rate of entrainment of particles into high-velocity regions, reduction of gas velocity in the bed and in downstream equipment and, in the case of catalysts, design of more attrition-resistant particles. See also Chapter 6.

Cyclones – High-speed impact of particles on surfaces in cyclone separators is a source for particle attrition and surface abrasion. Fines generated may not be collected and hence are lost from the process and/or lost to the environment. Where solids collected by the cyclone are returned to the process (e.g. in a fluidized bed) the cyclone may be responsible for increasing the fines content of the process.

Pneumatic transport – Particle velocities within pneumatic transport systems can be significant and this gives rise to particle–particle collisions and particle–surface collisions resulting in attrition and wear. This is especially the case around bends. Dense phase transport, where velocities are lower, is generally gentler on the particles compared with dilute phase transport. See also Chapter 7.

Mixing – This involves stirring of powders by some means, resulting in both particle–particle contacts and particle–surface contacts which result in attrition.

The consequences of attrition within a process are many. As mentioned earlier, material may be lost from the process creating the need for additional separation equipment and possibly economic and environmental concerns. Production of fine particles may give rise to significant changes in mean powder properties. Changes in particle size distribution, particle shape and surface properties will cause fluidization characteristics to change with time, changes in bulk density causing product packing problems, and adverse changes in powder flow properties in storage. Fines generation may increase the risk of dust explosion if the material is combustible (see Chapter 14). Catalyst properties may be changed as their size distribution, shape and surface properties are altered through attrition.

11.3.2 Measurement of the Degree of Attrition Within a Process

How the degree of attrition is measured will depend on the process. One could study the effect on individual particles to see how shape and size change. This could be extended to a population of particles, looking at changes in size, shape and surface area distribution. In some processes we may be more interested in the effect of attrition on the bulk properties such as bulk density, angle of internal friction and angle of wall friction (Chapter 9) and the way in which they influence powder behaviour in storage or packing. These methods are useful in comparing day-to-day changes within a given process but are not generally helpful in making comparisons between different processes and materials.

11.3.3   Measurement Methods

Attrition rates – There are various approaches to quantifying the rate of production of fines per unit time (Bemrose and Bridgwater, 1987). One approach is to measure the rate of production of particles less than a certain size – 45 μm is commonly used, but larger sizes may be applicable for certain processes (e.g. Amblard et al., 2015). The Hardgrove grindability index (Hardgrove, 1932) based on the amount of new surface produced, used primarily in comminution, has also been applied to the assessment of attrition. The Bond Work Index (Section 11.2.1) has also been used and attempts have been made to relate it to the Hardgrove index. The selection and breakage function approach (Section 11.2.2) has been applied with some success to the measurement of attrition. Although not offering a single number or index to describe attrition, this has the advantage of giving a more realistic description of the attrition process.

Attrition tests – Many standardized tests have been developed for assessing the grindability and attrition tendency of particulate solids (Bemrose and Bridgwater, 1987). Most use standardized test equipment with standard operating conditions and apply stress to the particles in a controlled way. Examples include exposing particles to a high-velocity air jet, a fluidized bed environment, shearing in a modified shear cell, and rolling in a drum or a small ball mill. Some tests have also been developed to assess the susceptibility of particulates to thermal shock, for example by adding a sample of cold test particles to a fluidized bed already at high temperature.

Tests on single particles have been available for many years. Early tests required particles of millimetre size, but more recently tests for smaller particles have become available. Two common approaches are the single particle crush test and the single particle impact test. The single particle crush test involves the slow application of a crushing force on a particle positioned between two flat surfaces or 'platens' (see Figure 11.3b). The impact test may involve particle–particle collision or particle–surface collision. Single particle tests can provide information on the energy requirements and the product size distribution upon fragmentation of the particles. However, abrasion, which may supply a significant amount of attrition in a process, is not specifically assessed in such tests. Boerefijn *et al.* (2000) found that a single particle impact test coupled with good hydrodynamic models of the fluidized bed was reliable in predicting attrition in the FCC process. This approach has been used to improve the attrition resistance of catalysts.

11.3.4   Energy Requirement for Attrition

Some assessment of energy requirement for size reduction by attrition is useful to determine which operations within a process might provide sufficient energy to give rise to attrition of a particular material. Predictions of the energy requirement might be done using, for example, the approaches given in Section 11.2.1.

11.4 TYPES OF COMMINUTION EQUIPMENT

The choice of machine selected for a particular comminution operation will depend on the following variables:

- Stressing mechanism.

- Size of feed and product.

- Material properties.

- Carrier medium.

- Mode of operation.

- Capacity or volumetric feed rate.

- Combination with other unit operations.

11.4.1 Stressing Mechanism

An initial classification of comminution equipment can be made according to the stressing mechanisms employed (see Section 11.1.3 above) as follows.

Machines using mainly mechanism 1, crushing

The *jaw crusher* behaves like a giant pair of nutcrackers (Figure 11.7). One jaw is fixed and the other, which is hinged at its upper end, is moved towards and away from the fixed jaw by means of toggles driven by an eccentric gear. The lumps of material are crushed between the jaws and leave the crusher when they are able to pass through a grid at the bottom.

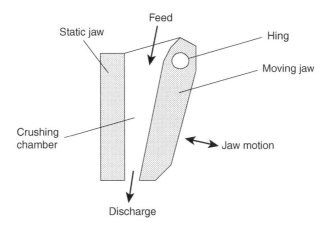

Figure 11.7 Schematic diagram of a jaw crusher

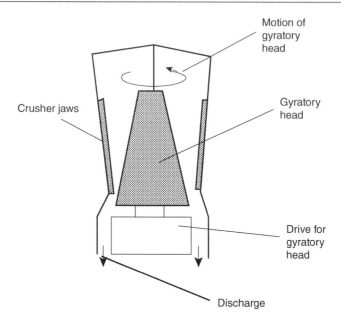

Figure 11.8   Schematic diagram of a gyratory crusher

The *gyratory crusher*, shown in Figure 11.8, has a fixed jaw in the form of a truncated cone. The other jaw is a cone which rotates inside the fixed jaw on an eccentric mounting. Material is discharged when it is small enough to pass through the gap between the jaws.

In the *crushing roll* machine, two cylindrical rolls rotate in opposite directions, horizontally and side by side with an adjustable gap between them (Figure 11.9). As the rolls rotate, they drag in material, which is choke-fed by gravity so that particle fracture occurs as the material passes through the gap between the rolls. These may be ribbed to give improved purchase between the material and rolls.

In the *horizontal table mill,* shown in Figure 11.10, the feed material falls on to the centre of a circular rotating table and is thrown out by centrifugal force. In moving outwards the material passes under a roller and is crushed.

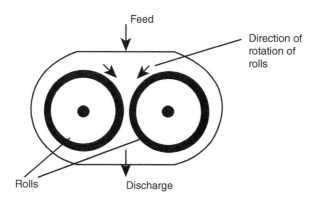

Figure 11.9   Schematic diagram of crushing rolls

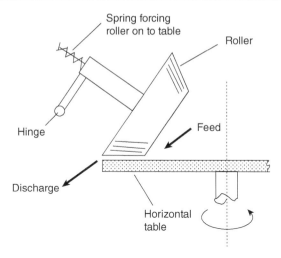

Figure 11.10   Schematic diagram of a horizontal table mill

Machines using mainly mechanism 2, high-velocity impact

The *hammer mill*, shown in Figure 11.11, consists of a rotating shaft to which are attached fixed or pivoted hammers. This device rotates inside a cylinder. The particles are fed into the cylinder either by gravity or by a gas stream. In the gravity-fed version the particles leave the chamber when they are small enough to pass through a grid at the bottom.

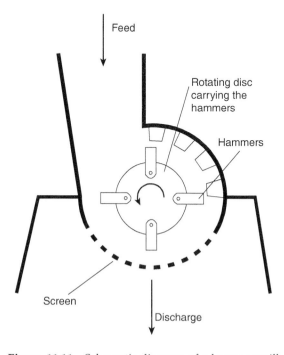

Figure 11.11   Schematic diagram of a hammer mill

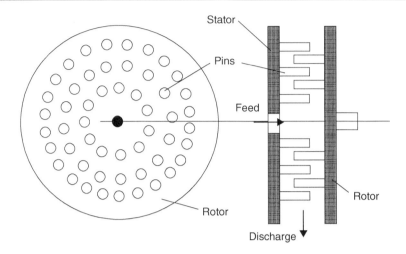

Figure 11.12   Schematic diagram of a pin mill

A *pin mill* consists of two parallel circular discs each carrying a set of projecting pins (Figure 11.12). One disc is fixed and the other rotates at high speed so that its pins pass close to those on the fixed disc. Particles are carried in air into the centre and as they move radially outwards are fractured by impact or by abrasion.

The *fluid energy mill* relies on the turbulence created in high-velocity jets of air or steam to produce conditions for interparticle collisions which bring about particle fracture. A common form of fluid energy mill is the loop or oval jet mill shown in Figure 11.13. Material is conveyed from the grinding area near the jets at the base of the loop to the classifier and exits at the top of the loop. These mills have a very high specific energy consumption and are subject to extreme wear when handling abrasive materials. These problems have been overcome to a certain extent in the fluidized bed jet mill in which the bed is used to absorb the energy from the high-speed particles ejected from the grinding zone.

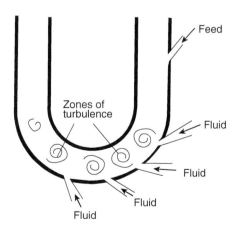

Figure 11.13   Schematic diagram of a fluid energy mill

Machines using a combination of mechanisms 1 and 2, crushing and impact with abrasion

The *media mill*, shown in Figure 11.14, is a cylinder, which may be vertical or horizontal, containing a stirred bed of a grinding medium, such as metals, glass, ceramics or polymer offering a wide range of properties and sizes chosen to suit the application. The grinding medium is agitated by discs or pins on a central rotor. In another design the rotor itself is a smooth cylinder, which fits into the mill vessel with only a small annular gap of size appropriate for the application. The feed, in the form of a slurry, is pumped into one end of the vessel and the product passes out of the other end through a screen, which retains the grinding medium.

In the *colloid mill*, the feed in the form of a slurry passes through the gap between a male, ribbed cone rotating at high speed and a female static cone (Figure 11.15).

The *ball mill*, shown in Figure 11.16, is a rotating cylindrical or cylindrical–conical shell about half filled with balls of steel or ceramic. The speed of rotation of the cylinder is such that the balls are caused to tumble over one another without causing cascading. This speed is usually less than 80% of the critical speed which would just cause the charge

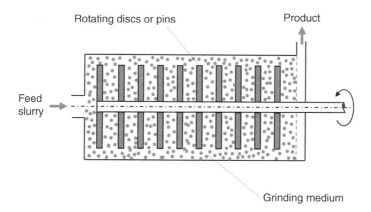

Figure 11.14 Schematic diagram of a media mill

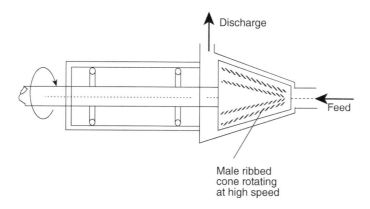

Figure 11.15 Schematic diagram of a colloid mill

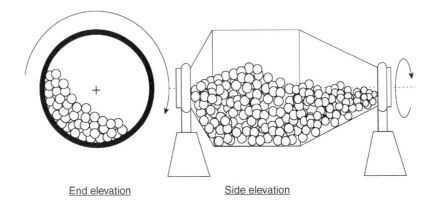

End elevation   Side elevation

Figure 11.16   Schematic diagram of a ball mill

of balls and feed material to be pushed to the wall by centrifugal forces. In continuous milling the carrier medium is air, which may be heated to avoid moisture which tends to cause clogging. Ball mills may also be used for wet grinding with water being used as the carrier medium, sometimes with the addition of surface-active chemicals. The size of balls is chosen to suit the desired product size. The conical section of the mill shown in Figure 11.16 causes the smaller balls to move towards the discharge end and accomplish the finer grinding.

Tube mills are very long ball mills, which are often compartmented by diaphragms, with balls graded along the length from large at the feed end to small at the discharge end. *Rod mills* use long steel rods instead of balls as the grinding medium. *Vibration mills* are like ball mills in that particles of the materials are crushed between ceramic or metal balls and the mill body. However, in the case of vibration mills the motion of the balls is generated by the vibration of the mill body.

11.4.2   Particle Size

Although it is technically interesting to classify mills according to the stressing mechanisms, it is the size of the feed and the product size distribution required, which in most cases determine the choice of a suitable mill. Generally, the terminology shown in Table 11.2 is used.

Table 11.3 indicates how the product size determines the type of mill to be used.

Table 11.2   Terminology used in comminution

Size range of product	Term used
1000–100 mm	Primary crushing
100–10 mm	Secondary crushing
10–1 mm	Coarse grinding
1 mm to 100 μm	Fine grinding
100–10 μm	Ultrafine grinding

Table 11.3 Categorizing comminution equipment according to product size

Down to 3 mm	3 mm–50 μm	<50 μm
Crushers	Ball mills	Ball mills
Table mills	Rod mills	Vibration mills
Edge runner mills	Pin mills	Media mills
	Tube mills	Fluid energy mills
	Vibration mills	Colloid mills

11.4.3 Material Properties

Material properties affect the selection of mill type, but to a lesser extent than feed and product particle size. The following material properties may need to be considered when selecting a mill:

- *Hardness*. Hardness is usually measured on the Mohs scale of hardness where graphite is ranked 1 and diamond is ranked 10. The property 'hardness' is a measure of the resistance to abrasion. This is measured by indentation (Figure 11.3).

- *Abrasiveness*. This is linked closely to hardness and is considered by some to be the most important factor in the selection of commercial mills. Very abrasive materials must generally be ground in mills operating at low speeds to reduce the wear of machine parts in contact with the material (e.g. ball mills).

- *Toughness*. This is the property which quantifies the material resistance to the propagation of cracks. It is defined as the *stress intensity factor* and is measured in a fracture mechanics test such as those shown in Figure 11.3. In tough materials, excess strain energy brings about plastic deformation rather than the propagation of new cracks. Brittleness is the opposite of toughness. Tough materials present problems in grinding, although in some cases it is possible to reduce the temperature of the material, as in *cryogrinding*, thereby reducing the propensity to plastic flow and rendering the material more brittle.

- *Cohesivity/adhesivity*. These properties control whether particles of material stick together and to other surfaces, as described in Section 2.3. Cohesivity and adhesivity are related to moisture content and particle size. A decrease of particle size or an increase in moisture content increases the cohesivity and adhesivity of the material. Problems caused by cohesivity/adhesivity due to particle size may be overcome by wet grinding.

- *Fibrous nature*. Materials of a fibrous nature are a special case and must be comminuted in shredders or cutters which are based on the hammer mill design.

- *Low melting point*. The heat generated in a mill may be sufficient to cause the melting of some materials, causing problems of increased toughness and increased cohesivity and adhesivity. In some cases, the problem may be overcome by using cold air as the carrier medium.

- *Other special properties.* Materials which are thermally sensitive and have a tendency to spontaneous combustion or high inflammability must be ground using an inert carrier medium (e.g. nitrogen). Toxic or radioactive materials must be ground using a carrier medium operating on a closed circuit.

11.4.4   Carrier Medium

The carrier medium may be a gas or a liquid. Although the most common gas used is air, inert gases may be used in some cases as indicated above. The most common liquid used in wet grinding is water although oils are sometimes used. The carrier medium not only serves to transport the material through the mill but, in general, transmits forces to the particles, influences friction and hence abrasion, affects crack formation and cohesivity/adhesivity. The carrier medium can also influence the electrostatic charging and the flammability of the material.

11.4.5   Mode of Operation

Mills operate in either batch or continuous mode. The choice between modes will be based on throughput, the process and economics. The capacity of batch mills varies from a few grams on the laboratory scale to a few tonnes on a commercial scale. The throughput of continuous milling systems may vary from several hundred grams per hour at laboratory scale to several thousand tonnes per hour at industrial scale.

11.4.6   Combination with Other Operations

Some mills have a dual purpose and thus may bring about drying, mixing or classification of the material in addition to its size reduction.

11.4.7   Types of Milling Circuit

Milling circuits are either 'open circuit' or 'closed circuit'. In open-circuit milling (Figure 11.17) the material passes only once through the mill, and so the main controllable

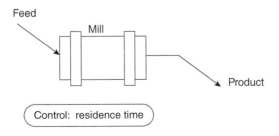

Figure 11.17   Open-circuit milling

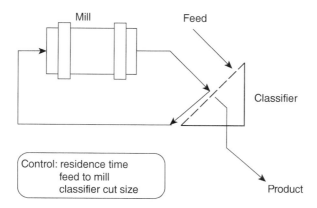

Figure 11.18   Closed-circuit milling

variable is the residence time of the material in the mill. Thus, the product size and distribution may be controlled over a certain range by varying the material residence time (throughput), i.e. feed rate governs product size and so the system is inflexible.

In closed-circuit milling (Figure 11.18) the material leaving the mill is subjected to some form of classification (separation according to particle size) with the oversize being returned to the mill with the feed material. Such a system is far more flexible since both product mean size and size distribution may be controlled.

Figures 11.19 and 11.20 show the equipment necessary for feeding material into the mill, removing material from the mill, classifying, recycling oversize material and removing product in the case of dry and wet closed milling circuits, respectively.

All of these examples can be reproduced in suitable flowsheeting packages, in steady state or in some cases set up to show the dynamics of changes in conditions.

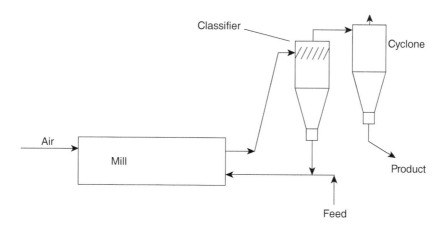

Figure 11.19   Dry milling: closed-circuit operation

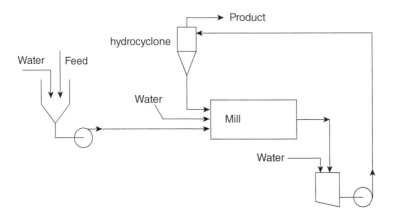

Figure 11.20   Wet milling: closed-circuit operation

11.5   WORKED EXAMPLES

WORKED EXAMPLE 11.1

A material consisting originally of 25 mm particles is crushed to an average size of 7 mm and requires 20 kJ/kg for this size reduction. Determine the energy required to crush the material from 25 to 3.5 mm assuming (a) Rittinger's law, (b) Kick's law and (c) Bond's law.

Solution

(a) Applying Rittinger's law as expressed by Equation (11.6):

$$20 = C_R \left(\frac{1}{7} - \frac{1}{25} \right)$$

Hence $C_R = 194.4$ and so with $x_2 = 3.5$ mm,

$$E = 194.4 \left(\frac{1}{3.5} - \frac{1}{25} \right)$$

Hence $E = 47.8$ kJ/kg.

(b) Applying Kick's law as expressed by Equation (11.10):

$$20 = -C_K \ln \left(\frac{7}{25} \right)$$

Hence $C_K = 15.7$ and so with $x_2 = 3.5$ mm,

$$E = -15.7 \ln \left(\frac{3.5}{25} \right)$$

Hence $E = 30.9$ kJ/kg.

(c) Applying Bond's law as expressed by Equation (11.11a):

$$20 = C_B \left(\frac{1}{\sqrt{7}} - \frac{1}{\sqrt{25}} \right)$$

Hence $C_B = 112.4$ and so with $x_2 = 3.5\,\text{mm}$,

$$E = 112.4 \left(\frac{1}{\sqrt{3.5}} - \frac{1}{25} \right)$$

Hence $E = 37.6\,\text{kJ/kg}$.

WORKED EXAMPLE 11.2

Values of breakage distribution function $b(i, j)$ and specific rates of breakage S_j for a particular material in a ball mill are shown in Table 11.W2.1. To test the validity of these values, a sample of the material with the size distribution indicated in Table 11.W2.2 is to be ground in a ball mill. Use the information in these tables to predict the size distribution of the product after one minute in the mill. (*Note:* S_j values in Table 11.W2.1 are based on one minute grinding time.)

Table 11.W2.1 Specific rates of breakage and breakage distribution function for the ball mill

Size interval (μm)	212–150	150–106	106–75	75–53	53–37	37–0
Interval no.	1	2	3	4	5	6
S_j	0.7	0.6	0.5	0.35	0.3	0
$b(1, j)$	0	0	0	0	0	0
$b(2, j)$	0.32	0	0	0	0	0
$b(3, j)$	0.3	0.4	0	0	0	0
$b(4, j)$	0.14	0.2	0.5	0	0	0
$b(5, j)$	0.12	0.2	0.25	0.6	0	0
$b(6, j)$	0.12	0.2	0.25	0.4	1.0	0

Table 11.W2.2 Feed size distribution

Interval no. (j)	1	2	3	4	5	6
Fraction	0.2	0.4	0.3	0.06	0.04	0

Solution

Applying Equation (11.18):

Change of fraction in interval 1:

$$\frac{dy_1}{dt} = 0 - S_1 y_1 = 0 - 0.7 \times 0.2$$
$$= -0.14$$

Hence, new $y_1 = 0.2 - 0.14 = 0.06$.

Change of fraction in interval 2:

$$\frac{dy_2}{dt} = b(2,1)S_1 y_1 - S_2 y_2$$
$$= (0.32 \times 0.7 \times 0.2) - (0.6 \times 0.4)$$
$$= -0.1952$$

Hence new $y_2 = 0.4 - 0.1952 = 0.2048$.

Change of fraction in interval 3:

$$\frac{dy_3}{dt} = \left[b(3,1)S_1 y_1 + b(3,2)S_2 y_2\right] - S_3 y_3$$
$$= \left[(0.3 \times 0.7 \times 0.2) + (0.4 \times 0.6 \times 0.4)\right] - (0.5 \times 0.3)$$
$$= -0.012$$

Hence, new $y_3 = 0.3 - 0.012 = 0.288$.

Similarly for intervals 4, 5 and 6:

$$new\ y_4 = 0.1816$$
$$new\ y_5 = 0.1429$$
$$new\ y_6 = 0.1227$$

Checking:

Sum of predicted product interval mass fractions $= y_1 + y_2 + y_3 + y_4 + y_5 + y_6 = 1.000$.

Hence product size distribution:

Interval no. (j)	1	2	3	4	5	6
Fraction	0.06	0.2048	0.288	0.1816	0.1429	0.1227

TEST YOURSELF

11.1 Explain why in brittle materials multiple fractures are common.

11.2 Using the concept of failure by crack propagation, explain why small particles are more difficult to break than large particles.

11.3 Summarize three different models for predicting the energy requirement associated with particle size reduction. Over what size ranges might each model be most appropriately applied?

11.4 Define *specific rate of breakage* and *breakage distribution function,* used in the prediction of product size distribution in a size reduction process.

11.5 Explain in words the meaning for the following equation.

$$\frac{\mathrm{d}y_i}{\mathrm{d}t} = \sum_{j=1}^{j=i-1} \left[b(i,j) S_j y_j \right] - S_i y_i$$

11.6 Describe the operation of the *hammer mill*, the *fluid energy mill* and the *ball mill*. In each case identify the dominant stressing mechanism responsible for particle breakage.

11.7 Under what conditions might wet grinding be used?

11.8 List five material properties that would influence selection of a mill type.

11.9 In processes handling and processing particulate solids, which operations are likely to give rise to attrition?

11.10 What problems might excessive attrition give rise to in a process?

EXERCISES

11.1

(a) Rittinger's energy law postulated that the energy expended in crushing is proportional to the area of new surface created. Derive an expression relating the specific energy consumption in reducing the size of particles from x_1 to x_2 according to this law.

(b) Table 11.E1.1 gives values of specific rates of breakage and breakage distribution functions for the grinding of limestone in a hammer mill. Given that values of specific rates of breakage are based on 30 s in the mill at a particular speed, determine the size distribution of the product resulting from the feed described in Table 11.E1.2 after 30 s in the mill at this speed.

[Answer: 0.12, 0.322, 0.314, 0.244.]

Table 11.E1.1 Specific rates of breakage and breakage distribution function for the hammer mill

Interval (μm)	106–75	75–53	53–37	37–0
Interval no.j	1	2	3	4
S_j	0.6	0.5	0.45	0
$b(1, j)$	0	0	0	0
$b(2, j)$	0.4	0	0	0
$b(3, j)$	0.3	0.6	0	0
$b(4, j)$	0.3	0.4	1.0	0

Table 11.E1.2 Feed size distribution

Interval	1	2	3	4
Fraction	0.3	0.5	0.2	0

11.2 Table 11.E2.1 gives information gathered from tests on the size reduction of a material in a ball mill. Assuming that the values of specific rates of breakage, S_j, are based on 25 revolutions of the mill at a particular speed, predict the product size distribution resulting from the feed material, details of which are given in Table 11.E2.2 after 25 revolutions in the mill at that speed.

[Answer: 0.125, 0.2787, 0.2047, 0.1661, 0.0987, 0.0779, 0.04878.]

Table 11.E2.1   Results of ball mill tests

Interval (μm)	300–212	212–150	150–106	106–75	75–53	53–37	37–0
Interval no. j	1	2	3	4	5	6	7
S_j	0.5	0.45	0.42	0.4	0.38	0.25	0
$b(1, j)$	0	0	0	0	0	0	0
$b(2, j)$	0.25	0	0	0	0	0	0
$b(3, j)$	0.24	0.29	0	0	0	0	0
$b(4, j)$	0.19	0.27	0.33	0	0	0	0
$b(5, j)$	0.12	0.2	0.3	0.45	0	0	0
$b(6, j)$	0.1	0.16	0.25	0.3	0.6	0	0
$b(7, j)$	0.1	0.08	0.12	0.25	0.4	1.0	0

Table 11.E2.2   Feed size distribution

Interval	1	2	3	4	5	6	7
Fraction	0.25	0.45	0.2	0.1	0	0	0

11.3 Table 11.E3.1 gives information on the size reduction of a sand-like material in a ball mill. Given that the values of specific rates of breakage S_j are based on five revolutions of the mill, determine the size distribution of the feed materials shown in Table 11.E3.2 after five revolutions of the mill.

[Answer: 0.0875, 0.2369, 0.2596, 0.2115, 0.2045.]

Table 11.E3.1   Results of ball mill tests

Interval (μm)	150–106	106–75	75–53	53–37	37–0
Interval no. (j)	1	2	3	4	5
S_j	0.65	0.55	0.4	0.35	0
$b(1, j)$	0	0	0	0	0
$b(2, j)$	0.35	0	0	0	0
$b(3, j)$	0.25	0.45	0	0	0
$b(4, j)$	0.2	0.3	0.6	0	0
$b(5, j)$	0.2	0.25	0.4	1.0	0

Table 11.E3.2 Feed size distribution

Interval	1	2	3	4	5
Fraction	0.25	0.4	0.2	0.1	0.05

11.4 Comminution processes are generally less than 1% efficient. Where does all the wasted energy go?

12

Size Enlargement

Size enlargement is the process by which particle size is increased, either by growth from solution for example, or by bringing smaller particles together to form larger ones. Size enlargement is one of the single most important process steps involving particulate solids in the process industries. It is mainly associated with the pharmaceutical, agricultural and food industries, but also plays an important role in other industries including processing of minerals, metals and ceramics.

Some of the particulate materials considered in this book are naturally occurring minerals but many are initially formed as particles by *crystallization* processes. This chapter includes a brief introduction to crystallization as a growth method.

We then consider processes in which smaller particles are formed into larger ones by introduction of a liquid *binder* phase or by direct compression, drawing on the background introduced in Chapter 2. The design of such processes requires an understanding of rate processes and population balances, extending the principles already introduced in Chapter 11. Some of the most important of the many industrial processes that are used for these operations are then described.

12.1 INTRODUCTION

There are many reasons why we may wish to increase the mean size of a product or intermediate. These include reduction of a dust hazard (explosion hazard or health hazard), to reduce caking and lump formation, to improve flow properties, to increase bulk density for storage, to create non-segregating mixtures of ingredients of differing original size, to provide a defined metered quantity of active ingredient (e.g. in pharmaceutical drug formulations), and to control the surface to volume ratio (e.g. in catalyst supports).

Adapted from an original chapter for the 2nd Edition by Karen Hapgood, now of Swinburne University, Melbourne, Australia.

Introduction to Particle Technology, Third Edition. Martin Rhodes and Jonathan Seville.
© 2024 John Wiley & Sons Ltd. Published 2024 by John Wiley & Sons Ltd.
Website: www.wiley.com/go/rhodes/particle3e

Methods by which size enlargement is brought about may be divided into:

(a) Wet granulation – consisting particularly of mixing processes in which the particles are brought into contact so that they stick together, aided by the addition of a liquid binder; here we use the term *granulation* to include all these types of process, although the term *agglomeration* is also used. Wet processes also include those in which a liquid is removed from a suspension in order to leave larger agglomerates; these include *extrusion* and some types of *spray drying*.

(b) Dry granulation–particularly by compaction processes such as *tableting* and *roll-pressing*.

It is not possible in the confines of a single chapter to cover the wide range of size enlargement methods in any detail, but examples of each major type will be given.

The starting point for many manufacturing processes is the creation of particles by crystallization, which is introduced first.

12.2   CRYSTALLIZATION

Crystallization is a widely used method for forming particles from chemicals in solution. It is a growth process in its own right but is also used to form particles which are then processed into larger entities by the methods introduced later in this chapter.

Crystallization is a very wide subject and only the basic principles will be covered here. For a more detailed summary of the subject, readers are referred to Davey and Garside (2000).

12.2.1   Crystal Structure

Materials in their solid state are either *crystalline* or *amorphous* or a combination of both. Crystalline materials are characterized by *ordered* molecular structures which repeat throughout the crystal, forming a crystal *lattice*, as in the example of sodium chloride shown in Figure 12.1. They also have a precise melting point, which relates to their ordered structure in which there are similar bond energies throughout. Amorphous materials have no such order and therefore no precise melting point. Many materials, such as polymers, show regions of order or crystallinity. Whether the structure is crystalline, amorphous or semi-crystalline has profound effects on its mechanical and chemical properties and therefore on its function in a product.

Some crystalline materials can exist in several *polymorphs*, which are different crystal structures resulting from alternative molecular packings. In general, different polymorphs show different densities and other physical and chemical properties, which can affect their function, such as bioavailability of a drug substance. Polymorphs have different melting points (because their lattice energies are different); the most stable form has the highest melting point. The other forms are termed *metastable* and will revert to the stable form, although the rate at which they do this is highly variable, from seconds to years.

From a particle technology point of view, the most obvious feature of a crystal is its distinctive shape or *habit*, which is again related to its molecular packing arrangement.

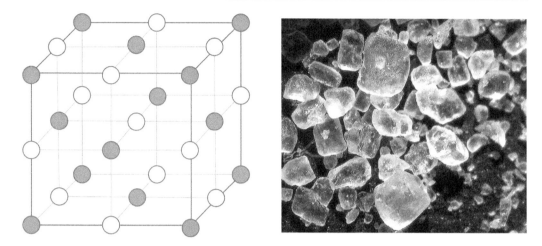

Figure 12.1 Sodium chloride crystal structure (face-centred cubic; blue = sodium, white = chlorine) and table salt crystals

This can have important effects in its subsequent processing, since particle shape affects behaviour in packing, flow and many other areas. For example, plate-like particles form structures which are very resistant to through-flow of fluids, so that they are difficult to filter and wash after they have formed. Shape also affects surface–volume ratio and therefore surface availability for chemical reaction and other purposes.

For crystals to grow, free molecules from the surrounding melt or solution must be attracted and incorporated. This process depends on the type of molecular packing at each face or *facet* of the crystal and factors such as its polarity. Facets with polar surfaces will be more attractive to molecules in polar solutions, for example. For similar reasons, undesired impurities or additives which are deliberately added to the crystallization process may bind to a surface and preferentially reduce or block the growth of one facet.

It is not unexpected, then, that different facets will grow at different rates. In general, the largest facet is the one with the slowest growth rate. This is explained by consideration of Figure 12.2. If the facets labelled 1 and 4 are the fastest growing, the initial hexagonal shape rapidly changes to a diamond. The slow-growing facets extend and the faster-growing ones effectively 'grow out'. If growth is fast enough and process conditions permit, the slower-growing facets take over the determination of the final shape.

12.2.2   Solubility and Saturation

The starting point for production of crystalline material is usually some form of reaction which results in the formation of the chemical constituent of interest as a liquid above its melting point or in solution, usually with other materials also present. It is then necessary to produce the solid. One way of doing this is to cool a melt to below its melting point, as in *prilling* of fertilizer granules; the other is to exceed the saturation limit of a solution so that crystals are formed. Both approaches are in common use but crystallization from solution is more usual in cases of fine chemical production for the pharmaceutical industry, for example.

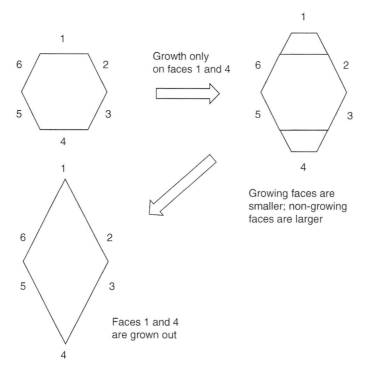

Figure 12.2 Preferential growth of a hexagonal shape produces a diamond. Taylor and Aulton (2022)/with permission of Elsevier

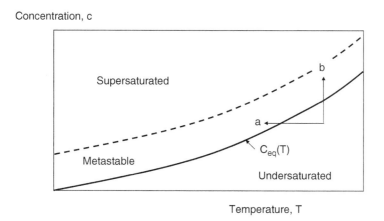

Figure 12.3 The solubility/supersolubility diagram. Davey and Garside (2000)/with permission of Oxford University Press

Pure chemicals have a limited solubility in a given solvent, the *equilibrium saturation*, as shown in Figure 12.3, and this increases with temperature. Solutions with concentrations below the solubility curve are *undersaturated*; those above it are *supersaturated*. If crystallization is to occur, saturation must be exceeded. For an initial starting point in the undersaturated region, Figure 12.3 shows two routes to doing this: to point (a) by decreasing the temperature, or to point (b) by increasing the concentration, through evaporation of the solvent, for example. It is the *concentration driving force* $\Delta c = c - c_{eq}$, which drives the crystallization process. Figure 12.3 also shows a *metastable* region in which the concentration is above the equilibrium saturation value. In the metastable region, existing crystals will grow but new crystals will not easily form; the width of this region depends on the kinetics of formation and growth.

12.2.3 Nucleation and Growth

The first step in crystallization is *nucleation* of a small number of molecules to form an ordered cluster around which further growth can take place. For thermodynamic reasons, such a cluster has to be of a certain critical size in order to survive. There is an energy barrier to spontaneous nucleation; as supersaturation increases, this energy barrier and the critical size of the nucleus both decrease. This dependence of nucleation rate on supersaturation is reflected in an equation for nucleation rate, J, which has units of number/s m^3:

$$J = K_J \exp\left(-B_J \gamma^3 / T^3 f^2\right) \tag{12.1}$$

where K_J and B_J are constants, γ is the surface energy of the cluster with respect to the surrounding solution (see Chapter 2), T is temperature and f is the fractional supersaturation, which is defined in terms of differences in chemical potential, but for an ideal solution can be thought of as proportional to the concentration driving force Δc. It is apparent from Equation (12.1) that J increases very rapidly in response to an increase in supersaturation, as shown schematically in Figure 12.4.

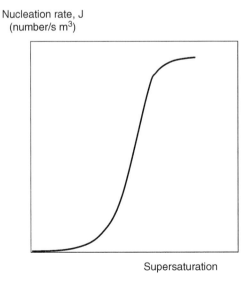

Figure 12.4   Nucleation rate as a function of supersaturation. Davey and Garside (2000)/Oxford University Press

If the materials used are free from contamination and the equipment used is clean, it is possible to cool a solution well into the supersaturated zone without spontaneous nucleation taking place, but if there is dust present, or surface imperfections on equipment, nucleation can be triggered there. In some crystallization processes, fine product is recycled in order to provide nucleation sites; this is known as *seeding* the crystallization. Sometimes attrition of crystals provides such fines *in situ*.

Having formed a nucleus, growth can then occur into the final crystal form. The mode of growth depends on the crystal structure, and as discussed earlier, different facets of a crystal may grow at different rates. The growth rate is dependent on the degree of supersaturation, but is not as sensitive to this as is the rate of nucleation. As in chemical reactions, the growth rate is also affected by mass transfer limitations, which can be reduced by stirring the crystallizer vessel, for example. In general, the mass-based growth rate R_G, is given by

$$R_G = k_r(\Delta c)^r \tag{12.2}$$

where k_r is a rate constant and the value of r is normally between 1 and 2, but may not be constant over the entire growth process. Growth is structure-sensitive: there are a number of possible modes, as shown schematically in Figure 12.5, and crystals of the same substance may grow at different rates.

Considering nucleation and growth together, both proceed faster at higher supersaturation, but with different dependencies. At low values of supersaturation, growth can proceed faster than nucleation and larger crystals result. At higher values, nucleation becomes very fast, resulting in smaller crystals because the available material in solution is divided over more possible sites.

An additional factor which affects the crystal size distribution is *Ostwald ripening*, which arises from the fact that small crystals are more soluble than large ones. If the solution is in equilibrium with large crystals then it is undersaturated with respect to the small ones, which will then tend to dissolve, providing more material for incorporation in the large ones. (This can be a problem in storage, leading to gradual growth in crystal sizes in some foods, for example, and a resulting gritty mouth feel.)

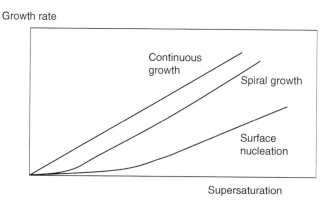

Figure 12.5 Crystal growth rate by various mechanisms. Davey and Garside (2000)/Oxford University Press

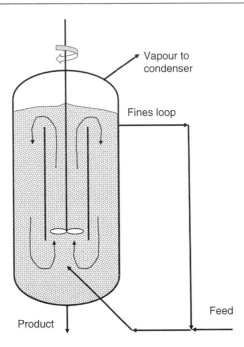

Figure 12.6 Schematic arrangement of a stirred tank crystallizer

12.2.4   Crystallizer Design

In a real crystallizer, nucleation and growth may be accompanied by crystal attrition and/or agglomeration and calculations of size distributions are best carried out using a population balance model such as that introduced in Section 12.3.3.

Crystallizer designs are very varied and include both batch and continuous modes and both stirred tanks and plug-flow-type devices. Figure 12.6 is a schematic example of a stirred-tank crystallizer, featuring an up-pumping impeller and concentric draft tube to promote circulation. The feed is introduced to the area around the base of the draft tube while the product is withdrawn from the lowest point, probably with the aid of some form of quiescent settling zone (not shown). In this example, a suspension of fines is withdrawn from a high point, again benefitting from gravity-separation from the larger crystals, and recycled with the feed solution. Close process control is necessary, particularly of the degree of supersaturation and therefore of temperature.

12.3   *GRANULATION*

12.3.1   Introduction

Granulation is particle size enlargement by combining smaller particles. The most common type of method is *wet granulation*, in which a liquid *binder* is distributed over the bed to initiate granule formation and some form of agitation is applied in order to bring the particles into contact so that the binder can act. The resulting assembly of particles is called a *granule* and consists of the primary particles arranged into a three-dimensional porous structure. Important properties of granules include their size, shape, porosity and composition uniformity, on which depend many other important characteristics. These properties are discussed further in Chapter 13 from the point of view of what is desirable in a product.

Interparticle forces were introduced in Chapter 2, including those due to van der Waals interactions and liquid bridges. The behaviour of particles which are surrounded, or partially surrounded, by liquid is very important to the understanding of wet granulation. As explained in Section 2.6 and shown in Figure 12.7, a small amount of liquid at the contact points between particles results in the formation of *pendular* liquid bridges, which can result in strong forces of attraction. The liquid within these isolated bridges is relatively immobile. As the proportion of liquid to particles is increased, the bridges become large enough to become connected, the liquid becomes free to move and the attractive force between particles decreases. This is the *funicular* region. When there is sufficient liquid to fill the interstitial pores between the particles completely, this is termed the *capillary* region and the granule strength falls further as there are fewer curved liquid surfaces and fewer boundaries for surface tension forces to act on. When the particles are completely dispersed in the liquid, which is the *droplet* state, the strength of the structure is very low.

In the pendular state, increasing the amount of liquid present has little effect on the strength of the bond between the particles until the funicular state is achieved. However, increasing the proportion of liquid enables the particles to be pulled further apart without bridge rupture. This has a practical implication for granulation processes: pendular bridges give rise to strong granules in which the quantity of liquid is not critical but should be less than that required to move into the funicular and capillary regimes.

The ratio of volume of liquid to interparticle pore volume is known as the *degree of saturation*. (Note that we are not concerned here with any internal or *intra*particle porosity.) Clearly, the degree of saturation is increased by increasing the liquid content of a granule while keeping the pore volume constant. The degree of saturation can also be increased by reducing the void fraction and moving the particles closer together (i.e. *densifying* or *consolidating* the granule). This reduces the open pore space available to

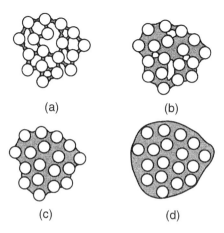

Figure 12.7 Liquid bonding between particles: (a) pendular; (b) funicular; (c) capillary and (d) droplet

the liquid, and the granule saturation then increases from the funicular state through to the droplet state.

We can see that there are a number of sub-processes which may be taking place simultaneously in a granulator:

(a) Individual particles and growing granules are in motion, resulting in collisions between them, each of which may result in further growth.

(b) Liquid is being introduced to the agitated bed, continuously or discontinuously, and is being distributed by the mixing of the particles, increasing their degree of saturation.

(c) Granules may be compacting under the action of repeated collisions so that their degree of saturation is also increased in this way.

(d) Liquid may be simultaneously evaporating, so reducing the degree of saturation.

The nature of the liquid binder varies according to the specific process. Commonly, the binder is a polymeric solution which acts as a glue and strengthens to a solid on drying. A hybrid process is melt granulation, where a polymeric binder is heated and sprayed in liquid form onto the powder, or introduced in powder form and melted *in situ*. The bed is then cooled during granulation to form solid bonds between the particles. Alternatively, a solvent may be used to induce dissolution and recrystallization of the material of which the particles are made.

12.3.2   Granulation Rate Processes

The formation of granules is controlled by three rate processes, illustrated in Figure 12.8. These are (i) *wetting and nucleation* of the original particles by the binding liquid, (ii) *coalescence* or *growth* to form granules plus *consolidation* of the granule and (iii) *breakage* of the granules. Nucleation is the term used, by analogy with crystallization, to describe the initial process of combining primary solid particles with a liquid drop to form new granules or nuclei. (Note, however, that there is no energy barrier to 'nucleation' in the granulation process, so that the analogy is not a perfect one.) Coalescence is the joining together of two granules (or a granule plus a primary particle) to form a larger granule. These processes combine to determine the properties of the product granule (size distribution, porosity, strength, dispersibility, etc.). The final granule size in a granulation process is controlled by the competing mechanisms of growth, breakage and consolidation.

Wetting and nucleation

Wetting is the process by which air (or other gas) within the voids between particles is replaced by liquid. Ennis and Litster (1997) stress the important influence which the extent and rate of wetting have on product quality in a granulation process. For example, poor wetting can result in much material being left ungranulated and requiring recycling. When granulation is used to combine ingredients, account must be taken of the different wetting properties which components of the final granule may have.

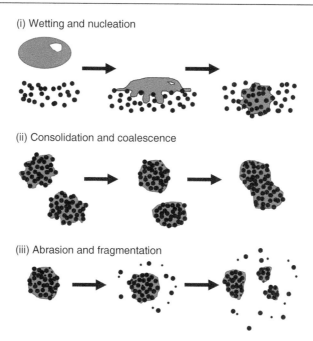

(i) Wetting and nucleation

(ii) Consolidation and coalescence

(iii) Abrasion and fragmentation

Figure 12.8 Summary of the three granulation mechanisms. Adapted from Iveson et al. (2001)

Wetting is governed by the surface tension of the liquid and the contact angle it forms with the material of the particles (see Chapter 2). The rate at which wetting occurs is important in granulation. An impression of this rate is given by the Washburn equation [Equation (12.3)] for the rate of penetration of liquid of viscosity μ and surface tension γ into a bed of powder when the effect of gravity is not significant:

$$\frac{\mathrm{d}z}{\mathrm{d}t} = \frac{R_p \gamma \cos\theta}{4\mu z} \tag{12.3}$$

where t is the time, z is the penetration distance of the liquid into the powder and θ is the dynamic contact angle of the liquid with the solid of the powder. R_p is the average pore radius, which is related to the packing density and the size distribution of the powder. Thus in granulation, the factors controlling the rate of wetting are surface tension, liquid viscosity, packing density and the size distribution.

The Washburn test requires some specialized testing equipment to perform; an alternative test is the drop penetration time test which is more directly related to the wetting of spray drops into a granulating powder (Hapgood et al., 2003). A drop of known volume V_d is gently placed onto a small powder bed with void fraction ε_b and the time taken for the drop to sink completely into the powder bed is measured. The drop penetration time t_p, is given by:

$$t_p = 1.35 \frac{V_d^{2/3} \mu}{\varepsilon_b^2 R \gamma \cos\theta} \tag{12.4}$$

The Washburn test and drop penetration time test are closely related, but the latter is simpler to perform as a screening and investigation test when developing or trouble-shooting a granulation process.

In general, good wetting is desirable. It gives a narrower granule size distribution and improved product quality through better control over the granulation process. In practice, the rate of wetting will significantly influence the extent of wetting of the powder mass, especially where evaporation of binder solvent takes place simultaneously with wetting. The brief analysis above shows us that the rate of wetting is increased by reducing viscosity, increasing surface tension, minimizing contact angle and increasing the size of pores within the powder. Viscosity is in turn determined by the binder concentration and the operating temperature. As concentration changes with solvent evaporation, the binder viscosity will increase. Small particles give small pores and large particles give large pores. Also, a wider particle size distribution will give rise to smaller pores. Large pores ensure a high rate of liquid penetration but give rise to a lower extent of wetting.

In addition to the wetting characteristics, the drop size and overall distribution of liquid are crucial parameters in granulation. If one drop is added to a granulator, only one nucleus granule will be formed, and the size of the nucleus granule will be proportional to the size of the drop. During atomization of the fluid onto the powder, it is important that the conditions in the spray zone balance the rate of incoming fluid drops with the rate of penetration into the powder and/or removal of wet powder from the zone. Ideally, each spray droplet should land on the powder without touching other droplets and sink quickly into the powder to form a new nucleus granule. These ideal conditions are called the *drop-controlled nucleation regime* and occur at low penetration time (described above) and low dimensionless spray flux Ψ_a, which is given by (Litster et al., 2001) as:

$$\Psi_a = \frac{3Q}{2vwd} \tag{12.5}$$

where Q is the solution flow rate, v is the powder velocity in the spray zone, d is the average drop diameter and w is the width of the spray. The dimensionless spray flux is a measure of the density of drops falling on the powder surface. At low spray flux ($\Psi_a \leq 1$) drop footprints will not overlap and each drop will form a separate nucleus granule. At high spray flux ($\Psi_a \approx 1$) there will be significant overlap of drops hitting the powder bed. Nuclei granules formed will be much larger and their size will no longer be a simple function of the original drop size. At a given spray flux value, the fraction of the powder surface that is wetted by spray drops as it passes beneath the spray zone (f_{wet}) is given by (Hapgood et al., 2004) as:

$$f_{wet} = 1 - \exp(-\Psi_a) \tag{12.6}$$

and the fraction of nuclei f_n formed by n drops can be calculated using (Hapgood et al., 2004):

$$f_n = \exp(-4\Psi_a)\left[\frac{(4\Psi_a)^{n-1}}{(n-1)!}\right] \tag{12.7}$$

The dimensionless spray flux parameter can be used both for scale-up and to estimate nuclei starting sizes for population balance modelling (see Section 12.3.3).

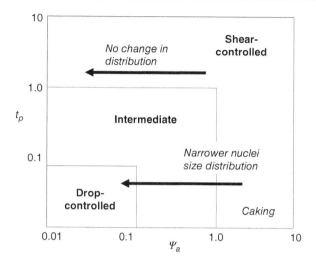

Figure 12.9 Nucleation regime map. For ideal nucleation in the drop-controlled regime, low Ψ_a and low t_p are required. After Hapgood et al. (2003)/John Wiley & Sons

When combined with the drop penetration time, Ψ_a forms part of a nucleation regime map (see Figure 12.9) (Hapgood et al., 2003). Three nucleation regimes have been defined: *drop controlled, shear controlled,* and an *intermediate* zone. Drop-controlled nucleation occurs when one drop forms one nucleus and is expected to occur when both the following conditions hold:

(1) Low spray flux Ψ_a – the spray density is low and relatively few drops overlap.

(2) Fast penetration time t_p – the drop must penetrate completely into the powder bed before it touches either other drops on the powder surface or new drops arriving from the spray.

If *either* criterion is not met, powder mixing and shear characteristics will dominate: this is the *mechanical dispersion regime.* Viscous or poorly wetting binders are slow to flow through the powder pores and form nuclei. Drop coalescence on the powder surface (also known as 'pooling') may occur and create a very broad nuclei size distribution. In the mechanical dispersion regime, the liquid binder can then only be dispersed by powder shear and agitation.

Granule consolidation

Consolidation is the term used to describe the increase in granule density caused by closer packing of primary particles as liquid is squeezed out as a result of collisions. Consolidation can only occur whilst the binder is still liquid. Consolidation determines the void fraction and density of the final granules. Factors influencing the rate and degree of consolidation include particle size, size distribution and binder viscosity. The granule void fraction ε, and the mass ratio of liquid to solid w, control the granule saturation s, which is the fraction of pore space filled with liquid:

$$s = \frac{w\rho_s(1-\varepsilon)}{\rho_l\varepsilon} \tag{12.8}$$

where ρ_s is the solid density and ρ_l is the liquid density. The saturation increases as the porosity decreases, and once the saturation exceeds 100%, further consolidation pushes liquid to the granule surface, making the surface wet. Surface wetness causes dramatic changes in granule growth rates (see below).

Growth

For two colliding primary granules to coalesce, their kinetic energy must be dissipated and the strength of the resulting bond must be able to resist the external forces exerted by the agitation of the powder mass in the granulator. Granules which are able to deform readily will absorb the collisional energy and create increased surface area for bonding. As granules grow so do the internal forces trying to pull the granule apart. It is possible to predict a critical maximum size of granule beyond which coalescence is not possible during collision (Figure 12.10).

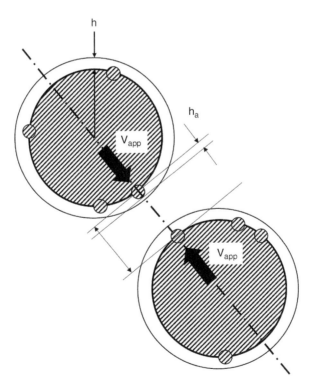

Figure 12.10 Two spherical particles with thin liquid layers in a normal collision. Adapted from Ennis et al. (1991)

Ennis et al. (1991) suggested a rationale for interpreting observed granule growth regimes in terms of the physics of collisions. Consider a collision between two rigid granules (assumed to be of low deformability) of density ρ_g, each coated with a layer of thickness h of liquid of viscosity μ, each having a diameter x and an approach velocity V_{app}. The parameter which determines whether coalescence will occur is a collision Stokes number[1] Stk:

$$Stk = \frac{4\rho_g V_{app} x}{9\mu} \tag{12.9}$$

The Stokes number is a measure of the ratio of collisional kinetic energy to energy dissipated through viscous dissipation. For coalescence to occur the Stokes number must be less than a critical value Stk^*, given by:

$$Stk^* = \left(1 + \frac{1}{e}\right) \ln\left(\frac{h}{h_a}\right) \tag{12.10}$$

where e is the coefficient of restitution for the collision and h_a is the dimension of the surface roughness of the granule.

Based on this criterion, three regimes of granule growth are identified for batch systems with relatively low agitation intensity, as shown in Figure 12.11. These are the *non-inertial*, *inertial* and *coating* regimes. Within the granulator at any time there will be a distribution of granule sizes and velocities which gives rise to a distribution of Stokes numbers for collisions. In the non-inertial regime, the Stokes number is less than Stk^* for all granules and primary particles and practically all collisions result in coalescence, i.e. the particle impact energy is too small to escape capture on contact and all collisions are 'successful'. In this regime, therefore, the growth rate is largely independent of liquid viscosity, granule or primary particle size and kinetic energy of collision. The rate of wetting of the particles controls the rate of growth in this regime; wherever binder liquid exists, capture will occur.

As granules grow, some collisions will occur for which the Stokes number exceeds the critical value, i.e. the collision energy is enough to escape capture. We now enter the

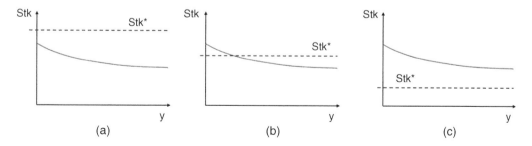

Figure 12.11 Stokes numbers in granulation processes (y indicates a space coordinate): (a) non-inertial, (b) inertial and (c) coating. Adapted from Ennis et al. (1991)

---

[1] Note that this is of a similar form, but different from, the Stokes numbers used elsewhere in this book, such as the Stokes numbers associated with gas–solid separation.

inertial regime [Figure 12.11(b)] in which the rate of growth is dependent on liquid viscosity, granule size and collision energy. The proportion of collisions for which the Stokes number exceeds the critical value increases throughout this regime and the proportion of successful collisions decreases. Once the average Stokes number for the powder mass in the granulator is comparable with the critical value, granule growth is balanced by breakage. Growth then continues by coating of primary particles onto existing granules, since these are the only possible successful collisions according to our criterion.

Further increases in the collision Stokes numbers, so that every collision has enough energy for the particles to escape capture [Figure 12.11(c)], results in no growth except by coating, which is a slow process.

Comparing the three scenarios in Figure 12.11, (a) and (c) are both difficult conditions in which to operate because they result in uncontrolled growth and virtually no growth respectively. Scenario (b) is more promising, because a granulator operating in this way will be responsive to changes in operating conditions and there may be a natural limit to growth.

This simple analysis breaks down when granule deformation is substantial, as in some high-shear, high-intensity granulators. The two types of growth behaviour that occur depending on granule deformation are shown in Figure 12.12. *Steady growth* [Figure 12.12(a)] occurs when the granule size increase is roughly proportional to granulation time i.e. a plot of granule size versus time is linear. *Induction growth* [Figure 12.12(b)] occurs when there is an initial period during which no increase in size occurs, followed by very rapid growth. During the induction period, the granules form and consolidate, but do not grow further until the granule void fraction is reduced enough to squeeze liquid to the surface. This excess free

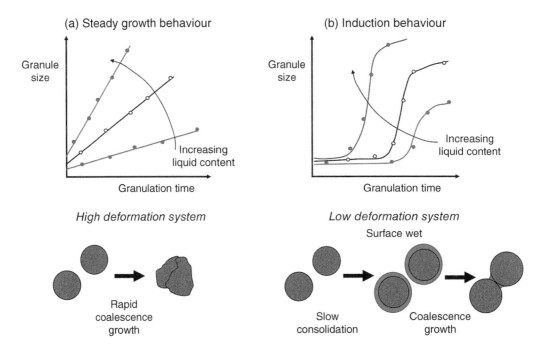

Figure 12.12 Schematic representation of the two main different types of granule growth and the way that they depend on the deformability of the granules: (a) steady growth and (b) induction growth. Hapgood et al. (2007)/With permission of Elsevier

liquid on the granules causes sudden coalescence of many granules and a rapid increase in granule size.

Deformation during granule collisions can be characterized by a Stokes deformation number Stk_{def}, relating the kinetic energy of collision to the energy dissipated during granule deformation (Tardos et al., 1997):

$$Stk_{\text{def}} = \frac{\rho_g U_c^2}{2Y_d} \qquad (12.11)$$

where U_c is the representative collision velocity in the granulator, ρ_g is the average granule density and Y_d is the dynamic yield stress (see Chapter 2) of the granule, respectively. Both Y_d and ρ_g are strong functions of granule void fraction and the granulation formulation properties, and are often evaluated at the maximum granule density, when the granules are strongest. This occurs when the granule void fraction reaches a minimum value ε_{min}, after which the granule density remains constant.

All types of granule growth can be described using the saturation and deformation Stk_{def} and the granule growth regime map (Hapgood et al., 2007) shown in Figure 12.13. At very low liquid contents, the product is similar to a dry powder. At slightly higher granule saturation, granule nuclei will form, but there is insufficient moisture for these nuclei to grow any further.

For systems with high liquid content, the behaviour depends on the granule strength and Stk_{def}. A weak system will form a slurry, an intermediate strength system will display steady growth, and a strong system (low Stk_{def}) will exhibit an induction time before rapid growth. At very high liquid saturations, the granules grow in size extremely quickly and any induction time is reduced to zero. The granule growth regime map has been successfully validated for several formulations in mixers, fluidized beds and tumbling granulators (Hapgood et al., 2007).

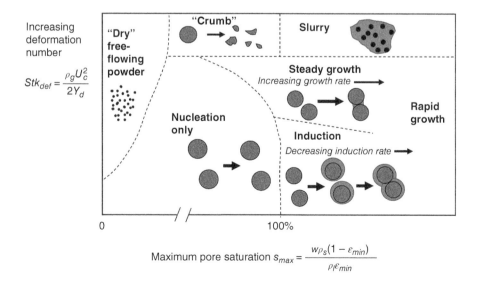

Figure 12.13 Granule growth regime map. Hapgood et al. (2007)/With permission of Elsevier

Granule breakage

Granule *breakage* also takes place during the granulation process. As introduced in Chapter 11 on size reduction, breakage (also called *fragmentation*) is the fracture of a particle – in this case, a granule – to form two or more pieces. *Abrasion* is the reduction in size of a granule by loss of primary particles from its surface. Empirical and theoretical approaches exist for modelling the different breakage mechanisms. In practice, breakage may be controlled by altering the granule properties (e.g. increase fracture toughness) and by making changes to the process (e.g. reduce agitation intensity).

Batch versus Continuous Granulation

Many granulation processes, particularly in the pharmaceutical industry, have traditionally been carried out on a batch basis. Such processes can be visualized in Figure 12.13 as moving from one part of the diagram to another – through nucleation to steady or rapid growth, for example. (Moving from nucleation to crumb or slurry formation would imply that the process has not been carried out correctly.) The alternative to carrying on all these sub-processes sequentially in time within a single batch vessel is to carry them out sequentially in space in a continuous process. This has many potential advantages, particularly in that it can enable better process control and reduced product variability. Typical equipment for carrying out both batch and continuous granulation is considered in Section 12.3.4.

12.3.3   Simulation of the Granulation Process

As in the processes of comminution and crystallization, the simulation of granulation hinges on the population balance. The population balance tracks the size distribution (by number, volume or mass) with time as the process progresses. It is a statement of the material balance for the process at a given instant. In the case of granulation, the instantaneous population balance equation is often written in terms of the number distribution of the volume of granules $n(v, t)$ (rather than granule diameter since volume is assumed to be conserved in any coalescence). $n(v, t)$ is the number frequency distribution of granule volume at time t. Its units are the number of granules per unit volume size increment v, per unit volume of granulator V. In words it is written as:

$$
\begin{pmatrix}
\text{Rate of increase} \\
\text{of number} \\
\text{of granules} \\
\text{in size interval} \\
v \text{ to } v + dv \\
\text{\small 1}
\end{pmatrix}
=
\begin{pmatrix}
\text{Net rate} \\
\text{of FLOW} \\
\text{of granules} \\
\text{into this} \\
\text{size interval} \\
\text{\small 2}
\end{pmatrix}
+
\begin{pmatrix}
\text{Rate at which} \\
\text{granules} \\
\text{ENTER this size} \\
\text{interval} \\
\text{by GROWTH} \\
\text{\small 3}
\end{pmatrix}
+
\begin{Bmatrix}
\text{Rate at which} & \text{Rate at which} \\
\text{granules ENTER} & \text{granules LEAVE} \\
\text{this size interval} - & \text{this size interval} \\
\text{by BREAKAGE} & \text{by BREAKAGE}
\end{Bmatrix}_{\small 4}
$$

$$(12.12)$$

Term (3) may be expanded to account for the different growth mechanisms and term (4) may be expanded to include the different mechanisms by which breakage occurs.

For a constant volume granulator the terms in the population balance equation become:

Term (1)

$$
\frac{\partial n(v, t)}{\partial t}
$$

$$(12.13)$$

Term (2)

$$\frac{Q_{in}}{V} n_{in}(v) - \frac{Q_{out}}{V} n_{out}(v) \tag{12.14}$$

Term (3)

$$\begin{matrix} \text{Net rate of growth} \\ \text{by COATING} \end{matrix} + \begin{matrix} \text{Net rate of growth} \\ \text{by NUCLEATION} \end{matrix} + \begin{matrix} \text{Net rate of growth by} \\ \text{COALESCENCE} \end{matrix}$$

Growth by coating causes granules to grow into and out of the size range v to $v + dv$.

$$\text{Hence, the net rate of growth by coating} = \frac{\partial G(v)n(v, t)}{\partial v} \tag{12.15}$$

$$\text{The rate of growth by nucleation} = B_{nuc}(v) \tag{12.16}$$

$G(v)$ is the volumetric growth rate constant for coating, $B_{nuc}(v)$ is the rate constant for nucleation. It is often acceptable to assume that $G(v)$ is proportional to the available granule surface area; this is equivalent to assuming a constant linear growth rate $G(x)$.

The rate of growth of granules by coalescence may be written as (Randolph and Larson, 1971):

$$\left[\underbrace{\frac{1}{2} \int_0^v \beta(u, v - u, t)n(u, t)n(v - u, t)du}_{(i)} \right] + \left[\underbrace{\int_0^\infty \beta(u, v, t)n(u, t)n(v, t)du}_{(ii)} \right] \tag{12.17a}$$

Term (i) is the rate of formation of granules of size v by coalescence of smaller granules. Term (ii) is the rate at which granules of size v are lost by coalescence to form larger granules. β is called the coalescence kernel. The rate of coalescence of two granules of volume u and $(v - u)$ to form a new granule of volume v is assumed to be directly proportional to the product of the number densities of the starting granules:

$$\begin{bmatrix} \text{collision rate of granules} \\ \text{of volume } u \text{ and } (v - u) \end{bmatrix} \propto \begin{bmatrix} \text{number density of} \\ \text{granules of volume } u \end{bmatrix} \times \begin{bmatrix} \text{number density of} \\ \text{granules of volume} \\ v - u \end{bmatrix}$$

The number densities are time dependent in general and so $n(u, t)$ is the number density of granules of volume u at time t and $n(v - u, t)$ is the number density of granules of volume $(v - u)$ at time t. The constant in this proportionality is β, the coalescence kernel or coalescence rate constant, which is in general assumed to be dependent on the volumes of the colliding granules. Hence, $\beta(u, v - u, t)$ is the coalescence rate constant for collision between granules of volume u and $(u - v)$ at time t.

The above assumes a pseudo-second-order process of coalescence in which all granules have an equal opportunity to collide with all other particles. In real granulation systems this assumption does not hold and collision opportunities are limited to local granules. Sastry and Fuerstenau (1970) suggested that for a batch granulation system, which is effectively restricted in space, the appropriate form for terms (i) and (ii) is:

$$\left[\underbrace{\frac{1}{2N(t)} \int_0^v \beta(u, v - u, t)n(u, t)n(v - u, t)du}_{(i)} \right] + \left[\underbrace{\frac{1}{N(t)} \int_0^\infty \beta(u, v, t)n(u, t)n(v, t)du}_{(ii)} \right] \tag{12.17b}$$

where $N(t)$ is the total number of granules in the system at time t.

The integrals in Equation (12.17b) account for all the possible collisions and the $\frac{1}{2}$ in term (i) of this equation ensures that collisions are only counted once.

In practice, many coalescence kernels are determined empirically and based on laboratory or plant data specific to the granulation process and the product.

According to Sastry (1975) the coalescence kernel is best expressed in two parts:

$$\beta(u, v) = \beta_0 \beta_1(u, v) \tag{12.18}$$

β_0 is the coalescence rate constant which determines the rate at which successful collisions occur and hence governs average granule size. It depends on solid and liquid properties and agitation intensity. $\beta_1(u, v)$ governs the functional dependency of the kernel on the sizes of the coalescing granules, u and v. $\beta_1(u, v)$ determines the shape of the size distribution of granules. Various forms of β have been published; Ennis and Litster (1997) suggest the form shown in Equation (12.19), which is consistent with the granulation regime analysis described above:

$$\beta(u, v) = \begin{cases} \beta_0, w < w^* \\ 0, w > w^* \end{cases} \quad \text{where } w = \frac{(uv)^a}{(u + v)^b} \tag{12.19}$$

w^* is the critical average granule volume in a collision and corresponds to the critical Stokes number value Stk^*. From the definition of the Stokes number given in Equation (12.9), the critical diameter x^* would be:

$$x^* = \frac{9\mu Stk^*}{4\rho_g V_{\text{app}}} \tag{12.20}$$

and so, assuming spherical granules,

$$w^* = \frac{\pi}{6} \left[\frac{9\mu Stk^*}{4\rho_g V_{\text{app}}} \right]^3 \tag{12.21}$$

The exponents a and b in Equation (12.19) are dependent on granule deformability and on the granule volumes u and v. In the case of small feed particles in the non-inertial regime, β reduces to the size-independent rate constant β_0 and the coalescence rate is independent of granule size. Under these conditions the mean granule size increases exponentially with time. Coalescence stops ($\beta = 0$) when the critical Stokes number is reached.

Using this approach, Adetayo and Ennis (1997) were able to demonstrate the three regimes of granulation (nucleation, transition and coating) traditionally observed in drum granulation and to model a variety of apparently contradictory observations.

Term (4)

The rate of breakage into and out of the size interval may be accounted for by the use of the specific rate of breakage and breakage distribution function as used in the simulation of population balances in comminution (see Chapter 11).

12.3.4 Granulation Equipment

Some of the features of the most commonly used granulators are summarized in Table 12.1.

Tumbling granulators

In tumbling granulators a tumbling motion is imparted to the particles in an inclined cylinder (drum granulator) or pan (disc granulator, Figure 12.14). Tumbling granulators operate in continuous mode and are able to deal with large throughputs (see Table 12.1). Solids and liquid feeds are delivered continuously to the granulator. In the case of the disc granulator the tumbling action gives rise to a natural classification of the contents according to size. Advantage is taken of this effect and the result is a product with a narrow size distribution.

Table 12.1 Types of granulator, their features and typical applications

Method	Product granule size (mm)	Granule density	Scale/ throughput	Comments	Typical applications
Tumbling (disc, drum)	0.5–20	Moderate	1–1000 t/h	Produces spherical granules	Fertilizers, iron ore, agricultural chemicals
Mixer-granulators (continuous and batch)	0.1–2	Low-moderate-high	<50 t/h or <500 kg/ batch	Handles a wide variety of materials, including cohesive ones	Chemicals, detergents, pharmaceuticals, ceramics
Fluidized (bubbling beds, spouted beds)	0.1–2	Low-moderate	<50 t/h or <500 kg/ batch	Easy to scale up. Less tolerant of cohesive materials. Good for coating. Drying is possible in the same vessel	Continuous (fertilizers, detergents); batch (pharmaceuticals, agricultural chemicals)
Spray drying	0.005–0.2	Low	1 kg/h to 30 t/h	Spray-drying can produce very porous 'instant' products. Drying in the same vessel	Milk, instant foods, detergents, pharmaceuticals
Pressure methods (extrusion, roll pressing, tableting)	0.5–10	High	1–10 t/h	Size is determined directly by the dimensions of the equipment	Pharmaceuticals, catalysts, chemicals, metals, ceramics

Adapted from Ennis and Litster (2018).

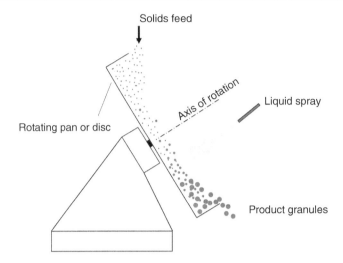

Figure 12.14   Schematic representation of disc granulator

Batch mixer granulators

In mixer granulators the motion of the particles is brought about by some form of agitator rotating at low or high speed on a vertical or horizontal axis. Rotation speeds vary from 50 revolutions per minute (rpm) in the case of the horizontal *pug mixers* used for fertilizer granulation to over 3000 rpm in the case of the vertical high shear continuous granulator used for detergents and agricultural chemicals. For vertical axis mixers used by the pharmaceutical industry (Figure 12.15), impeller speeds range from 500 to 1500 rpm for mixers less than 30 cm in diameter, and decreasing to 50–200 rpm for mixers larger than 1 m in diameter. In general, the agitator speed decreases as mixer scale increases, in order to

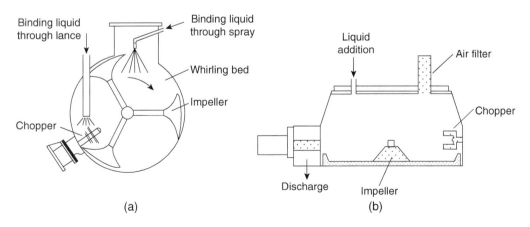

Figure 12.15   Schematic representation of a horizontal axis (a) and vertical axis (b) mixer granulator. Reprinted from *Perry Chemical Engineers' Handbook*, 7th Edition, 'Section 20: size enlargement', Ennis and Litster, pp. 20–77. Copyright (1977) with permission of The McGraw-Hill Companies

maintain either (a) constant maximum velocity at the blade tip or (b) constant mixing patterns and Froude number $\omega^2 D/2g$, where ω is the angular velocity of the blade, D is the diameter and g is acceleration due to gravity. This is equivalent to $u^2/(D/2)g$, where u is the blade tip speed.

Continuous mixer granulators

As discussed earlier, it is sometimes advantageous to operate granulation processes continuously. Particularly when combined with automated process control, this can result in reduced product variability, reduced equipment size and product hold-up or inventory, reduced need for manual intervention and therefore better overall economic performance. In the manufacture of detergent powders, for example, it is common practice to couple two horizontal axis mixers together, as shown in Figure 12.16. The granulation binder is added to the first machine, which operates at high impeller speed with a residence time of seconds, and additional densification and addition of secondary ingredients takes place in a second, low-speed machine with a residence time of minutes, before the product drops into a fluidized bed drier/cooler.

Extrusion granulators

Extrusion granulation is a multi-step process used in a number of industries, which starts by mixing a dense wet mass or paste of the material to be granulated. This is then extruded to produce a uniform product which can be cut into short rods or blocks for further processing (Figure 12.17). Some ceramics and building materials are formed in this way. In the pharmaceutical industry, extrusion is used to produce rod-shaped particles of uniform diameter, which are then formed into spheres on a rotating disc, termed a *spheronizer*. This method has the advantage that it is able to produce uniform spherical particles with high levels of active ingredients, i.e. without the addition of large amounts of other non-active material.

Twin-screw granulators

A major advance in pharmaceutical granulation technology has been the development and widespread adoption of continuous *twin-screw granulation*. Although this has some

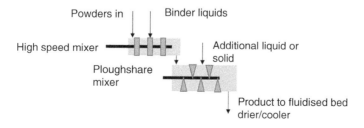

Figure 12.16 Continuous granulation using mixers in series. Adapted from patents from Bayer, Unilever, P&G, 1970s onwards

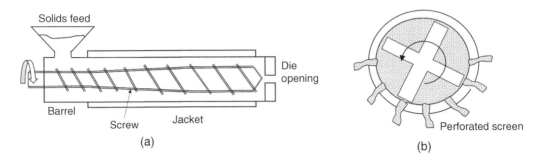

Figure 12.17   Types of extruder (a) single screw and (b) radial perforated screen

superficial similarities with extrusion granulation, it is in fact quite different because the machine does not run full of material, as an extruder does, and there is no die at the end through which the material is forced to flow. In this type of granulator, the solids to be granulated are first dry-mixed, the binder and other constituents are added and work is done on the resulting mixture at various points along the two co-rotating screws. Semi-wet granules are discharged from the end into some form of drier, typically a fluidized bed. Figure 12.18 shows an example of the configuration of the twin-screws. Elements of various kinds and with particular functions are slotted onto the two shafts so as to produce zones for conveying, mixing and other purposes. Liquids and other solids can be added at desired points along the length of the machine.

Experiments have shown that, as for all granulators, there is an operating window on axes of the degree of agitation (in this case, the screw speed) versus the liquid–solid ratio

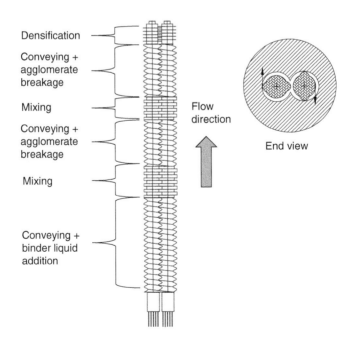

Figure 12.18   Example of a twin-screw granulator configuration. Adapted from Keleb et al. (2004)

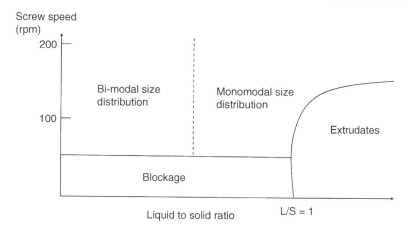

Figure 12.19 Window of operation for a laboratory twin-screw granulator. Adapted from Tu et al. (2009)

in which a good unimodal granule size distribution results, as shown schematically in Figure 12.19. Liquid–solid ratios which are too high result in wet extruded material, while speeds which are too low lead to blockage.

Figure 12.20 shows a photograph of the twin-screw granulator in place above a fluidized bed drier; note its very small size, which is a consequence of its high throughput. This type of granulator is frequently used as part of a tablet manufacturing line, as shown schematically in Figure 12.21, which also indicates the options for online process measurement.

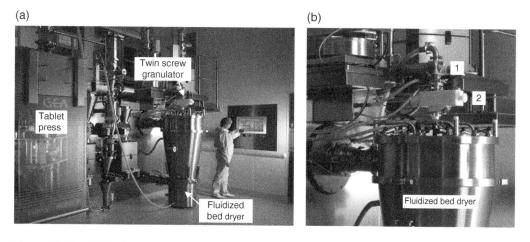

Figure 12.20 (a) Twin-screw granulator as part of a tablet manufacturing line. (b) Close-up of the twin-screw granulator mounted above fluidized bed dryer. [1] is the twin-screw granulator and [2] is a granule sampling device. Courtesy of GEA Group

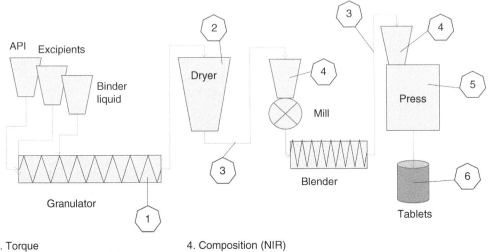

1. Torque
2. Temperature and humidity
3. Particle size distribution
4. Composition (NIR)
5. Press characteristics
6. Final tablet weight and composition

Figure 12.21 Schematic representation of a complete continuous tableting line, indicating process monitoring options (API = active pharmaceutical ingredient; NIR = near infra-red)

Fluidized bed granulators

In fluidized bed granulators the particles are set in motion by fluidizing air. The fluidized bed may be either a bubbling or a spouted bed (see Chapter 6) and may operate in batch or continuous mode. Liquid binder and wetting agents are sprayed in fine droplet form above or within the bed. The advantages which this granulator has over others include good heat and mass transfer, mechanical simplicity, ability to combine the drying stage with the granulation stage and ability to produce small granules from powder feeds. However, running costs and breakage rates can be high compared with other devices.

Two schematic arrangements of fluidized bed granulators are shown in Figure 12.22. Both types have a conical section at the base to promote circulation of particles, which is essential in an application where the particles will be cohesive. The bottom spray arrangement is here shown with a draft tube, again to promote particle circulation and reduce the distribution of particle cycle times. This is sometimes known as a *Wurster coater*, after one of its originators

For further details of industrial equipment for granulation and other means of size enlargement the reader is referred to Salman et al. (2007).

Roll pressing

All of the granulation methods listed above involve the addition of some form of wet binder to the powder to be granulated. Roll pressing is a dry technique which therefore presents advantages in that there is no need for a subsequent drying step.

In roll pressing, powder is fed between two counter-rotating rolls, as shown in Figure 12.23, where the friction against the roll surface, which may be suitably machined,

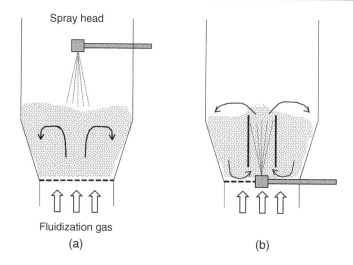

Figure 12.22 Fluidized bed granulators (a) top spray and (b) bottom spray, with draft tube. Particle circulation indicated

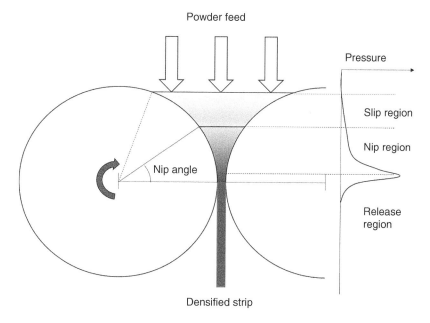

Figure 12.23 Roll press (schematic)

drags the powder into the narrow gap between them. The pressure which is generated in this way is typically hundreds of bars and acts to densify and compact the powder into a continuous *strip* or discrete *briquettes*, if the rolls are shaped to enable this. In the pharmaceutical industry, continuous strip is usually produced, which is then reduced in size in a mill and screened to produce a free-flowing intermediate product suitable for feeding to a tableting press.

Figure 12.23 shows a vertical feed, although other angles of feed are possible. The feed may be under gravity or via a screw. Powder first enters a *slip region*, so called because there is relative movement between the roll surface and the particles in this region. At a certain angle – the *nip angle* – the powder velocity at the wall equals the roll surface velocity and the powder is 'nipped'. As it continues into the gap it is densified under rapidly increasing pressure up to the point of closest approach. The maximum pressure does not necessarily coincide exactly with this point because of wall slip and other factors. The nip angle, as shown in Figure 12.23, is typically less than 10° and depends on the material characteristics and operating parameters. Clearly, a larger nip angle will result in a greater maximum pressure. Guigon et al. (2007) discuss how it may be calculated, as well as other aspects of roll press design and operation.

The pressure is rapidly reduced after the maximum, the strain energy in the compacted strip is released and the strip is ejected.

12.4   TABLETING

The tablet is the most common pharmaceutical dosage form. They are made in high-speed rotary tableting machines which can produce more than 10 000 tablets per minute. This explains the importance of pre-granulation into a free-flowing powder feed material, as described in the previous sections.

Figure 12.24 shows schematically the sequence of operations in a tablet press. Between (a) and (b), the bottom punch moves down and powder flows into the tablet die. Between (b) and (c), the top punch moves down at the same time as the bottom punch moves up, the powder is compressed and the tablet is formed. Between (c) and (d) the top punch returns to its original position and the bottom punch moves up to eject the finished tablet. A high-speed press contains multiple punches of this kind, arranged in a circle.

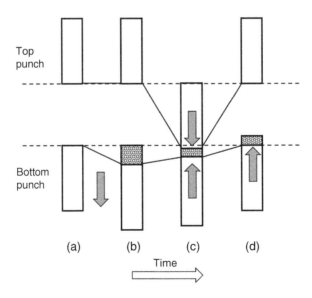

Figure 12.24   Tablet pressing sequence (simplified): (a) → (b): Bottom punch moves down to accept powder fill. (b) → (c): The top punch moves down while bottom punch moves up, to compress tablet. (c) → (d): The top punch returns to position while bottom punch moves up to eject the finished tablet

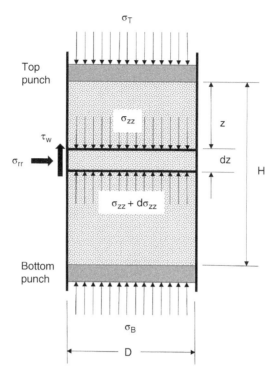

Figure 12.25 Force balance on an infinitesimal slice for a powder under compression in a die. Seville and Wu (2016)/With permission of Elsevier

Compression of a tablet can be analysed in a similar way to the analysis of stresses in hoppers described in Section 9.7. Consider the compaction of a powder in a cylindrical die of diameter D. Figure 12.25 shows a force balance on an infinitesimal slice of powder a distance z below the surface of the top punch. In this case, the weight of the powder is small compared with the forces of compression. Neglecting this, a force balance on the element gives

$$\frac{\pi D^2}{4} \sigma_{zz} = \frac{\pi D^2}{4}(\sigma_{zz} + d\sigma_{zz}) + \tau_W \pi D dz \qquad (12.22)$$

which can be simplified to

$$\frac{d\sigma_{zz}}{dz} + \frac{4\tau_W}{D} = 0 \qquad (12.23)$$

From the earlier discussion about stresses in hoppers, we know that a simple approximation in the case of cohesionless powders is Janssen's assumption that the radial stress acting normal to the wall σ_{rr}, is proportional to the vertical stress σ_{zz} as in Equation (9.13):

$$\frac{\sigma_{rr}}{\sigma_{zz}} = k \qquad (12.24)$$

and the wall shear stress τ_W, is related to the normal stress on the wall σ_{rr} by

$$\tau_W = \mu_W \sigma_{rr} \tag{12.25}$$

where μ_W is the coefficient of friction at the die wall.

Incorporating these into Equation (12.23) gives:

$$\frac{d\sigma_{zz}}{dz} + \frac{4\mu_W k}{D} = 0 \tag{12.26}$$

At $z = 0$, the applied stress at the top punch is σ_T. Solving Equation (12.26) with this boundary condition gives

$$\sigma_{zz} = \sigma_T \left[\exp\left(-\frac{4\mu_W kz}{D} \right) \right] \tag{12.27}$$

Figure 12.26 shows how the stress transmission ratio σ_{zz}/σ_T, varies with distance z below the top punch, for a given set of dimensions which are typical of pharmaceutical tablets and an example set of material properties.

It is clear from Figure 12.26 that the local stress (in this case the vertical stress is shown) falls by a substantial fraction over a few mm of tablet depth. This would have consequences for the local density and therefore the mechanical properties of the tablet and the rate at which it might be expected to disintegrate and dissolve in use (see Chapter 13). The reason why the local stress falls in this way is, of course, that part of the stress applied by the punch is supported by friction at the walls, which suggests a way of countering the effect by reducing the wall friction coefficient. This is done in practice by addition of a fine lubricant powder such as magnesium stearate. This analysis also explains why double-ended compaction is usually used, with top and bottom punches moving together simultaneously; this minimizes the reduction in local stress. Finally, of course, tablets are usually not perfect cylinders but more rounded shapes, partly because rounding of the edges removes the parts of the tablet which are difficult to compress, and therefore reduces the tendency towards attrition.

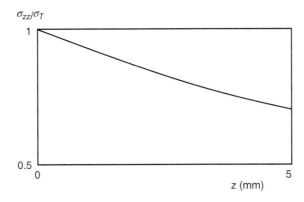

Figure 12.26 Stress transmission ratio as a function of distance below top punch, z ($D = 10$ mm; material properties: $\mu_W = 0.4$; $k = 0.33$). Seville and Wu (2016)/With permission of Elsevier

After formation by compression, the tablet is then ejected by being pushed upwards by the bottom punch, as in Figure 12.23. In both machine design and product formulation it is important to consider the stress relaxation which then occurs, so as to avoid delamination of the tablet caused by release of its stored elastic energy.

FURTHER READING

For general reading on the topic of size enlargement in particle technology, readers are referred to: Salman et al. (2007); Davey and Garside (2000).

12.5   WORKED EXAMPLES

WORKED EXAMPLE 12.1

A pharmaceutical product is being scaled up from a pilot scale mixer granulator with a batch size of 15 kg to a full-scale mixer granulator with a 75 kg batch size. In the pilot scale mixer, 3 kg of water is added to the mixer over six minutes through a nozzle producing 200 µm diameter spray drops across a 0.2 m wide spray. During scale-up, the ratio of liquid to dry powder is kept constant but the solution flowrate can be scaled to maintain constant spray time or constant spray rate through the nozzle. If the flow rate is increased to maintain constant spray time, the new nozzle produces 400 µm drops over a 0.3 m wide spray zone. Powder velocity in the spray zone is currently 0.7 m/s at pilot scale. At full scale, the powder velocities are 0.55 m/s and 1 m/s at the 'low' and 'high' impeller speeds, respectively. Calculate the change in dimensionless spray flux for the following cases:
(a) base case at pilot scale;

(b) scale-up to full scale using spray time of six minutes and low impeller speed;

(c) full scale using constant spray rate and low impeller speed;

(d) full scale using constant spray rate and high impeller speed.

Solution

Using Equation (12.5) and ensuring consistent units, the calculations are summarized in Table 12.W1.1.

Table 12.W1.1

Scale-up approach	(a) Pilot scale base case	(b) Constants spray time, low impeller	(c) Constant spray rate, low impeller	(d) Constant spray rate, high impeller
Batch size (kg)	15	75	75	75
Spray amount (kg)	3	15	15	15
Spray time (min)	6	6	30	30
Flow rate (kg/min)	0.5	2.5	0.5	0.5

(continued overleaf)

Table 12.W1.1   *(continued)*

Scale-up approach	(a) Pilot scale base case	(b) Constants spray time, low impeller	(c) Constant spray rate, low impeller	(d) Constant spray rate, high impeller
Drop size (μm)	200	400	200	200
Spray width (m)	0.2	0.3	0.2	0.2
Impeller speed (rpm)	216	108	108	220
Powder velocity (m/s)	0.7	0.55	0.55	1
Spray flux Ψ_a	0.45	0.95	0.57	0.31

WORKED EXAMPLE 12.2

When the spray flux is 0.1, 0.2, 0.5 and 1.0:

(a) Calculate the fraction of the spray zone wetted.

(b) Calculate the number of nucleus formed from only one drop (f_1).

(c) If the drop-controlled regime was defined as the spray flux at which 50% or more of the nuclei are formed from a single drop, what would be the critical value of spray flux and what fraction of the spray zone would be wetted?

Solution

Using Equation (12.6) for (a) and (c), solutions are shown in Table 12W2.1. To calculate the fraction of nuclei formed from a single drop for (b), use Equation (12.7) with $n = 1$, which simplifies to $f_1 = \exp(-4\Psi_a)$.

Table 12.W2.1

Ψ_a	f_{wet}(%)	f_1(%)
0.1	10	67
0.17	16	51
0.2	18	45
0.5	39	14
1	63	2

TEST YOURSELF

12.1 When and why is it necessary to increase particle size?

12.2 Sketch the solubility diagram (concentration versus temperature). Starting in the undersaturated region, what are the two approaches to precipitation of crystals?

12.3 How do nucleation and growth depend upon supersaturation? What is then your strategy for producing (a) large crystals, (b) small crystals?

12.4 What is Ostwald ripening and why does it occur?

12.5 Redraw Figure 12.7 showing how saturation increases as the void fraction of the granule decreases.

12.6 How do the rate of wetting and drop penetration time change with (a) increase in viscosity, (b) decrease in surface tension, (c) increase in contact angle?

12.7 If the flow rate of solution to be added to a granulator doubles, how much wider does the spray zone need to be to maintain constant spray flux? If the spray cannot be adjusted, what does the bed velocity need to be to maintain constant spray flux?

12.8 At a spray flux of 0.1, calculate the number of nuclei formed from one drop, two drops and three drops, etc. What fraction of the powder surface will be wetted by the spray?

12.9 Explain the non-inertial, inertial, and coating regimes of granule growth. What happens to the maximum granule size as (a) the collision velocity between particles increases, (b) the viscosity increases?

12.10 Explain the difference between the steady growth and the induction growth regimes on the granule growth regime map.

12.11 Explain how induction growth is linked to granule void fraction and saturation.

12.12 Explain the five terms in the granulation population balance.

12.13 What are the possible advantages of continuous granulation?

12.14 Explain the sequence of operations in a tableting press.

12.15 Why do tablets show an internal density variation and what can be done to reduce it?

EXERCISES

12.1 In the pharmaceutical industry, any batch that deviates from the set parameters is designated as an 'atypical' batch and must be investigated before the product can be released. You are a pharmaceutical process engineer, responsible for granulating a pharmaceutical product in a 600 l mixer containing 150 kg of dry powders. It was noticed while manufacturing a new batch that the liquid delivery stage ended earlier than usual and the batch contained larger granules than normal. During normal production, the impeller speed is set to 90 rpm and water is added at a flowrate of 2 l/min through a nozzle producing an average drop size of 400 μm. Due to an incorrect setting, the actual flow rate used in the atypical batch was 3.5 l/min and the actual drop size was estimated at 250 μm. The spray width and powder surface velocity were unaffected and remained constant at 40 and 60 cm/s, respectively. Calculate the dimensionless spray flux for the normal case and the atypical batch, and explain why this would have created larger granules.

[Answer: normal 0.52; atypical, 1.46]

12.2 Calculate the fraction or percentage of nuclei formed from one drop, two drops and three drops, etc., at spray flux values of (a) 0.05, (b) 0.3, (c) 0.8. What fraction of the powder surface will be wetted by the spray?

[Answer: (a) 0.82, 0.16, 0.02, 0.0, 0.0; (b) 0.30, 0.36, 0.22, 0.09, 0.03; (c) 0.04, 0.13, 0.21, 0.22, 0.18.]

12.3 A tableting press is to be selected for production of a tablet of diameter 12 mm and thickness 4 mm. The material properties are: $\mu_W = 0.4$; Janssen constant $k = 0.33$. Find the maximum variation in local stress within the tablet in the cases (a) of single punch compression, (b) double punch compression. (Hint: treat the tablet in two equal halves, top and bottom.) If a lubricant is available so that the wall friction coefficient is reduced to 0.35, what difference does this make?

12.4 Explain how you would monitor and control a continuous twin-screw granulation process feeding a tableting press.

13

Particulate Products

The range of commercial products which incorporate particles in some form is vast and new examples appear every day. This chapter introduces some of the principles of design of such products. In all cases, the product is designed to perform some specified function or set of functions. Many particulate products must remain stable until they are needed and then undergo a transition into a final form, such as a pharmaceutical tablet which must dissolve in order to make a drug available to the body. We introduce here the concepts behind dispersion and dissolution of such products. Many particulate products exist as suspensions in a liquid; we discuss the stability of such suspensions using the theory behind interparticle forces introduced in Chapter 5. Examples of product development are given from a number of industrial sectors.

13.1 INTRODUCTION

This chapter concerns particles as *products*, and it is important to appreciate that products are sold on the basis of their *performance* – how well they work; how closely they meet the customer's needs – rather than on their scientific description. The performance of particulate products depends on their *chemical* form (the chemical ingredients that are used to make them, sometimes called their *formulation*) but also on their *physical* form (sometimes referred to as *structure*). Important contributors to their physical form include particle size, shape and density and the *distributions* of each of these. Some examples of particulate products and their structural forms are given in Table 13.1.

One of the most important classes of particulate products is the category of *fast-moving consumer goods* (FMCG), which include cosmetics and toiletries, paints, household surface cleaners, fabric washing and dish washing products, as well as a wide range of foods such as margarines and spreads, ice cream, chocolate, instant drinks and soups. These are all

Introduction to Particle Technology, Third Edition. Martin Rhodes and Jonathan Seville.
© 2024 John Wiley & Sons Ltd. Published 2024 by John Wiley & Sons Ltd.
Website: www.wiley.com/go/rhodes/particle3e

Table 13.1 Examples of particulate products and their structures

Sector	Product examples	Structures
Food and drink	Ice cream, chocolate, margarine, spreads, tea and coffee, instant drinks and soups	Dispersions, soft solids, agglomerates
Home care products	Fabric washing and conditioning products, dishwashing products, surface cleaners	Dispersions, creams, agglomerates, tablets
Personal care products	Soaps, shampoos, skin creams, toothpaste	Soft solids, creams, pastes
Pharmaceuticals	Drug delivery products	Dispersions, tablets, creams
Coatings and fibres	Paints, coatings, synthetic fibres	Dispersions, fibres
Chemicals	Polymers, pigments, catalysts, sorbents	Agglomerates, tablets, extrudates
Agrochemicals	Animal feed, fertilizers, nutrients, weedkillers, insecticides	Dispersions, agglomerates, extrudates, pellets
Energy	Solid biofuels, heat-carriers and phase-change materials, fuel cell components	Agglomerates, extrudates, pellets

high-added-value products which can command a substantial market price as a result of 'smart' formulation. Such formulations frequently change in order to meet new needs and so obtain market advantage.

FMCG products have some common features (Edwards and Instone, 2001):

(a) They are made up from a combination of raw materials which are used to form a complex multiphase structured mixture (e.g. an *agglomerate*, *suspension*, *gel*); the mixture consists of an *active ingredient* (*AI*), which gives the product its primary function, and a number of *structurants*, which help to build and maintain the structure.

(b) They are formed in such a way that they are structured on a scale of 1–100 μm, meaning that sub-units of the structure such as suspended particles or droplets are of this scale.

(c) It is this *microstructure* which largely determines the product properties in use, including their appearance.

(d) The microstructure must remain stable as they pass along the supply chain, until the consumer uses the product.

(e) The microstructure is designed to be destroyed during product usage, as it is sheared, melted, dissolved or otherwise transformed. The way in which this happens – for example, the sequence of dissolution of components in a laundry *tablet* – is another important way of controlling the performance in use.

It is not only the ingredients which create the product structure, but also the manufacturing process. The *processing conditions* (order of addition of the ingredients, temperature, shear and flow environment, etc.) are very important in determining the product microstructure, so that *product quality* is determined by a combination of formulation and processing conditions.

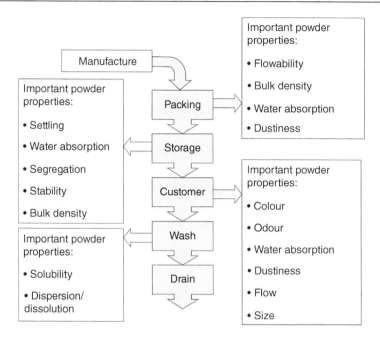

Figure 13.1 Powder properties at various stages in the packing, distribution and usage of a detergent powder product (University of Birmingham student project)

The list of customer requirements for a particular product can be extensive. Figure 13.1, which is taken from a student project, shows the powder properties which are required at each stage of delivery of a detergent powder for fabric cleaning, from manufacture to its eventual disposal. These stages include packing into containers, storage before and after it reaches the point of sale, customer usage and the disposal of the wash water into the domestic drainage system. Most of these powder properties will be considered separately in this chapter. The most important point to note here is that the manufacture and usage of all products carries with it an associated environmental burden. Calculation of that burden, its usage in *Life Cycle Analysis* and the resulting decision-making about product development strategies is beyond the scope of this book; interested readers are referred to Baumann and Tillman (2015).

13.2   STRUCTURE AND PROPERTIES

Structure of a product can have many purposes: structural changes can make products more (or less) easily dispersed and ingested (in a food or a pharmaceutical tablet), more visually attractive (in a paint or coating), more absorbent (in a fabric, for example), more reactive (in a catalyst), stiffer and/or stronger (in a building material), more thermally, electrically or acoustically insulating (in multiple applications), and there are many other examples. Figure 13.2 shows some of them.

The following example (taken from Knight, 2001) shows how the structure of a common particulate product can affect its performance.

(a)   (b)

(c)   (d)

Figure 13.2 Examples of particulate structures (a) detergent powder; (b) broken surface of a detergent tablet, showing particles embedded in a continuous matrix; (c) single sugar crystals and (d) microcapsules of melamine-formaldehyde designed to deliver an active ingredient to its destination. Professor Zhibing Zhang / with permission of University of Birmingham

WORKED EXAMPLE 13.1 Agglomerates – Illustrating a hierarchy of structures

What principles can we use to design an agglomerated product for dissolution in a liquid?

The agglomerate (otherwise known as a granule) is an important and widespread type of particulate product, in which fine ingredient particles are bound together either by compaction alone, using a tableting press, for example, or by adding a liquid *binder* to stick them together. (Examples of such equipment are shown in Chapter 12.) Here we consider the design of an agglomerated product such as that shown in Figure 13.2a.

Following the earlier principle of identifying the important product attributes, we might identify the reasons why agglomerated products are preferred to other options as:

- To improve the product's dispersion and dissolution characteristics.

- To reduce dustiness, because agglomerated particles are less likely to become airborne[1].

- To enhance flow properties, because agglomerates are less likely to be impeded by cohesive forces – think about an agglomerated powder flowing into a container.

- To make the product more attractive, e.g. by improving appearance or feel.

- To reduce or to increase the bulk density.

- To reduce the tendency for the particles to cohere into a *cake* during storage[2].

An agglomerated product must be strong enough to resist breaking down in transport to its point of use (no one wants to buy a box of very fine washing powder, for example) but not so strong that it then resists dispersion and dissolution in a washing machine. Design of an agglomerated product is therefore a compromise which best meets the range of desirable features.

Bulk density is an important product property. As the bulk density decreases, the size of the package for a given weight goes up, and larger packages are more expensive to transport and store, more costly to make and less convenient for the customer. The bulk density ρ_B is simply given by:

$$\rho_B = \rho_{abs}(1 - \varepsilon_B)(1 - \varepsilon_a) \tag{13.1}$$

where ρ_{abs} is the density of the material making up the particles (absolute or skeletal density), ε_B is the void fraction of the volume *between* the agglomerates and ε_a is the void fraction of volume *within* the agglomerates (Figure 13.3). In effect, the agglomerated product shows a hierarchy of structure, from the individual particle of size (in this example) of order 100 μm to the single agglomerate of order 1 mm to the multi-agglomerate dosage scale of order 1 cm (one standard teaspoon[3] = 5 ml = 5 cm^3). Typically, for a narrow size distribution of agglomerates which are not too extreme in shape, ε_B is in the range 0.4–0.5, while ε_a may be much less, depending on how the agglomerates are made, such as 0.05–0.2. For a detergent powder, a typical constituent material density might be 1650 kg/m^3. Taking a value for ε_B as 0.45 and a range for ε_a of 0.05–0.2 gives a range for ρ_B of about 725–860 kg/m^3. The differences in bulk density here may seem small, but they can result in different packaging possibilities and potentially large changes in the resulting market share.

Consider now the rate at which liquid can penetrate into a mass of particles; the practical application might be in dispersing a washing powder or a pharmaceutical powder or an 'instant' drink or food product. Figure 13.4 shows an idealized situation in which liquid is free to rise into an assembly of agglomerates. The liquid is pulled up into the assembly by surface tension forces (see Chapter 2) – it is said to be 'capillary-driven' – and in this example air is completely displaced

---

[1] This also enhances product safety, because airborne fine particles can be an inhalation hazard. See Chapter 14.
[2] *Caking* occurs when bridges grow between particles, sometimes due to some of the product dissolving in moisture which is present in the product, and resulting in a solid mass of material which is then difficult to handle.
[3] Real teaspoons often hold rather less than 5 ml – try it!

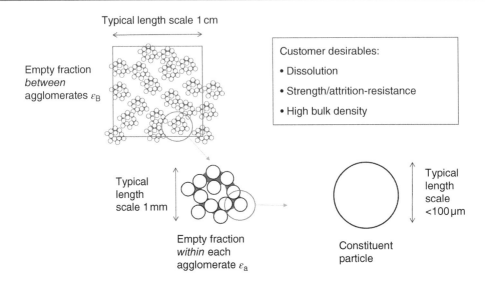

Typical length scale 1 cm

Empty fraction *between* agglomerates ε_B

Customer desirables:
• Dissolution
• Strength/attrition-resistance
• High bulk density

Typical length scale 1 mm

Empty fraction *within* each agglomerate ε_a

Constituent particle

Typical length scale <100 µm

Figure 13.3   Structured agglomerated products

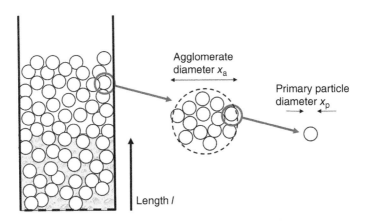

Agglomerate diameter x_a

Primary particle diameter x_p

Length l

Figure 13.4   Liquid penetration into an assembly of agglomerates

as the liquid rises. Liquid is also being pulled into each agglomerate, which is a slower process because the void fraction in each agglomerate is much less.

The rate at which the liquid rises through the column of agglomerates is given by the equation:

$$l^2 = \frac{1}{6}\left[\frac{\varepsilon_B}{1-\varepsilon_B}\right]\frac{x_a\gamma}{\mu}t \qquad (13.2)$$

where l is the height to which the liquid has risen over time t, ε_B is the void fraction of the assembly of agglomerates, (i.e. *between* the agglomerates), as defined earlier, x_a is the agglomerate diameter and γ and μ are the liquid surface tension[4] and viscosity respectively. This equation is derived from a combination of the Poiseuille and Kozeny equation introduced in Chapter 6, and is one form of the *Washburn equation* (see Knight 2001 for the derivation). With the variables arranged as shown here it is evident that the penetration length for a given time depends on both the void fraction between agglomerates and a dimensionless group containing the particle size, surface tension and viscosity. It may be helpful to think of the liquid penetration being *driven* by the surface tension and *opposed* by its viscosity.

Table 13.2 shows how far liquids of different viscosities can penetrate in 1 s in an example where the agglomerates are of diameter 1 mm.

Whether this rate of penetration of liquid into the bed is acceptable depends on the application. If it is not acceptable then one answer might be to agitate the particles by stirring, as in some forms of granulation (see Chapter 12).

Note that this simple analysis assumes that the viscosity and surface tension remain constant; if the solid partially dissolves into the liquid then the liquid properties will change. Most probably, the effective liquid viscosity will increase, slowing down the rate of penetration into the bed.

Now consider the liquid uptake into each agglomerate, as shown in Figure 13.5.

Table 13.2 Predicted effect of liquid viscosity on penetration distance in 1 s ($x_a = 1$ mm; $\gamma = 50$ mJ/m^2; void fraction of assembly of agglomerates, $\varepsilon_B = 0.5$)

Viscosity (Pa s)	Penetration distance (mm)
0.001 (e.g. water)	100
1 (e.g. glycerol, paint)	3
1000 (e.g. polymer melt)	0.1

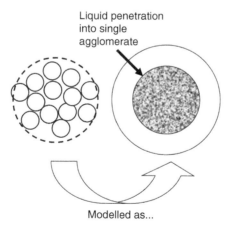

Liquid penetration into single agglomerate

Modelled as...

Figure 13.5 Liquid penetration into a single agglomerate

---

[4] Strictly, the surface energy of the liquid-solid interface.

Table 13.3 Predicted time for complete penetration of a liquid into an agglomerate (spherical agglomerate $x_a = 1$ mm, made from constituent particles of diameter $x_p = 1$ μm; liquid $\gamma = 50$ mJ/m^2; agglomerate void fraction, $\varepsilon_a = 0.5$)

Viscosity (Pa s)	Penetration time (s)
0.001 (e.g. water)	0.4
1 (e.g. glycerol, paint)	40
1000 (e.g. polymer melt)	4×10^4 (>11 h)

Assuming that the agglomerate is spherical, of diameter x_a and of uniform void fraction ε_a, and that the air within the pores can escape, the time t_C taken for the liquid to penetrate completely is given by:

$$t_C = \frac{25}{24}\left[\frac{1-\varepsilon_a}{\varepsilon_a^2}\right]\frac{\mu}{x_p\gamma}x_a^2 \tag{13.3}$$

where x_p is here the diameter of the particles making up the agglomerates. Again, the dimensionless group containing surface tension and viscosity appears, with the liquid penetration into the agglomerate being driven by surface tension and opposed by viscosity.

Table 13.3 shows the predicted time for complete penetration, for some example values of the important variables.

As before, whether these timescales are appropriate depends upon the particular application. Even from this idealized example it is possible to see how the choice of size of particle and size of agglomerate can determine whether its performance as a product is going to be acceptable in meeting consumer needs and desirables.

13.3   DISSOLUTION AND DISPERSION

As mentioned earlier, many particulate products are designed to dissolve in a liquid. The example above concerned the rate at which a liquid could penetrate into a mass of particles or a single porous particle. The next stage in dissolution after the two phases – the solid particle and the surrounding liquid – are brought into contact, is that the *solute* substance is liberated from the solid phase and moves into the liquid phase, termed the *solvent*. The solute, which is usually in molecular form, is at a high concentration close to the solid surface and the concentration decreases with distance away from it, as shown in Figure 13.6.

As dissolution proceeds, the solute molecules must *diffuse* through the boundary layer and into the bulk liquid. It is usually this diffusion process which is the rate-limiting step in dissolution, in which case dissolution is said to be *diffusion-controlled*. The movement of solute will then commonly obey a simplified version of Fick's law of diffusion:

$$\frac{dC}{dt} = k\Delta C \tag{13.4}$$

where C is the solute concentration at any position at time t, $\Delta C = C - C_2$ where C_2 is the concentration in the bulk, and k is the diffusional rate constant (s^{-1}).

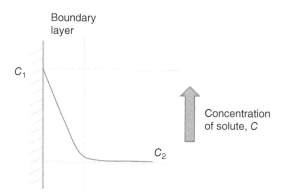

Figure 13.6 Solute concentration close to the solid surface

A development of this approach is the *Noyes-Whitney equation*, which describes the rate of dissolution from a single particle (see Taylor and Aulton, 2022). The rate of diffusional mass transfer dm/dt, through the boundary layer shown in Figure 13.6 is given by:

$$\frac{dm}{dt} = k_1 \frac{A\Delta C}{h} \tag{13.5}$$

where A is the area available for diffusion and h is its thickness, ΔC is the concentration difference across the boundary layer (= $C_1 - C_2$) and k_1 is the diffusion coefficient, usually given the symbol D (m^2 s^{-1}). This equation can often be simplified since

(1) At equilibrium the region immediately next to the solid will be saturated with solute, so $C_1 = C_S$, the saturation solubility.

(2) If the bulk of liquid is large and/or the solute is removed at a faster rate than it enters solution, C_2 becomes small by comparison with C_S (this is known as the 'sink' condition).

Thus Equation (13.5) becomes:

$$\frac{dm}{dt} = \frac{DAC_S}{h} \tag{13.6}$$

WORKED EXAMPLE 13.2: Improving tablet dissolution rate

A particular drug substance is available in tablet form but shows poor *bioavailability* because its rate of dissolution is too low. What can be done to improve it? (Bioavailability is a measure of the extent to which an active substance reaches its target in the human body intact.) Equation (13.6) indicates some of the factors which determine the dissolution rate. This can be enhanced by increasing the area available for dissolution, either by making the solid available in smaller particles or by introducing porosity into the structure. However, in order to be effective, the solid needs to be wetted, and it becomes progressively more difficult for liquids to infiltrate fine particles, as shown in Example 13.1, so there is a limit to how much A in Equation (13.6) can be increased.

An option that might be considered here is to include in the structure a *disintegrant*: a component which acts to break the structure apart by swelling on contact with the liquid (starch is an example) or by some other mechanism, such as generation of a gas (e.g. sodium bicarbonate, which produces carbon dioxide).

A larger value of the saturation solubility C_S will increase the dissolution rate. C_S depends on the nature of the solid, or solids, since there will usually be several components present. Is it possible to modify the crystalline form of the active substance, for example, to increase C_S, or to reformulate it with different excipients so as to enhance C_S? Solubility is also a function of temperature and of the nature of the liquid, but if this is a dosage form for the human body, these are fixed. It is common, however, for the usage instructions to suggest taking the medicine with a glass of water. In addition to helping with swallowing, this also helps to ensure that C_2 in Figure 13.6 remains low and ΔC in Equation (13.5) is as large as possible.

The diffusion coefficient D again depends on the intrinsic properties of the chemical substances, including the viscosity of the liquid. In the case of delivery to the body, the properties of the liquid are largely fixed.

The thickness of the boundary layer h is affected by the flow conditions of the liquid. Again, within the body this may be impossible to alter, but it indicates an obvious way in which the dissolution rate can be increased in other applications, using apparatus such as a stirred tank.

Since bioavailability is strongly linked to dissolution, the measurement of dissolution rate is an important activity for the pharmaceutical industry, both in development of new drugs and dosage forms and for quality control in manufacture. In the human patient (*in vivo*), the rate of dissolution can be measured by taking samples of plasma or urine and measuring the drug concentration in them. However, this is not appropriate during drug discovery and development, where experiments must take place in laboratory glassware (*in vitro*).

A number of standard types of dissolution test are available, of which the two most common are shown in Figure 13.7, together with a typical dissolution curve. The drug dosage form is either (a) placed in a cage or (b) allowed to move freely within an agitated tank, typically of volume 1 L.

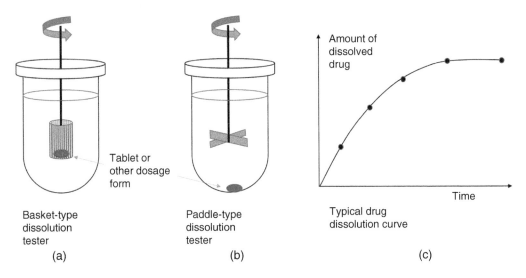

Figure 13.7 Common types of dissolution tester (a) basket type, (b) paddle type, and (c) a typical dissolution curve

Agitation of the liquid occurs due to rotation of the cage or impeller at 50–100 rpm. The liquid is made up to mimic the location within the body, such as the gastrointestinal tract, and held at the body temperature of 37°C. Small samples of the liquid are taken, typically at 5–15 minute intervals, allowing a dissolution curve such as (c) to be produced. Clearly, the types of apparatus shown in Figure 13.7 are very crude ways of mimicking the conditions in the body, which has led to the development of more sophisticated methods that more closely replicate those conditions, but these are not in common use.

13.4   SUSPENSIONS AND THEIR STABILITY

It is often convenient to prepare a product as a *suspension* of particles or droplets in a continuous liquid phase. Common examples include pharmaceutical preparations, foods and coatings such as paint. There are a number of natural examples such as milk – a suspension of micron-scale fat droplets in an aqueous liquid which also contains proteins and nanoparticles of calcium phosphate. Many suspensions are opaque because the suspended particles or drops scatter light. Milk is a dense white because the size of the droplets is comparable with the wavelength of light and they therefore scatter strongly. Skimmed milk appears more transparent because most of the light-scattering fat droplets have been removed.

Milk is an example of a two-phase oil-in-water liquid-liquid mixture in which the phases can be *inverted*. It is possible to break the fat droplets by mechanical action (*churning*) so that they *coalesce* and separate from the aqueous liquid, which can be drained off, eventually producing butter. This is a two-phase water-in-oil mixture in which the continuous fat phase is a solid at room temperature and can therefore be spread – an important property for the consumer. Suspensions of this kind also have an interesting mouthfeel because the fat phase will partially melt and flow at mouth temperature. A large number of consumer products mimic these oil-in-water and water-in-oil structures, such as soya- and oat-derived milks and margarine.

The most important property of suspensions is *stability*. In order for them to remain useful and attractive to the consumer they must not separate or settle. Particles or droplets which are denser than the continuous liquid phase will naturally tend to settle under gravity, and less-dense materials will tend to rise to the surface, in each case at a rate which depends on their size, density and the properties of the liquid. Settling as a separation process is considered in Chapter 5. Here the interest is in preventing or reducing separation. Clearly, one way of doing this is to choose a small particle size, but a problem which frequently arises is *aggregation* or *coalescence* or *flocculation*, in all of which particles combine into larger entities which then have an enhanced tendency towards settling. Particles or droplets tend to aggregate or coalesce because this minimizes their surface energy, and stabilization depends on putting some sort of energy barrier in opposition to this.

Methods for stabilizing dispersions include the following (Figure 13.8):

- Electrostatic repulsion: an example of this is by means of the electrical double-layer described in Chapter 5 and shown in Figure 13.8a, in which each particle is surrounded by a ring of ions of the same (usually negative) charge. Note that the effectiveness of the electrical double-layer depends on the concentration of free ions in the fluid and therefore on its pH. A high concentration of electrolyte compresses the double layer and

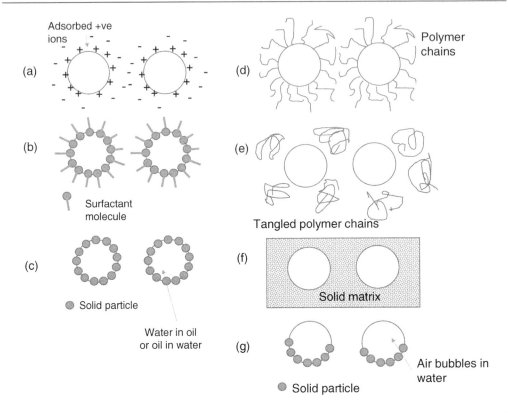

Figure 13.8 Methods for stabilizing dispersions: (a) Electrical double-layer repulsion; (b) surfactant stabilization; (c) Pickering emulsion; (d) steric stabilization; (e) depletion stabilization; (f) particles held in a solid matrix (liquid below its yield stress) and (g) particles attached to air bubbles in a foam or froth

makes it less effective. (Milk is stabilized by a naturally occurring double layer. Addition of a little acid – vinegar, for example – immediately causes flocculation and separation of the fat and aqueous phases.)

- Addition of surface-active *stabilizers* to the suspension (Figure 13.8b): these are usually molecules consisting of a polar part such as an ion or a hydroxyl group and a non-polar part such as a hydrocarbon chain; the polar part is attracted to the aqueous phase and the non-polar part to the oil or fat. Such molecules, which are described as *amphiphilic* (attracted to both oil and water phases), find it energetically most favourable to sit at the oil-water interface, thus reducing its interfacial energy and stabilizing the suspension. Particles themselves can also act in this way: a *Pickering emulsion* of oil and water is stabilized by the addition of fine particles which accumulate at the interface (Figure 13.8c). An everyday example of this is vinaigrette salad dressing, in which droplets of a vinegar-rich aqueous phase are stabilized in an oil continuous phase by the addition of a fine solid such as mustard powder.

- Addition of long-chain polymers to the suspension (Figure 13.8d): this approach can enhance stability in two different ways. The molecules may attach to the dispersed particles or droplets and their long tails may prevent particles coming into close proximity. This is called *steric* stabilization and in this case they are acting rather like the surface-active stabilizers described above. Alternatively, the polymer molecules may associate with each other, aggregating within the continuous phase and preventing relative movement, known as *depletion binding* (Figure 12.8e).

- Changing the viscosity of the continuous phase: increasing the viscosity of the fluid will slow settling (see Chapter 3) but not prevent it. However, it is also possible to make the fluid non-Newtonian (see Chapter 5), so that no flow occurs until a certain yield stress has been exceeded. In this case, the suspension will act as a solid at low stress levels (Figure 13.8f). An example of this is non-drip paint, which is a solid until stirred or spread with a brush, but then reverts to an immobile film as it dries. Paint is considered further in Section 13.5.3.

- Addition of a gaseous phase: some suspensions can be stabilized in a *froth* by use of a gaseous foaming agent which forms small bubbles throughout the structure, to which the particles or droplets attach (Figure 13.8g). Examples of this kind of structure are shaving foams, whipped cream and the froths on certain types of coffee.

It is important to note that multiple stabilization mechanisms can act simultaneously. Real products usually contain many ingredients, including several whose primary function is stabilization.

13.5   SOME INDUSTRIAL PRODUCT SECTORS

This section considers in more detail three industrial sectors from the list in Table 13.1.

13.5.1   Pharmaceutical Dosage Forms

Pharmaceutical *dosage forms* make up an important class of particulate products. The most common examples are *tablets* and *capsules* (Figure 13.9), but medicines are also

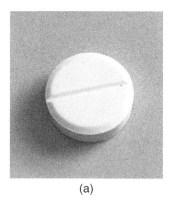
(a)
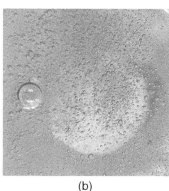
(b)
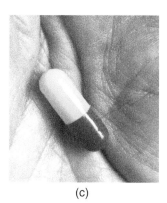
(c)

Figure 13.9 Pharmaceutical dosage forms: (a) tablets; (b) dispersible tablets, formulated so that they effervesce in water and (c) capsules containing pellets

prepared as *suspensions* of particles in a liquid, as particles to be dispersed into a gas for inhalation (see Chapter 14), and in many other ways.

The choice of delivery method depends on many factors, particularly the type of drug and the dosage rate required, and the patient/customer acceptability.

In addition to the AI (the drug), pharmaceutical dosage forms are almost always formulated from a mixture of other ingredients (perhaps as many as ten) which are collectively known as 'excipients' (Table 13.4). The mass fraction of the dosage form that is pharmaceutically 'active' varies considerably: for example, each aspirin tablet weighing 500 mg may typically contain 75 mg of AI. Vitamin C tablets typically contain more than 50% by weight of AI, while potent drugs are present at the 1–100 µg level per tablet, equivalent to less than 0.1% by weight. These extremes present different kinds of problems to the formulator and the manufacturer.

In principle, excipients do not have a therapeutic function in themselves and they are carefully selected so as not to interact chemically with the AI or each other. However, they clearly do have an effect on the rate at which the AI is made available in the body, as described later.

It is apparent from Table 13.4 that excipients can have two main functions: (i) to improve the performance of the product by ensuring effective delivery of the AI to the body, and (ii) to improve the 'processability' of the product so that it can be made reliably, reproducibly and efficiently. The reproducibility of manufacture through, for example,

Table 13.4 Excipients used in tablets and capsules

Excipient type	Function	Examples
Bulking agent or binder	Chemically inert, non-hygroscopic, hydrophilic, good compression properties	Lactose, cellulose and their derivatives, polyvinylpyrrolidone (PVP), polyethylene glycol (PEG)
Antiadherents	Reduce adhesion between tablet and metal surfaces during compression	Magnesium stearate
Glidants	Reduce interparticle cohesion and friction	Silica gel, fumed silica, talc, magnesium carbonate
Lubricants	Reduce particle–surface friction during processing	Talc, silica, magnesium stearate, stearic acid
Coatings	Protection from moisture, ease of swallowing, controlled release	Ether hydroxypropyl methylcellulose (HPMC)
Disintegrants	Expand and dissolve when wet causing the tablet to break apart, so releasing the active ingredients	Crosslinked polyvinylpyrrolidone (crospovidone), crosslinked sodium carboxymethyl cellulose (croscarmellose sodium)
Colours	Improve appearance and aid identification	Titanium dioxide
Flavours	Improve taste, masking unpleasant drug flavour	Fruit extracts
Sorbents	Limits fluid sorbing by adsorption or absorption in a dry state	Magnesium carbonate, kaolin
Preservatives	Enhance shelf life	Antioxidants such as vitamins A, E and C, amino acids cysteine and methionine, citric acid and derivatives

low variability of drug loading from tablet to tablet, is clearly of overriding importance in medicinal applications.

13.5.2   Food

Many foods come in a particulate form. Since the beginning of agriculture, grains and foods produced from them have made up a large part of the human diet. Cereal grains include wheat, oats, rice, corn (maize), barley, sorghum, rye and millet. An equally important class is that of beans and legumes, of which some of the most important are chickpeas, lentils, peas, kidney beans, soya beans, cocoa beans and coffee beans. Next come crystallized solids such as sugar and salt, which are naturally occurring components and also frequent additives to many foodstuffs. They are important contributors to taste and texture, for example in chocolate, itself the product of cocoa beans. Historically, the desirability of spices, very often found in particulate form, drove the establishment of global trading routes.

Naturally derived foods are invariably structured materials: both meats and vegetables consist of cellular structures, while milk-based foods may be emulsions or gels, as discussed in Section 13.4. Man-made foods are often designed to mimic naturally occurring structures, in order to deliver a similarly desirable texture and taste to the consumer.

Food processing challenges include the following (Seville *et al.*, 2007):

- *Structure and formation of food granules*: Example 13.1 showed how agglomerates with 'instant' properties might be constructed. These must be structured such that they hydrate and dissolve quickly but are also strong enough to withstand the stresses of transport through the supply chain.

- *Diffusion and reaction in food solids*: the chemical reactions which occur in cooking, such as the well-known *Maillard reaction* between amino acids and reducing sugars, result in the formation of hundreds of flavouring and colouring compounds which give cooked foods their distinct desirable (or undesirable) properties. Migration of species within food solids during their processing is an inevitable consequence of cooking and drying; their surface composition may be very different from their bulk, since active materials migrate to the surface during processing.

- *Structuring of liquids and soft solids*: suspensions of droplets and particles can be produced by stirring the mixture and these suspensions can form a solid gel as they are cooled. Cooling without stirring results in near-spherical droplets; simultaneous stirring and cooling can result in the formation of elongated particles and fibres like those shown in Figure 13.10, which illustrates the range of possibilities in the multiscale structuring of gelled particle structures.

13.5.3   Paint and Coatings

Paint is a universal material, usually supplied as a liquid and applied to surfaces in such a way as to form a thin coating which dries to a solid. It provides obvious decorative properties but is frequently also used to protect the coated surface from physical damage and

Quiescent cooled   Shear cooled

Gellan – k- carrageenan
(2%w/w – 2%w/w)

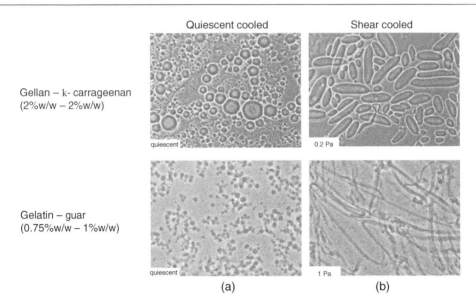

Gelatin – guar
(0.75%w/w – 1%w/w)

(a)   (b)

Figure 13.10 Examples of food gel particles obtained through (a) cooling through the gelation temperature, and (b) cooling through gelation while laminar flow processing at constant low shear stress. Wolf *et al.* (2000)/Reproduced from Oxford University Press

chemical corrosion. The latter category includes large-volume use for coating structures such as bridges, ships, aircraft and road vehicles.

Most paints are either oil-based or water-based, and typical ingredients are listed in Table 13.5. The colour may come from dissolved dyes or dispersed pigment particles. Paints and coatings also frequently contain other kinds of particles which give them special properties such as toughness or resistance to abrasion. Pigments may be synthetic molecules or natural materials such as clays and calcium carbonate. Pigments such as titanium dioxide may be added to make the paint more opaque and protect it from attack by ultraviolet light. The trend is towards increasing the use of water-based paints and addition of particulate fillers so as to reduce the use of volatile liquids. It is possible to formulate spray paint at >75% solids by weight and for brush-applied paint to be >85% solids by weight. Increasing use is being made of 100% powder coatings, in which the paint is entirely solid and applied by electrostatically enhanced spraying or by dipping the article to be coated into a fluidized bed of the powder particles.

In most paints and coatings, the size and concentration of suspended particles are both important. The colour properties of paint and its light-reflecting properties such as gloss depend strongly on the pigment size in relation to the wavelength of light. As noted elsewhere (Chapter 1), particles of sizes close to the wavelength of incident light scatter that light most strongly (which is why fog is opaque and rain is not). For this reason, pigments in high-quality paints are commonly of order 1–10 μm in size. (The visible spectrum covers a wavelength range of approximately 0.4–0.7 μm.) For example, latex wall paint contains pigment particles in the range 20–30 μm and has comparatively low binder content, while automotive paint contains finer particles, from about 2 to 10 μm in size, and a higher binder content. The difference is the gloss required of the paint finish, as shown in Figure 13.11. Domestic wall paint is said to be 'flat', with little light reflected, while automotive paint requires a defect-free finish with a particularly high gloss.

Table 13.5 Paint constituents and their roles

	Components	Typical function
Pigment (dispersed phase)	Primary pigment	Optical effects. Inhibition of corrosion
	Extender	To increase opacity and barrier properties
Fluid (continuous phase)	Polymer or resin binder	Provides the basis for a continuous film and surface seal
	Solvent or diluent	Provides the means by which the paint can be applied
	Additives	Enhancements of various kinds, including flow agents and reaction catalysts

Lambourne and Strivens (1999)/With permission of Elsevier.

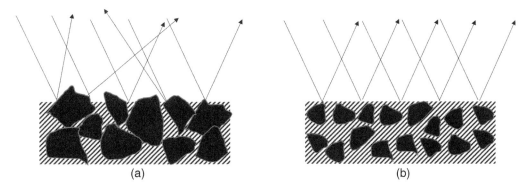

Figure 13.11 Light reflection from paint surfaces: (a) larger particles with a high proportion of pigment produce a matt finish, with some absorption of light and diffuse reflection and (b) smaller particles with a lower proportion of pigment result in a gloss finish and reflection at a single angle

The size and concentration of particles of pigment in paint also contribute to the rheology of the paint (see also Chapter 5). For example, high solids loading can create a *thixotropic* fluid: possessing time-dependent shear thinning properties, so that it is solid under static conditions but will flow over time when shaken or sheared. This is obviously a useful property in a 'non-drip' paint.

The important formulation variable in paint is the *pigment-volume concentration* (PVC), defined as the ratio of pigment volume to total paint volume:

$$\text{PVC} = \frac{V_{\text{pigment}}}{V_{\text{pigment}} + V_{\text{binder}}} \tag{13.7}$$

PVC affects both the physical and optical properties of paint. As indicated in Figure 13.11, less binder makes the pigment stand out from the surface, giving a matt finish which is also less resistant to damage. More binder provides a smoother and glossier surface. However, there is a critical value of this ratio, CPVC, below which the paint is saturated with binder and the surface becomes glossy. Further binder addition (lowering PVC)

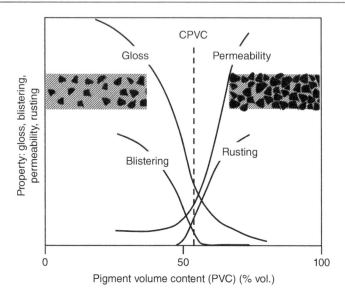

Figure 13.12   Effect of varying PVC on several paint properties. After Meyer et al. (1997)

does not improve glossiness but may affect other properties, such as mechanical strength and permeability. The value of CPVC depends on the combination of binder, pigment and other components but is generally 35–65%. Figure 13.12 shows how the important paint properties vary with PVC for a particular case in which CPVC = 55%.

FURTHER READING

The topic of Particulate Products is vast and so in a single chapter it is possible only to give an introduction. For more examples, the reader is referred to the following: Bröckel et al. (2007); Litster (2016); Norton et al. (2013); Taylor and Aulton (2022) and Wesselingh et al. (2007).

TEST YOURSELF

13.1 Why are particulate products often structured into agglomerates?

13.2 Why do some products take up liquid slowly and how might this process be accelerated?

13.3 What is Fick's first law of diffusion and how is this related to the Noyes-Whitney equation for the rate of dissolution from a single particle?

13.4 What are 'sink' conditions in dissolution?

13.5 How can you improve the dissolution rate for a given substance?

13.6 How can a suspension be stabilized against separation of the phases?

13.7 In pharmaceutical products, what are excipients and what are their functions?

13.8 What are the particular challenges associated with processing of particles for food?

13.9 What effect does the ratio of particles to binder have on the properties of paint and coatings?

13.10 In particulate products in general, what are the effects of particle size on product properties?

EXERCISES

13.1 A powder of particle size 100 μm is to be saturated with water by capillary uptake into a bed of height 10 cm and void fraction 0.5. How long will this take?

[Answer: 83 s]

13.2 A powder of size 10 μm is agglomerated into particles of diameter 1 mm and void fraction 0.4. How long will it take for water to completely saturate each agglomerate, assuming that air can easily escape?

[Answer: 5.4 s]

13.3 Make a list of the products in your home which contain particles. What features must these products have in order to make them desirable to the consumer? What are the active ingredients in each case? What are the other ingredients and what function does each of these ingredients have? What size are the particles in the product and why? How do you think these products have been made?

[Hint: Ingredient function can be difficult to determine. Try looking up patents for products like the ones you have.]

13.4 Design a suspension of particles of calcium carbonate for surface cleaning. What particle size would you choose, what are the other ingredients and what are their functions?

[Hint: Try looking up patents for similar products.]

14
Health and Safety

This chapter covers the fire and explosion hazards of fine powders and both the positive and negative health effects of fine powders. Although the hazards of fine, combustible powders (dusts) are well known and control measures well understood, fires and explosions in plants handling such materials are still common in the food, wood processing, metal handling and power generation industries and occasionally occur in other industries processing combustible powders (pharmaceuticals, plastics). In this chapter, we begin with combustion fundamentals and how they apply to fine powders dispersed in air. The characteristics of dust explosions are then discussed and the test methods used to measure these characteristics are described. This section concludes with a look at the measures available to control the hazards associated with the handling and processing of combustible fine powders.

The section on health effects begins with a description of the respiratory system and an analysis of the interaction of the system with fine particles. The devices available for the pulmonary delivery of drugs are then described and compared. This section concludes with a discussion of the negative health effects of fine powders and the control measures available to minimize exposure of personnel.

14.1 FIRE AND EXPLOSION HAZARDS OF FINE POWDERS

14.1.1 Introduction

Finely divided combustible solids, or dusts, dispersed in air can give rise to explosions in much the same way as flammable gases. In the case of flammable gases, fuel concentration, local heat transfer conditions, oxygen concentration and initial temperature all affect ignition and resulting explosion characteristics. In the case of dusts, however, more variables are involved (e.g. particle size distribution, moisture content) and so the analysis

Introduction to Particle Technology, Third Edition. Martin Rhodes and Jonathan Seville.
© 2024 John Wiley & Sons Ltd. Published 2024 by John Wiley & Sons Ltd.
Website: www.wiley.com/go/rhodes/particle3e

and prediction of dust explosion characteristics is more complex than for the flammable gases. Dust explosions have been known to give rise to serious property damage and loss of life. Most people are probably aware that dust explosions have occurred in grain silos, flour mills and in the processing of coal. However, explosions of dispersions of fine particles of metals (e.g. aluminium), plastics, sugar and pharmaceutical products can be particularly potent. Process industries where fine combustible powders are used and where particular attention must be directed towards control of dust explosion hazards include plastics, food processing, metal processing, pharmaceuticals, agricultural, chemicals and solid fuels. Process steps where fine powders are heated have a strong association with dust explosion; examples include dilute pneumatic conveying and spray drying, which involves heat and a dilute suspension.

Data collected by dustsafetyscience.com show that, in North America alone, over the period 2017–2021 there were an average of 33 dust explosions and 148 dust fires per year, resulting in an average of 44 injuries and 3 fatalities per year. The main industries involved were wood processing, food and agriculture, solid fuels, metals and power generation. Food and wood products made up nearly 75% of the combustible dust fires and explosions recorded during that period. Of the dust explosions recorded, the sources of the majority were storage silos, dust collection systems, dryers, conveyors and elevators.

For the purpose of education and training, reports and analyses of dust fires and explosions are provided online by organizations such as the US Chemical Safety Board (CSB) and the National Fire Protection Association (NFPA).

An example from the CSB is summarized here (www.csb.gov). The Imperial Sugar Company factory at Port Wentworth, Georgia, USA made and packaged granulated and powdered sugar. In February 2008, a huge explosion and fire destroyed the packing building causing 14 deaths and injuring 38 others.

In the factory, sugar was transported from large silos to the packing building by a complex system of bucket elevators, screw conveyors and conveyor belts. During transfer, sugar spilled onto the floor around the equipment – accumulating up to several centimetres deep in some areas. The sugar contained fine particles which became airborne. Hammer mills were used to break down the granulated sugar into powdered sugar, creating even more dust. The hammer mills were connected to a dust collection system – an exhaust system drawing dusty air from around the machine to a series of filters. This system was undersized and in disrepair. The dust collection system was not connected to the bucket elevators, screw conveyors and conveyor belts. Workers regularly used compressed air to clean machinery – dispersing more dust. Significant amounts of sugar dust had collected on elevated hard-to-clean surfaces, such as pipes, structural girders and light fittings, around the factory.

The conveyor belt, which was fed from the silos, had recently been completely enclosed to avoid contamination. The feed chutes from the silos occasionally became blocked, causing spillage from the belt and a build-up of dust in the air inside the enclosure. On the day of the incident, sugar dust in the enclosure reached flammable concentration and contacted an ignition source (likely an overheated bearing) and exploded. This primary explosion blew apart the conveyor enclosure and the resulting pressure wave and fireball entered the multi-storey packing building. Here, the pressure wave disturbed accumulated dust on surfaces creating large dust clouds, which were ignited by the flames from the primary explosion. This secondary explosion blew out walls and buckled concrete floors, disturbing more dust and giving rise to further secondary explosions and fires. Workers had difficulty escaping since the explosion had cut the electricity to the lights and blocked the stairwells.

Summarizing the faults that gave rise to this accident:

- Poor design of building – too many surfaces for dust to accumulate, building unable to safely vent.

- Poor housekeeping – dust allowed to accumulate on floors and surfaces.

- Poor worker training – used compressed air to clean equipment.

- Inadequate design of dust collection system – but note that the dust collection system could become the source of explosion if incorrectly designed and operated.

- Poor maintenance – overheating bearings, dust collection system in disrepair.

Secondary explosions cause the most damage and loss of life since they usually involve greater quantities of flammable dust than the primary explosion and occur outside the plant equipment but inside a building. In many dust explosion accidents, a small primary explosion, resulting in loss of containment, gives rise to massive secondary explosions. Secondary explosions usually occur because of poor housekeeping and poor building design (too many hard-to-clean surfaces where dust can accumulate).

14.1.2   Combustion Fundamentals

Flames

A *flame* is a gas rendered luminous by emission of energy produced by chemical reaction. In a stationary flame (for example, a candle flame or gas stove flame), unburned fuel and air flow into the flame front as combustion products flow away from the flame front. A stationary flame may arise from combustion of either premixed fuel and air, as observed in a Bunsen burner with the air hole open, or by diffusion of air into the combustion zone, as for a Bunsen burner with the air hole closed.

When the flame front is not stationary it is called an explosion flame. In this case the flame front passes through a premixed fuel–air mixture. The heat released and gases generated result in either an uncontrolled expansion effect or, if the expansion is restricted, a rapid build-up of pressure.

Explosions and detonations

Explosion flames travel through the fuel–air mixture at velocities ranging from a few metres per second to several hundreds of metres per second and this type of explosion is called a *deflagration. Flame speeds* are governed by many factors including the heat of combustion of the fuel, the degree of turbulence in the mixture and the amount of energy supplied to cause ignition. It is possible for flames to reach supersonic velocities under some circumstances. Such explosions are accompanied by pressure shock waves, are far more destructive and are called *detonations*. The increased velocities result from increased gas densities generated by pressure waves. In practice it is likely that all detonations begin as deflagrations.

Flammability limits

Within an elemental volume of fuel–air mixture, the heat generated by the combustion reaction increases exponentially with increasing temperature, whilst the heat lost from the element to the surrounding mixture increases linearly with temperature. Whether a flame is sustained (i.e. ignition occurs in the mixture) is dependent on the temperature and the rate of heat generation by the reaction per unit volume of mixture.

Below a certain fuel concentration, ignition will not occur since the rate of heat generation within the element is insufficient to match the rate of heat loss to the surroundings. This concentration is known as the lower flammability limit C_{fL} of the fuel–air mixture (Figure 14.1). It is generally measured under standard conditions in order to give reproducible heat transfer conditions. At C_{fL} the oxygen is in excess. As the fuel concentration is increased beyond C_{fL} the amount of fuel reacting per unit volume of mixture and the quantity of heat generated per unit volume by the reaction will increase until the stoichiometric ratio for the reaction is reached. For fuel concentration increase beyond the stoichiometric ratio, the oxygen is limiting and so the amount of fuel reacting per unit volume of mixture and the quantity of heat generated per unit volume decrease with fuel concentration. A point is reached when the heat release per unit volume of mixture is too low to sustain a flame. This is the upper flammability limit, C_{fU}. This is the concentration of fuel in the fuel–air mixture above which a flame cannot be propagated. For many fuels the amount of fuel reacting per unit volume of fuel–air mixture (and hence the heat generated per unit volume mixture) at C_{fL} is similar to that at C_{fU} showing that, for a fuel–air mixture, there is a minimum value of heat generated per unit volume required to propagate a flame. This value is exceeded between C_{FL} and C_{FU} (see Worked Example 14.5).

Thus, in general, there is a range of fuel concentration in air within which a flame can be propagated. From the analysis above it will be apparent that this range will widen (C_{fL} will decrease and C_{fU} will increase) as the initial temperature of the mixture is increased. In practice, therefore, flammability limits are measured and quoted at standard temperatures (usually 20 °C).

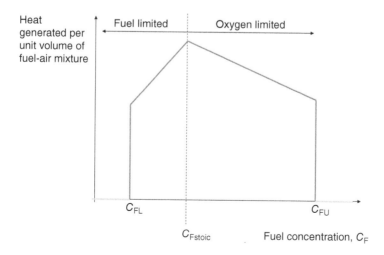

Figure 14.1 Heat generated per unit volume of fuel–air mixture versus fuel concentration

For mixtures of gaseous fuels in air the lower and upper flammability limits can be used as reference points to ensure safe operation of equipment and processes. Where necessary, flammable mixtures can be avoided by ensuring operation at fuel concentrations below C_{fL} or above C_{fU}.

Minimum oxygen for combustion

At the lower limit of flammability there is more oxygen available than is required for stoichiometric combustion of the fuel. For example, the lower flammability limit for propane in air at 20 °C is 2.2% by volume.

For complete combustion of propane according to the reaction:

$$C_3H_8 + 5O_2 \rightarrow 3CO_2 + 4H_2O$$

five volumes of oxygen are required per volume of fuel propane.

In a fuel–air mixture with 2.2% propane the ratio of air to propane is:

$$\frac{100 - 2.2}{2.2} = 44.45$$

And since air is approximately 21% oxygen, the ratio of oxygen to propane is 9.33. Thus, in the case of propane at the lower flammability limit oxygen is in excess by approximately 87%.

It is therefore possible to reduce the concentration of oxygen in the fuel–air mixture whilst still maintaining the ability to propagate a flame. If the oxygen is replaced by a gas which has similar physical properties (nitrogen for example) the effect on the ability of the mixture to maintain a flame is minimal until the stoichiometric ratio of oxygen to fuel is reached. The oxygen concentration in the mixture under these conditions is known as the minimum oxygen for combustion (MOC). MOC is therefore the stoichiometric oxygen equivalent to the lower flammability limit. Thus

$$MOC = C_{fL} \times \left(\frac{molO_2}{molfuel} \right)_{stoich}$$

For example, for propane, since under stoichiometric conditions five volumes of oxygen are required per volume of fuel propane,

$$MOC = 2.2 \times 5 = 11\% \text{ oxygen by volume}$$

14.1.3   Combustion in Dust Clouds

Fundamentals specific to dust cloud explosions

The combustion rate of a solid in air will in most cases be limited by the surface area of solid presented to the air. This is the case even if the particle size is quite large, say a few millimetres in size. However, if the particles of solid are small enough to be dispersed in air without too much propensity to settle, the reaction rate will be great enough to permit an explosion flame to propagate. For a dust explosion to occur, the solid material of which the particles are composed must be combustible, i.e. it must react exothermically with the oxygen in air. However, not all combustible solids give rise to dust explosions.

For the generalized conditions for flame propagation discussed above to apply to dust explosions we need only to add in the influence of particle size on reaction rate. The rate of heat generated per unit volume is now determined by the surface area of solid fuel particles per unit volume of the suspension, which is typically inversely proportional to the solid particle size. Thus, the likelihood of flame propagation and explosion will increase with decreasing particle size. Qualitatively, this is because finer fuel particles:

- More readily form a dispersion in air.

- Have a larger surface area per unit mass of fuel.

- Offer a greater surface area for reaction (higher reaction rate).

- Consequently generate more heat per unit mass of fuel.

- Heat up more rapidly.

Theoretically the lower and upper flammability limits apply to any fuel–air mixture. However, for dust suspensions the flammability limits cannot be used in any control strategy, since they are dependent on the nature of the solid, the size distribution and the surface area. An additional complication is that dispersions of dusts in air are often far from uniform.

Characteristics of dust explosions

Consider design engineers wishing to know the potential fire and explosion hazards associated with a particulate solid made or used in a plant which they are designing. They are faced with the same problem which they face in gathering any "property" data of particulate solids; unlike liquids and gases there are few published data, and what is available is unlikely to be relevant. The particle size distribution, surface properties and moisture content all influence the potential fire hazard of the powder, so unless the engineers can be sure that their powder is identical in every way to the powder used to obtain the published data, they must have the explosion characteristics of the powder determined by testing. Having made the decision, the engineers must ensure that the sample given to the test laboratory is truly representative of the material to be produced or used in the final plant.

Standards differ in different countries, but most tests include an assessment of some or all of the following explosion characteristics:

- Minimum dust concentration for explosion.

- Minimum energy for ignition – the minimum energy of an electric spark needed to initiate a dust explosion.

- Minimum ignition temperature – the minimum temperature of a hot surface able to ignite a dust cloud.

- Maximum explosion pressure.

- Maximum rate of pressure rise during explosion.

- Minimum oxygen for combustion.

An additional screening test is sometimes used. This is simply a test for *explosibility* in the test apparatus, classifying the dust as able or unable to ignite and propagate a flame in air at room temperature under test conditions. Another test which may be useful is the *minimum dust layer ignition temperature*, which determines the minimum temperature at which a dust deposit will self-ignite when exposed to a heated surface as a function of the deposit volume. This test is relevant to the design of storage volume, duration and conditions.

Apparatus for determination of dust explosion characteristics

There are several different devices for determination of dust explosion characteristics. All devices include a vessel which may be open or closed, an ignition source which may be an electrical spark, chemical igniter or electrically heated wire coil and a supply of air for dispersion of the dust. The simplest type is known as the vertical tube apparatus and is shown schematically in Figure 14.2. The sample dust is placed in the dispersion cup. Delivery of dispersion air to the cup is via a solenoid valve. Ignition may be either by electrical spark across electrodes or by heated coil. The vertical tube apparatus may be used for the screening test and for determination of minimum dust concentration for explosion, minimum energy for ignition and in a modified form for MOC.

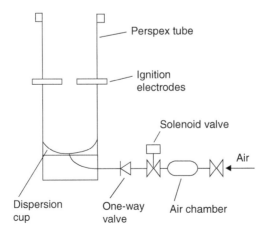

Figure 14.2   Vertical tube apparatus for determination of dust explosion characteristics

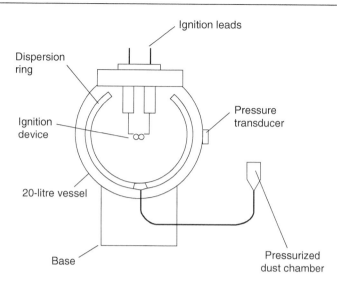

Figure 14.3 The 20-litre sphere apparatus for determination of dust explosion characteristics (after Lunn, 1992).

A second apparatus, known as the 20-litre sphere, is used for determination of maximum explosion pressure and maximum rate of pressure rise during explosion. These give an indication of the severity of explosion and enable the design of explosion protection equipment. This apparatus, which is shown schematically in Figure 14.3, is based on a spherical 20-litre pressure vessel fitted with a pressure transducer. The dust to be tested is first charged to a reservoir and then blown by air into the sphere via a perforated dispersion ring. The vessel pressure is reduced to about 0.4 bar before the test so that upon injection of the dust, the pressure rises to atmospheric. Ignition is by a pyrotechnical device with a standard total energy (typically 10 kJ) positioned at the centre of the sphere. Ignition is slightly delayed after air injection in order to allow the turbulence to reduce.

The third basic test device is the Godbert–Greenwald furnace apparatus (Figure 14.4), which is used to determine the minimum ignition temperature and the explosion characteristics at elevated temperatures. The apparatus includes a vertical electrically heated furnace tube which can be raised to controlled temperatures up to 1000 °C. The dust under test is charged to a reservoir and then dispersed through the tube. If ignition occurs, the furnace temperature is lowered in 10 °C steps until ignition does not occur. The lowest temperature at which ignition occurs is taken as the ignition temperature. Since the quantity of dust used and the pressure of the dispersion air both affect the result, these are varied to obtain a minimum ignition temperature.

Application of the test results

The *minimum dust concentration for explosion* is measured in the vertical tube apparatus and is used to give an indication of the quantities of air to be used in extraction systems for combustible dusts. Since dust concentrations can vary widely with time and location in a plant, it is not considered wise to use concentration control as the sole method of protection against dust explosion.

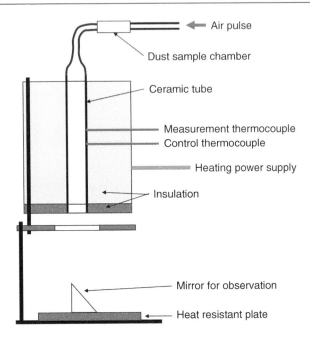

Figure 14.4 Schematic cross-section of the Godbert–Greenwald furnace apparatus

The *minimum energy for ignition* is measured primarily to determine whether the dust cloud could be ignited by an electrostatic spark. Ignition energies of dusts can be as low as 15 mJ; this quantity of energy can be supplied by an electrostatic discharge.

The *minimum ignition temperature* indicates the maximum temperature for equipment surfaces in contact with the powder. For new materials it also permits comparison with well-known dusts for design purposes. Table 14.1 gives some values of explosion parameters for common materials.

The *maximum explosion pressure* is usually in the range 8–13 bar and is used mainly to determine the design pressure for equipment when explosion containment or protection is opted for as the method of dust explosion control.

The *maximum rate of pressure rise* during an explosion is used in the design of explosion relief. It has been demonstrated that the maximum rate of pressure rise in a dust explosion is inversely proportional to the cube root of the vessel volume, i.e.

$$\left(\frac{\mathrm{d}P}{\mathrm{d}t}\right)_{max} = V^{1/3}K_{St} \qquad (14.1)$$

Table 14.1 Explosion parameters for some common materials (Schofield, 1985)

Dust	Mean particle size (μm)	Maximum explosion pressure (bar)	Maximum rate of pressure rise (bar/s)	K_{St}
Aluminium	17	7.0	572	155
Polyester	30	6.1	313	85
Polyethylene	14	5.9	494	134
Wheat	22	6.1	239	65
Zinc	17	4.7	131	35

Table 14.2 Dust explosion classes based on 1 m^3 test apparatus

Dust explosion class	K_{St} (bar m/s)	Comments
St 0	0	Non-explosible
St 1	0–200	Weak to moderately explosible
St 2	200–300	Strongly explosible
St 3	>300	Very strongly explosible

The value of K_{St} is found to be constant for a given powder. Typical values are given in Table 14.1. The severity of dust explosions is classified according to the St class based on the K_{St} value (see Table 14.2).

The minimum oxygen for combustion is used to determine the maximum permissible oxygen concentration when *inerting* is selected as the means of controlling the dust explosion, i.e. an inert gas is used to prevent combustion. Organic dusts have an MOC of about 11% if nitrogen is the diluent and 13% in the case of carbon dioxide. Inerting requirements for metal dusts are more stringent since MOC values for metals can be far lower.

14.1.4 Control of the Hazard

Introduction

As with the control of any process hazard, there is a hierarchy of approaches that can be taken to control dust explosion hazard. These range from the most desirable strategic approach of changing the process to eliminate the hazardous powder altogether to the merely tactical approach of avoiding ignition sources. The main approaches are listed below:

Preventative measures:

- Change the process to eliminate the dust.

- Minimize dust cloud formation by good housekeeping and design.

- Replace the oxygen with an inert gas.

- Reduce oxygen to below MOC.

- Add moisture to the dust.

- Add diluent powder to the dust.

- Exclude ignition sources.

- Control dust concentration to be outside flammability limits.

Protective measures:

- Vent the vessel to a safe place to relieve pressure generated by the explosion.

- Detect the start of an explosion and inject a suppressant.

- Design the plant to withstand the pressure generated by any explosion.

Good housekeeping and building design

Since the flammable dust may occasionally escape containment, any building housing the plant must be designed with a minimum of surfaces on which airborne dust might settle. Dust which does escape containment must be removed to avoid accumulation. A layer of fine dust, such as flour, 0.3 mm thick on the floor is sufficient to fill a room with an explosible dust cloud up to 3 m above floor level.

Ignition sources

Excluding ignition sources sounds a sensible policy. However, statistics of dust explosions indicate that in a large proportion of incidents the source of ignition was unknown. Thus, whilst it is good policy to avoid sources of ignition as far as possible, this should not be relied on as the sole protection mechanism. It is interesting to look briefly at the ignition sources which have been associated with dust explosions.

- *Flames*. Flames from the burning of gases, liquids or solids are effective sources of ignition for flammable dust clouds. Several sources of flames can be found in a process plant during normal operation (e.g. burners, pilot flames, etc.) and during maintenance (e.g. welding and cutting flames). These flames would usually be external to the vessels and equipment containing the dust. A good permit-to-work system should be in place to ensure a safe environment before any maintenance commences.

- *Hot surfaces*. Careful design is required to ensure that surfaces likely to be in contact with dust do not reach temperatures which can cause ignition. Attention to detail is important; for example, ledges inside equipment should be avoided to prevent settling of dust and possible self-ignition. Dust must not be able to build up on hot or heated surfaces, otherwise surface temperatures will rise as heat dissipation from the surface is reduced. Care must also be taken outside the vessel; for example, if dust is allowed to settle on electric motor housings, overheating and ignition may occur.

- *Electric sparks*. Sparks produced in the normal operation of electrical power sources (by switches, contact breakers and electric motors) can ignite dust clouds. Special electrical equipment is available for application in areas where there is a potential for dust explosion hazard. Sparks from electrostatic discharges are also able to ignite dust clouds. Electrostatic charges are developed in many processing operations (particularly those involving dry powders) and so care must be taken to ensure that such charges are led to earth to prevent accumulation and eventual discharge. Even the energy in the charge developed on a process operator can be sufficient to ignite a dust cloud.

- *Mechanical sparks and friction.* Sparks and local heating caused by friction or impact between two metal surfaces or between a metal surface and foreign objects inadvertently introduced into the plant have been known to ignite dust clouds.

Venting

If a dust explosion occurs in a closed vessel at atmospheric pressure, the pressure will rise rapidly (up to and sometimes beyond 600 bars/s) to a maximum of around 10 bars. If the vessel is not designed to withstand such a pressure, deformation and possible rupture will occur. The principle of explosion venting is to discharge the vessel contents through an opening or vent to prevent the pressure rising above the vessel design pressure. In this context, venting can include having a roof or section of roof which will lift in the event of an explosion. Venting is a relatively simple and inexpensive method of dust explosion control but cannot be used when the dust, gas or combustion products are toxic or in some other way hazardous, or when the rate of pressure rise is greater than 600 bar/s. The design of vents is best left to the expert although there are published guides (Lunn, 1992). The mass and type of the vent determine the pressure at which the vent opens and the delay before it is fully open. These factors together with the size of the vent determine the rate of pressure rise and the maximum pressure reached after the vent opens. Figure 14.5 shows typical pressure rise profiles for explosions in a vessel without venting and with vents of different sizes.

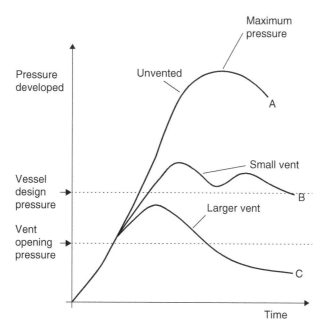

Figure 14.5 Pressure variation with time for dust explosions (A) unvented, (B) vented with inadequate vent area and (C) vented with adequate vent area (after Schofield, 1985)

Suppression

The pressure rise accompanying a dust explosion is rapid, but it can be detected in time to initiate some action to suppress the explosion. Suppression involves discharging a quantity of inert gas or powder into the vessel in which the explosion has commenced. Modern suppression systems triggered by the pressure rise accompanying the start of the explosion have response times of the order of a few milliseconds and are able to effectively extinguish the explosion. The fast-acting trigger device can also be used to vent the explosion, isolate the plant item, or shut down the plant if necessary.

Inerting

Nitrogen and carbon dioxide are commonly used to reduce the oxygen concentration of air to below the MOC. Even if the oxygen concentration is not reduced as far as the MOC value, the maximum explosion pressure and the maximum rate of pressure rise are much reduced (Palmer, 1990). Total replacement of oxygen is a more expensive option but provides an added degree of safety.

Minimize dust cloud formation

This cannot be relied on as the sole control measure but should be incorporated in the general design philosophy of a plant involving flammable dusts. Examples are (1) use of dense phase conveying as an alternative to dilute phase, (2) use of cyclone separators and filters instead of settling vessels for separation of conveyed powder from air, and (3) avoiding situations where a powder stream is allowed to fall freely through air (e.g. in charging a storage hopper). Outside the vessels of the process, good housekeeping practice should ensure that deposits of powder are not allowed to build up on ledges and surfaces within a building. This avoids secondary dust explosions caused when these deposits are disturbed and dispersed by a primary explosion or shock wave.

Containment

Where plant vessels are of small dimensions it may be economical to design them to withstand the maximum pressure generated by the dust explosion (Schofield and Abbott, 1988). The vessel may be designed to contain the explosion and be replaced afterwards or to withstand the explosion and be reusable. In both cases design of the vessel and its accompanying connections and ductwork is a specialist task. For large vessels the cost of design and construction to contain dust explosions is usually prohibitive.

14.1.5   Worked Examples

WORKED EXAMPLE 14.1

It is proposed to protect a section of duct used for pneumatically transporting a plastic powder in air by adding a stream of nitrogen. The air flow rate in the present system is $1.6\,\text{m}^3/\text{s}$ and the air carries 3% powder by volume. If the minimum oxygen for combustion (by replacement of oxygen with nitrogen) of the powder is 11% by volume, what is the minimum flow rate of nitrogen which must be added to ensure safe operation?

Solution

The current total air flow of $1.6 \, m^3/s$ includes 3% by volume of plastic powder and 97% air (made up of 21% oxygen and 79% nitrogen by volume). In this stream the flow rates are therefore:

powder: $0.048 \, m^3/s$

oxygen: $0.3259 \, m^3/s$

nitrogen: $1.226 \, m^3/s$

At the limit, the final concentration of the flowing mixture should be 11% by volume. Hence, using a simple mass balance assuming constant densities,

$$\frac{\text{volume flow } O_2}{\text{total volume flow}} = \frac{0.3259}{1.6 + n} = 0.11$$

from which, the minimum required flow rate of added nitrogen, $n = 1.36 \, m^3/s$.

WORKED EXAMPLE 14.2

A combustible dust has a lower flammability concentration limit in air at 20 °C of 0.9% by volume. A dust extraction system operating at $2 \, m^3/s$ is found to have a dust concentration of 2% by volume. What minimum flow rate of additional air must be introduced to ensure safe operation?

Solution

Assume that the dust explosion hazard will be reduced by bringing the dust concentration in the extract to below the lower flammability limit. In $2 \, m^3/s$ of extract, the flow rates of air and dust are:

air: $1.96 \, m^3/s$

dust: $0.04 \, m^3/s$

At the limit, the dust concentration after the addition of dilution air will be 0.9%, hence:

$$\frac{\text{volume of dust}}{\text{total volume}} = \frac{0.04}{2 + n} = 0.09$$

from which the minimum required flow rate of added dilution air, $n = 2.44 \, m^3/s$.

WORKED EXAMPLE 14.3

A flammable dust is suspended in air at a concentration within the flammable limits and with an oxygen concentration above the minimum oxygen for combustion. Sparks generated by a grinding wheel pass through the suspension at high speed, but no fire or explosion results. Explain why.

Solution

In this case it is likely that the temperature of the sparks will be above the measured minimum ignition temperature and the energy available is greater than the minimum ignition energy. However, an explanation might be that the heat transfer conditions are unfavourable. The high-speed sparks have insufficient contact time with any element of the fuel–air mixture to provide the energy required for ignition.

WORKED EXAMPLE 14.4

A fine flammable dust is leaking from a pressurized container at a rate of 2 litre/min into a room of volume $6\,m^3$ and forming a suspension in the air. The minimum explosible concentration of the dust in air at room temperature is 2.22% by volume. Assuming that the dust is fine enough to settle only very slowly from suspension, (a) what will be the time from the start of the leak before explosion occurs in the room if the air ventilation rate in the room is $4\,m^3/h$, and (b) what would be the minimum safe ventilation rate under these circumstances?

Solution

(a) Mass balance on the dust in the room:

$$\left(\begin{array}{c}\text{rate of}\\\text{accumulation}\end{array}\right)=\left(\begin{array}{c}\text{rate of flow}\\\text{into the room}\end{array}\right)-\left(\begin{array}{c}\text{rate of flow out}\\\text{of room with air}\end{array}\right)$$

assuming constant gas density,

$$V\frac{dC}{dt}=0.12-4C$$

where 0.12 is the leak rate in m^3/h, V is the volume of the room and C is the dust concentration in the room at time t.

Rearranging and integrating with the initial condition, $C = 0$ at $t = 0$,

$$t=-1.5\ln\left(\frac{0.12-4C}{0.12}\right)h$$

Assuming the explosion occurs when the dust concentration reaches the lower flammability limit, 2.22%:

$$\text{time required}=2.02\,h$$

(b) To ensure safety, the limiting ventilation rate is that which gives a room dust concentration of 2.22% at steady state (i.e. when $dC/dt = 0$). Under this condition,

$$0=0.12-FC_{fL}$$

hence, the minimum ventilation rate, $F = 5.4\,m^3/h$.

WORKED EXAMPLE 14.5

Using the information in Table 14.W5.1, calculate the heat generated per unit volume for each of the fuels at their lower and upper flammability limits. Comment on the results of your calculations.

Table 14.W5.1 Combustion data for various fuels

Substance	Lower flammability limit (volume percent of fuel in air) (20 °C and 1 bar)	Upper flammability limit (volume percent of fuel in air)	Standard enthalpy of reaction (MJ/kmol)
Cyclohexane	1.3	8.4	−3953
Benzene	1.4	8.0	−3302
Toluene	1.27	7	−3948
Ethanol	3.3	19	−1366
Methanol	6	36.5	−764
Methane	5.2	33	−890
Propane	2.2	14	−2220

Below C_{Fstoic}:

Heat release does not occur below the lower flammability limit C_{FL} since flame propagation does not occur (Table 14.W5.2).

Table 14.W5.2 Heat generated per m^3 fuel–air mixture at C_{FL} for various fuels

Fuel	C_{FL}	$-\Delta H_c$ (MJ/kmol)	Heat generated per m^3 mixture (MJ/m^3)
Cyclohexane	0.013	3953	2.14
Benzene	0.014	3302	1.92
Toluene	0.0127	2948	2.09
Ethanol	0.033	1366	1.87
Methanol	0.06	764	1.91
Methane	0.052	890	1.93
Propane	0.022	2220	2.03

At C_F the volume of fuel in $1\,m^3$ mixture = $C_F\,m^3$

Assuming ideal gas behaviour, molar density $= \dfrac{n}{V} = \dfrac{P}{RT}$

At 20 °C and 1 bar pressure, molar density = $0.0416\,\text{kmol}/m^3$

So, the heat released per m^3 mixture =

$$(-\Delta H_c) \times C_F \times 0.0416 \qquad (14.W5.1)$$

Hence, at C_{FL}: heat released per m^3 mixture = $(-\Delta H_c) \times C_{FL} \times 0.0416$

For all the fuels listed in Table 14.W5.1, at the lower flammability limit the heat generated per unit volume of mixture is around $2\,MJ/m^3$ $(1.98 \pm 8\%)$ demonstrating that it is the heat generated per unit volume which determines whether a flame will propagate in the fuel–air mixture.

Beyond C_{Fstoic}

When C_F increases beyond C_{Fstoic} oxygen is limiting. Therefore, the first step is to determine the oxygen concentration (kmol/m³ mixture) for a given C_F.

For 1 m³ of mixture: volume of fuel = C_F and volume of air = $1 - C_F$

Taking air as 21 vol% oxygen and 79 vol% nitrogen,

volume of oxygen = 0.21 $(1 - C_F)$

Assuming ideal gas behaviour, the molar density at 20 °C and 1 bar is 0.0416 kmol/m³ (see above),

Moles of oxygen per m³ in a fuel–air mixture at C_F, n_{O_2} = 0.0416 × 0.21$[1 - C_F]$ kmol/m³

Hence, moles of fuel reacting per m³ in a fuel–air mixture at C_F

$= n_{O_2}$ × moles fuel reacting with every mole of oxygen $= n_{O_2}$ × $\dfrac{\text{stoichiometric coefficient of fuel}}{\text{stoichiometric coefficient of oxygen}}$

Hence, beyond C_{Fstoic} the heat released per m³ fuel–air mixture:

$$= 0.0416 \times 0.21[1 - C_F] \times R_{ST} \times (-\Delta H_c) \qquad (14.W5.1)$$

$$\left[\text{where,} \quad R_{ST} = \frac{\text{stoichiometric coefficient of fuel}}{\text{stoichiometric coefficient of oxygen}} \right]$$

Stoichiometric coefficients:

Cyclohexane: $C_6H_{12} + 9O_2 \rightarrow 6CO_2 + 6H_2O$, so $R_{ST} = \dfrac{1}{9} = 0.1111$

Benzene: $C_6H_6 + 7.5O_2 \rightarrow 6CO_2 + 3H_2O$, so $R_{ST} = 0.1333$

Toluene: $C_7H_8 + 9O_2 \rightarrow 7CO_2 + 4H_2O$, so $R_{ST} = 0.1111$

Ethanol: $C_2H_5OH + 3O_2 \rightarrow 2CO_2 + 3H_2O$, so $R_{ST} = 0.3333$

Methanol: $CH_3OH + 1.5O_2 \rightarrow CO_2 + 2H_2O$, so $R_{ST} = 0.6666$

Methane: $CH_4 + 2O_2 \rightarrow CO_2 + 2H_2O$, so $R_{ST} = 0.5$

Propane: $C_3H_8 + 5O_2 \rightarrow 3CO_2 + 4H_2O$, so $R_{ST} = 0.2$

Table 14.W5.3 gives the calculated values of heat generated per m³ fuel–air mixture at the upper flammability limit ($C_F = C_{FU}$) for these fuels [based on Equation (14.W5.2)].

Amongst most of the fuels, there is little variation in the value for heat generated per m³ of fuel–air mixture at the upper flammability limit (3.43 ± 4% for those excluding methanol and methane,

Table 14.W5.3 Heat generated per m³ fuel–air mixture at C_{FU} for various fuels

Fuel	C_{FU}	$-\Delta H$ (MJ/kmol)	R_{ST}	Heat generated per m³ mixture (MJ/m³)
Cyclohexane	0.084	3953	0.1111	3.51
Benzene	0.08	3302	0.1333	3.54
Toluene	0.07	2948	0.1111	3.56
Ethanol	0.19	1366	0.3333	3.22
Methanol	0.365	764	0.6666	2.83
Methane	0.33	890	0.50	2.60
Propane	0.14	2220	0.20	3.34

which have slightly lower values). Also, we note that the values at the upper flammability limit are somewhat higher than (but certainly of the same order of magnitude as) the values at the lower flammability limit. These differences are likely to be due to the different physical properties (conductivity, specific heat capacity, for example) of the fuel–air mixture at *low fuel concentrations* compared with the physical properties at *higher fuel concentrations*.

WORKED EXAMPLE 14.6

Based on the information for methane in Table 14.W5.1, produce a plot of heat released per m³ of mixture as a function of fuel concentration for methane–air mixtures at atmospheric pressure and 20 °C. In the light of this, explain why fuels have an upper flammability limit.

Solution

For $C_F < C_{FL}$ and $C_F > C_{FU}$ no heat will be released since no combustion takes place.

For $C_{FL} < C_F < C_{Fstoic}$ (the fuel-limiting range) the heat released per m³ of fuel–air mixture will be described by (see Example 14.5):

$$(-\Delta H_C) \times C_F \times 0.0416 \qquad (14.W6.1)$$

For methane, the heat released per m³ of fuel–air mixture becomes: 37.02 C_F MJ/m³.

For $C_{Fstoic} < C_F < C_{FU}$ (the oxygen-limiting range) the heat released per m³ of fuel–air mixture will be described by (see Example 14.5):

$$0.0416 \times 0.21(1 - C_F) \times R_{ST} \times (-\Delta H_C) \qquad (14.W6.2)$$

For methane, the heat release per m³ of fuel–air mixture becomes: 3.89 $(1 - C_F)$ MJ/m³

Determine C_{Fstoic}:

$$\text{For } 1\,\text{m}^3 : C_F = \frac{\text{volume of fuel}}{\text{total volume of mixture}} = \frac{\text{vol fuel}}{\text{vol fuel} + \text{vol O}_2 + \text{vol N}_2}$$

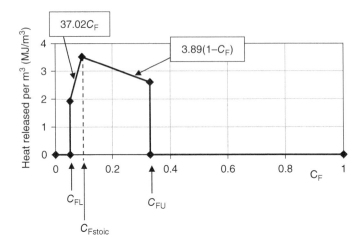

Figure 14.W6.1 The heat released per m^3 of methane–air mixtures as a function of methane concentration C_F

Taking air as 21 mol% oxygen and 79 mol% nitrogen,

$$C_F = \frac{\text{vol fuel}}{\text{vol fuel} + \text{vol O}_2 + \text{vol N}_2} = \frac{\text{vol fuel}}{\text{vol fuel} + 4.762\,\text{vol O}_2}$$

At stoichiometric conditions (assuming ideal gas behaviour),

$$\frac{\text{vol fuel}}{\text{vol O}_2} = \frac{\text{mol fuel}}{\text{mol O}_2} = R_{ST}$$

$$\text{So}: C_{Fstoic} = \frac{\text{vol fuel}}{\text{vol fuel} + 4.762\left(\frac{\text{vol fuel}}{R_{ST}}\right)} = \frac{R_{ST}}{R_{ST} + 4.762}$$

For methane this gives $C_{Fstoic} = 0.095$

Plotting these gives Figure 14.W6.1.

TEST YOURSELF ON FIRE AND EXPLOSION HAZARDS

14.1 List five process industries which process combustible particulate materials and in which there is, therefore, the potential for dust explosion.

14.2 What is meant by the terms *lower flammability limit* and *upper flammability limit*?

14.3 The lower flammability limits for benzene, methanol and methane are 1.4, 6.0 and 5.2 vol%, respectively. However, for each of these fuels, the heat generated per unit volume of fuel–air mixture at the lower flammability limit is approximately 1.92 MJ/m^3. Explain the significance of these statements.

14.4   Explain why, for a suspension of combustible dust in air, the likelihood of flame propagation and explosion increases with decreasing particle size.

14.5   List and define five explosion characteristics that are determined experimentally in the 20-litre sphere test apparatus.

14.6   In decreasing order of desirability, list five approaches to reducing the risk of dust cloud explosion.

14.7   Explain why a policy of eliminating ignition sources from a process plant handling combustible powders cannot be relied upon as the sole measure taken to prevent a dust explosion.

14.8   What factors must be considered when designing a vent to protect a vessel from the effects of a dust explosion?

14.2   HEALTH EFFECTS OF FINE POWDERS

14.2.1   Introduction

When we think of the health effects of fine powders, we might think first of the negative effects related to inhalation of particles which have acute or chronic effects on the lungs and the body, e.g. silica, coal dust. However, with the invention of the metered dose inhaler, the widespread use of fine particle drugs delivered directly to the lungs for the treatment of asthma began. Pulmonary delivery, as this method is called, is now a widespread method of drug delivery. In this section, therefore, we will look at both the negative and positive health effects of fine powders, beginning with a description of the respiratory system and an analysis of the interaction of the system with fine particles.

14.2.2   The Human Respiratory System

Operation

The requirement for the human body to exchange oxygen and carbon dioxide with the environment is fulfilled by the respiratory system. Air is delivered to the lungs via the nose and mouth, the *pharynx* and *larynx* and the *trachea*. Beyond the trachea, the single airway branches to produce the two main *bronchi* which deliver air to each of the two lungs. Within each lung the bronchi divide repeatedly to produce many smaller airways called *bronchioles*, giving an inverted tree-like structure (Figure 14.6). The upper bronchioles and bronchi are lined with specialized cells some of which secrete mucus whilst others have hairs or cilia which beat causing an upward flow of mucus along the walls. The bronchioles lead to alveolar ducts (or *alveoli*) which terminate in *alveolar sacs*. In the adult male there are about 300 million alveoli, each approximately 0.2 mm in diameter. The walls of the alveoli have a rich supply of blood vessels. Oxygen from the air diffuses across a thin membrane around the alveolar sacs into the blood and carbon dioxide from the blood diffuses in the reverse direction.

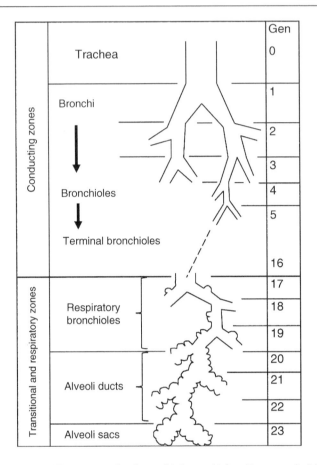

Figure 14.6   Human tracheobronchial tree (After Zeng et al., 2001)

Air is forced in and out of the lungs by the movement of the *diaphragm* muscle beneath the *thoracic cage* or chest cavity. The lungs are sealed within the chest cavity, so that as the diaphragm muscle moves down, the pressure within the cavity falls below atmospheric, causing air to be drawn into the lungs. The lungs expand, filling the chest cavity. As the diaphragm muscle moves up, the lungs are squeezed and the air within them is expelled to the environment.

Dimensions and flows

A systematic description commonly used is that of Weibel (1963). In this treatment the respiratory system is considered as a sequence of regular branches or dichotomies. The tree starts with the trachea, as *generation* 0, leading to the two main bronchi, as 'generation' 1. Within the lungs the branching continues until generation 23 representing the alveolar sacs. According to Weibel's model, at generation n there are 2^n tubes. So, for example, at generation 16 (the terminal bronchioles) there are 65 536 tubes of diameter 0.6 mm.

Table 14.3 Characteristics of the respiratory tract, based on a steady flow of 60 litre/min (Zeng et al., 2001)

Part	Number	Diameter (mm)	Length (mm)	Typical air velocity (m/s)	Typical residence time (s)
Nasal airways		5–9		9	
Mouth	1	20	70	3.2	0.022
Pharynx	1	30	30	1.4	0.021
Trachea	1	18	120	4.4	0.027
Two main bronchi	2	13	37	3.7	0.01
Lobar bronchi	5	8	28	4.0	0.007
Segmental bronchi	18	5	60	2.9	0.021
Bronchioles	504	2	20	0.6	0.032
Secondary bronchioles	3024	1	15	0.4	0.036
Terminal bronchioles	12 100	0.7	5	0.2	0.023
Alveolar ducts	8.5×10^5	0.8	1	0.0023	0.44
Alveolar sacs	2.1×10^7	0.3	0.5	0.0007	0.75
Alveoli	5.3×10^8	0.15	0.15	0.00004	4

Table 14.3 is based more on direct measurement and gives the typical dimensions of the components of the airways together with typical air velocities and residence times.

The nasal airways are quite tortuous and they change diameter several times in their path. The lower section is lined with hairs. The narrowest section is the nasal valve, which has a cross-sectional area of 20–60 mm^2 (equivalent to 5–9 mm diameter) and typically accounts for 50% of the resistance to flow in the nasal airways. Typical air flow rates in the adult nasal airways range from 180 ml/s during normal breathing to 1000 ml/s during a strong sniff. This gives velocities in the nasal airways as high as 9 m/s during normal breathing and 50 m/s during a strong sniff.

The mouth leading to the pharynx and larynx presents a far smoother path for the flow of air and offers lower resistance. However, whilst moving through the pharynx and larynx the air stream is subject to some sharp changes in direction. A typical air velocity in the mouth during normal breathing is 3 m/s.

We see from Table 14.3 that due mainly to the continuous branching, the air velocity in the airways decreases at the start of the bronchioles. The result is that the residence time in the different sections of the airways is of the same order until the alveolar region is reached, where the residence time increases significantly.

14.2.3 Interaction of Fine Powders with the Respiratory System

Airborne particles entering the respiratory tract will be deposited on the surface of any part of the airways with which they come into contact. Because the surfaces of the airways are always moist, there is a negligible chance of a particle becoming re-entrained in the air after once contacting a surface. Particles which do not make contact with the airway surfaces will be exhaled. The deposition of particles in the airways is a complicated process. Zeng et al. (2001) identify five possible mechanisms contributing to the deposition of particles carried into the respiratory tract. These are sedimentation, inertial impaction, diffusion, interception and electrostatic precipitation, and are, as might be expected, the same mechanisms responsible for particle collection in filters (see Chapter 8). After introducing each mechanism, their relative importance in each part of the respiratory tract will be discussed.

Sedimentation

Particles sediment in air under the influence of gravity. As discussed in Chapter 3, a particle in a static fluid of infinite extent will accelerate from rest under the influence of the gravity and buoyancy forces (which are constant) and the fluid drag force, which increases with relative velocity between the particle and the fluid. When the upward-acting drag and buoyancy forces balance the gravity force, a constant, terminal velocity is reached. For particles of the size of relevance here (less than 40 µm), falling in air, the drag force will be given by Stokes' law [Equation (3.3)] and the terminal velocity will be given by Equation (3.15):

$$U_{\mathrm{T}} = \left[\frac{x^2 \left(\rho_{\mathrm{p}} - \rho_{\mathrm{f}} \right) g}{18 \mu} \right] \tag{3.15}$$

Particles in the size range under consideration accelerate rapidly to their terminal velocity in air. For example, a 40 µm particle of density 1000 kg/m^3 accelerates to 99% of its terminal velocity (47 mm/s) in 20 ms and travels a distance less than 1 mm in doing so. So, for these particles in air, we can take the sedimentation velocity as equivalent to the terminal velocity. The above analysis is relevant to particles falling in stagnant air or air in laminar flow. If the air flow becomes turbulent, as it may do in certain airways at higher breathing rates, the characteristic velocity fluctuations increase the tendency of particles to deposit.

Inertial impaction

Inhaled air follows a tortuous path as it makes its way through the airways. Whether the airborne particles follow the path taken by the air at each turn depends on the balance between the force required to cause the particle to change direction and the fluid drag available to provide that force. If the drag force is sufficient to cause the required change in direction, the particle will follow the gas and will not be deposited.

Consider a particle of diameter x travelling at a velocity U_P in air of viscosity μ within an airway of diameter D. Let us consider the extreme case where this particle is required to make a 90° change in direction. The necessary force F_{R} is that required to cause the particle to stop and then be re-accelerated to velocity U_P. The distance within which the particle must stop is of the order of the airway diameter, D, and so:

work done = force × distance = particle kinetic energy

$$F_{\mathrm{R}} D = 2 \left(\frac{1}{2} m U_{\mathrm{p}}^2 \right) \tag{14.2}$$

(The 2 on the right-hand side is there because we need to re-accelerate the particle.)

$$\text{Therefore, required force, } F_R = \frac{\pi}{6}x^3\rho_p U_p^2 \frac{1}{D} \tag{14.3}$$

The force available is the fluid drag. Stokes' law [Equation (3.3)] will apply to the particles under consideration and so:

$$\text{Available drag force, } F_D = 3\pi\mu x U_p$$

The rationale for using U_p as the relevant relative velocity is that this represents the maximum relative velocity that would be experienced by the particle as it attempts to continue in a straight line.

The ratio of the force required F_r to the force available F_d is then:

$$\frac{x^2\rho_p U_p}{18\mu D} \tag{14.4}$$

The greater the value of this ratio is above unity, the greater will be the tendency for particles to impact with the airway walls and so deposit. The further the value is below unity, the greater will be the tendency for the particles to follow the gas. This dimensionless ratio is known as the Stokes number:

$$Stk = \frac{x^2\rho_p U_p}{18\mu D} \tag{14.5}$$

(We also met the Stokes number in Chapter 8, where it was one of the dimensionless numbers used in the scale-up of gas cyclones for the separation of particles from gases, and for inertial collection of particles in filters. There are obvious similarities between the collection of particles in a gas cyclone or filter and 'collection' of particles in the airways of the respiratory system. Note that this Stokes number is not to be confused with the one we met in Chapter 12, describing collision between granules.)

Diffusion

The motion of smaller airborne particles is influenced by the bombardment of air molecules. This is known as Brownian motion and results in a random motion of the particles (see also Chapter 5). Since the motion is random, the particle displacement is expressed as a root mean square:

$$\text{Root mean square displacement in time } t, L = \sqrt{6\alpha t} \tag{14.6}$$

where α is the diffusion coefficient given by:

$$\alpha = \frac{kT}{3\pi\mu x} \tag{14.7}$$

for a particle of diameter x in a fluid of viscosity g at a temperature T. k is the Boltzmann constant, which has the value 1.3805×10^{-23} J/K.

Interception

Interception is the deposition of the particle by reason of its size and shape compared with the size of the airway.

Electrostatic precipitation

Particles and droplets may become electrostatically charged, particularly during the dispersion stage, by interaction with each other or with nearby surfaces. It has been speculated that charged particles could induce opposite charges on the walls of some airways, resulting in the particles being attracted to the walls and deposited.

Relative importance of these mechanisms within the respiratory tract

Under the humid conditions found with the respiratory tract, it is most likely that any charges on particles will be quickly dissipated, and so electrostatic precipitation is unlikely to play a significant role in particle deposition anywhere in the airways.

Deposition by interception is also not significant, since the particles of interest are far smaller than the airways.

We will now consider the relative importance of the other three mechanisms, *sedimentation, inertial impaction* and *diffusion,* in the various parts of the respiratory tract. First, it is interesting to compare, for different particle sizes, the displacements due to sedimentation and diffusion under the conditions typically found within the respiratory tract. Table 14.4 makes this comparison for particles of density $1000\,kg/m^3$ in air at a temperature of $30\,°C$.

From Table 14.4, we can see that diffusion does not become significant compared with sedimentation until particle size falls below $1\,\mu m$. In the case where Brownian motion were to act downwards with the sedimentation, the minimum displacement would occur

Table 14.4   A comparison of displacements due to sedimentation and diffusion for particles of density $1000\,kg/m^3$ in air at $30\,°C$, with density $1.21\,kg/m^3$ and viscosity $1.81 \times 10^{-5}\,Pa\,s$

Particle diameter (μm)	Particle terminal velocity (m/s)	Displacement in 1 s due to sedimentation (μm)	Displacement in 1 s due to Brownian motion (μm)
50	7.5×10^{-2}	75 000	1.7
30	2.7×10^{-2}	27 000	2.2
20	1.2×10^{-2}	12 000	2.7
10	3.0×10^{-3}	3000	3.8
5	7.5×10^{-4}	750	5.4
2	1.2×10^{-4}	120	8.5
1	3.0×10^{-5}	30	12.0
0.5	7.5×10^{-6}	7.5	17.0
0.3	2.7×10^{-6}	2.7	21.9
0.2	1.2×10^{-6}	1.2	26.8
0.1	3.0×10^{-7}	0.3	37.9

Table 14.5 Time required for particles of different diameters to be displaced a distance equivalent to the airway diameter, due to the combined effect of sedimentation and diffusion

Airway component	Diameter (mm)	Typical air residence time (s)	Required particle component residence time (s)		
			5 μm	1 μm	0.5 μm
Two main bronchi	13	0.01	17.2	309	580
Bronchioles	2	0.032	2.6	48	90
Secondary bronchioles	1	0.036	1.3	24	45
Terminal bronchioles	0.7	0.023	1	17	30
Alveolar ducts	0.8	0.44	1	19	35
Alveolar sacs	0.3	0.75	0.4	7	13
Alveoli	0.15	4	0.2	3.5	6.6

for particles of around 0.5 μm in diameter. This would suggest that particles around this diameter would have the least chance of being deposited by these mechanisms. We see that particles smaller than about 10 μm would require significant residence times to travel far enough to be deposited. For example, in Table 14.5, we look at the time required for particles to be displaced a distance equivalent to the airway diameter, due to the combined effect of sedimentation and diffusion. In practice, sedimentation becomes a significant mechanism for deposition only in the smaller airways and in the alveolar region, where air velocities are low, airway dimensions are small and air residence times are relatively high.

Now we will consider the importance of inertial impaction as a mechanism for causing deposition. In Table 14.6 we have calculated [using Equation (14.5)] the Stokes numbers for particles of different sizes in the various areas of the respiratory duct, based on the information provided in Table 14.3. This calculation assumes that the particle velocity is the same as the air velocity in the relevant part of the respiratory tract.

Table 14.6 Stokes number for a range of different size particles, and typical Reynolds numbers in each section of the respiratory tract

Region	Stokes number for particles of different sizes						Re flow
	1 μm	5 μm	10 μm	20 μm	50 μm	100 μm	
Nose	3.1×10^{-3}	7.7×10^{-4}	0.31	1.2	7.7	31	5415
Mouth	4.9×10^{-4}	1.2×10^{-4}	0.05	0.2	1.2	5	4254
Pharynx	1.5×10^{-4}	3.7×10^{-5}	0.01	0.1	0.4	1	2865
Trachea	7.6×10^{-4}	1.9×10^{-4}	0.08	0.3	1.9	8	5348
2 main bronchi	8.7×10^{-4}	2.2×10^{-4}	0.09	0.3	2.2	9	3216
Lobar bronchi	1.5×10^{-3}	3.8×10^{-4}	0.15	0.6	3.8	15	2139
Segmental bronchi	1.8×10^{-3}	4.4×10^{-4}	0.18	0.7	4.4	18	955
Bronchioles	9.6×10^{-4}	2.4×10^{-4}	0.10	0.4	2.4	10	84
Secondary bronchioles	1.3×10^{-3}	3.2×10^{-4}	0.13	0.5	3.2	13	28
Terminal bronchioles	9.5×10^{-4}	2.4×10^{-4}	0.10	0.4	2.4	10	10
Alveolar ducts	8.7×10^{-6}	2.2×10^{-6}	0.0009	0.003	0.022	0	0
Alveolar sacs	7.7×10^{-6}	1.7×10^{-6}	0.0007	0.003	0.017	0	0
Alveoli	7.7×10^{-7}	1.9×10^{-7}	0.0001	0.003	0.002	0	0

Recalling that the larger the value of the Stokes number is above unity, the greater will be the tendency for particles to impact with the airway walls and so deposit, we see that particles as small as 50 µm are likely to be deposited in the mouth or nose, pharynx, larynx and trachea. These figures are for steady breathing; higher rates will give higher Stokes numbers and so would cause smaller particles to be deposited at a given part of the tract.

Whilst moving through the mouth or nose, pharynx and larynx, the air stream is subject to some sharp changes in direction. This induces turbulence and instabilities which increase the chances of particle deposition by inertial impaction.

In summary, inertial impaction is responsible for the deposition of the larger airborne particles and this occurs mainly in the upper airways. In practice, therefore, we find that only those particles smaller than about 10 µm will travel beyond the main bronchi. Such particles have a decreasing tendency to be deposited by inertial impaction the further they travel into the lungs but are more likely to be deposited by the action of sedimentation and diffusion as they reach the smaller airways and the alveolar region, where air velocities are low, airway dimensions are small and air residence times are relatively high.

These conclusions are very similar to those introduced in the discussion on gas–solid separation in Chapter 5, and particularly with Figure 5.16, which shows schematically the way that the dominant mechanism of collection changes from inertial to diffusional as particle size decreases, with a minimum collection at a size of order 1 µm.

14.2.4 Pulmonary Delivery of Drugs

Target areas: The delivery of drugs for the treatment of lung diseases (asthma, bronchitis, etc.) via aerosols direct to the lungs is attractive for a number of reasons. Compared with other methods of drug delivery (oral, injection), pulmonary delivery gives a rapid, predictable onset of drug action with minimum dose and minimum side effects. The adult human lung has a very large surface area for drug absorption (typically 120 m^2). The alveoli wall membrane is permeable to many drug molecules and has rich blood supply. This makes *aerosol delivery* attractive for the delivery of drugs for the treatment of other illnesses. (As discussed in Chapter 5, aerosols are suspensions of liquid drops or solid particles in air or other gas.) In pulmonary delivery using aerosols, the prime targets for the aerosol particles are the alveoli, where the conditions for absorption are best. In practice, particles larger than about 2 µm rarely reach the alveoli and particles smaller than about 0.5 µm may reach the alveoli but are exhaled without deposition (Smith and Bernstein, 1996; see also analysis above).

The technologies in common use for the delivery of drugs to the lungs are described here.

Nebulizers – devices which generate a fine mist of aqueous drug solution – have been used in hospitals for over 100 years, although the modern nebulizer bears little resemblance to the earlier devices. The modern nebulizer (Figure 14.7) is used where patients cannot use other devices or when large doses of drugs are required. Portable nebulizers have been developed, although these devices still require a source of power for air compression.

The *metered-dose inhaler (MDI)* has been used widely by asthma sufferers for many years – despite its disadvantages (Zeng et al., 2001). In the MDI, the drug, which is dispersed or dissolved in a liquid propellant, is held in a small, pressurized container (Figure 14.8). Each activation of the device releases a metered quantity of propellant

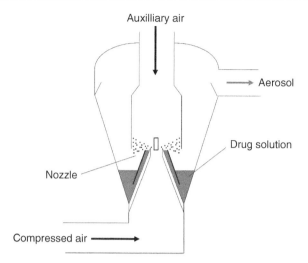

Figure 14.7 Schematic representation of a nebulizer. Compressed air expands as it leaves the nozzle. This causes reduced pressure which induces the drug solution to flow up and out of the nozzle where it is atomized by contact with the air stream (After Dalby et al., 1996).

carrying a predetermined dose of drug. The liquid in the droplets rapidly vaporizes leaving solid particles. The high velocity of discharge from the container means that many particles impact on the back of the throat and are therefore ineffective. Another disadvantage is that the patient, who may be in a stressed condition, must coordinate the activation of the MDI with *inspiration* (taking an inward breath).

   The third common type of device for pulmonary drug delivery is the *dry powder inhaler (DPI)*. This device now has many forms, some of which appear remarkably simple in their design (Figure 14.9). Particular effort has gone into the formulation of the powder. In most cases the drug is of the order a few micrometres in size and adheres – usually by natural forces – to inert "carrier particles", which are much larger in size (100 μm or

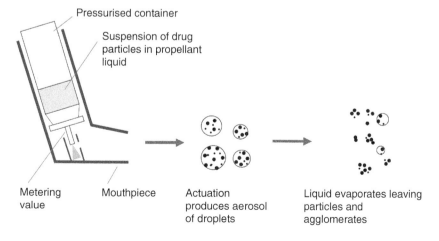

Figure 14.8   Metered-dose inhaler. After Dalby et al., 1996.

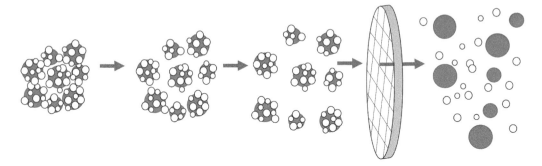

Figure 14.9 Carrier particles in a dry powder inhaler. The powder may be initially loosely compacted, but by the shearing action of the air stream and impingement on a screen the agglomerates are dispersed and the drug particles dislodged from the carrier particles. After Dalby et al., 1996.

more). The carrier particles are required for two reasons: because the required quantity of drug is so small that it would be difficult to package; and because such a fine powder would be very cohesive – therefore difficult to handle and unlikely to easily disperse in air. The intention is that the carrier particles should be left in the device, or in the back of the mouth or the upper airways, and that the drug particles should detach during inspiration and travel to the lower airways before deposition.

14.2.5   Harmful Effects of Fine Powders

Particles which find their way into the lungs can also have a negative effect on health. In the workplace and in our everyday lives, exposure to fine particle aerosols should be avoided. History has shown that exposure to coal dust, silica dust and asbestos dust, for example, can have disastrous effects on the health of workers, even many years after exposure. Many other workplace dusts, less well known, have been found to have negative health effects.

As discussed above, particles smaller than 10 μm present the greatest risk, since they can penetrate deep into the lungs – offering the greatest chance of chemical absorption into the blood as well as physical interaction with the lungs. Less is known about the effects of ultrafine particles below 1 μm in size, but it has been observed that some particles below 200 nm (0.2 μm) can pass through the blood–brain barrier, which is both a cause for concern and an opportunity for devising appropriate therapeutic strategies. Nano-sized drug formulations are in increasing use for targeted delivery of relatively insoluble drugs to various parts of the body.

If a dust hazard is suspected in the workplace, the first step is to monitor the working environment to determine the exposure of the worker. One of the better methods of achieving this is for the worker to wear a portable sampling device, which gives a measure of the type of particles and their size distribution in the air immediately around the worker. Such devices usually sample at a typical respiration rate and velocity and some devices are designed to capture directly only the *respirable* particles (particles capable of reaching the alveoli) or the *inhalable* particles (particles capable of being inhaled).

The second step in dealing with the potential dust hazard is for the results of the monitoring process to be compared with the accepted workplace standards for the particulate materials in question. If the concentration of respirable particles is found to exceed the accepted limit for that material, then we proceed to the third stage, namely control.

When dealing with any hazard, there is a hierarchy of control measures that may be put in place. In the modern workplace, the aim is to produce a safe environment rather than a safe person. To understand what is meant by this, consider the hierarchy of control measures:

- Specification.

- Substitution.

- Containment.

- Ventilation.

- Reduced time of exposure.

- Protective equipment.

In designing a process, an engineer or scientist should aim for control measures at the top of the hierarchy. Only as a last resort should the measures lower down in the hierarchy be used.

Specification: Devising an alternative process which does not include this hazard. In the case of a dust hazard, this may mean using a completely different process: for example using wet milling rather than dry milling. Granulation may be an option – with loose granules, the large surface area provided by fine particles can still be made available, but without the associated dust hazard.

Substitution: Replacing the hazardous material with a non-hazardous material.

Containment: Designing the process using equipment which ensures that hazardous materials are contained and do not, under normal operation, escape into the environment.

An example is using pneumatic transport rather than conveyor belts or other mechanical conveyors in transporting powders within the workplace, or using fully enclosed conveyor belts.

Ventilation: Accepting that the hazardous material is present in the workplace environment and creating air flows to draw the material away from workers or reduce its concentration in the environment.

Reduced time of exposure: Accepting that the hazardous material is present in the workplace environment and reducing the time spent by each worker in that environment.

Protective equipment: Accepting that the worker must work in an environment where the hazardous material is present and providing suitable protective equipment for the worker to wear. Examples for controlling dust hazard (in order of decreasing efficacy) are: air-line helmets – clean air is provided under pressure via flexible tubing to a full headset worn by the worker; positive pressure sets – a pump and filter worn by the worker provides air to a headset which may partially or fully enclose the worker's head; airstream helmets – a pump and filter fitted to a hard hat with a visor such that the filtered air stream is blown across the worker's face; ori-nasal respirators – a well-fitting rubber or plastic mask covering the nose and mouth and fitted with efficient filters suitable for the material in question; disposable face masks – masks made of filter material covering the nose and mouth are usually not well-fitting.

TEST YOURSELF ON HEALTH EFFECTS OF FINE POWDERS

14.9 Make a diagrammatic sketch of the human lung, labelling the following regions: alveoli sacs, bronchi, trachea, respiratory bronchioles.

14.10 In the human respiratory systems what are the typical diameters and air velocities in the trachea, terminal bronchioles, main bronchi?

14.11 List the mechanisms by which inhaled particles may become deposited on the walls of the airways in the respiratory systems. Which mechanism dominates for particles smaller than around 10 μm in the upper airways, and why?

14.12 What is the Stokes number and what does it tell us about the likelihood of particles being deposited on the walls on the respiratory system?

14.13 In which part of the respiratory system is sedimentation important as a mechanism for depositing particles?

14.14 For pulmonary delivery of drugs, what size range of particles is desirable and why?

14.15 Describe the construction and operation of three types of devices for the delivery of respirable drugs.

14.16 What is a carrier particle and why is it needed?

14.17 What steps should be taken to determine whether dust in the workplace has the potential to cause a health hazard through inhalation?

14.18 Explain what is meant by the hierarchy of control measures when applied to the control of a dust hazard.

EXERCISES – FIRE AND EXPLOSION HAZARDS

14.1 It is proposed to protect a section of the duct used for pneumatically transporting a food product in powder form in air by adding a stream of carbon dioxide. The air flow rate in the present system is $3 \, m^3/s$ and the air carries 2% powder by volume. If the minimum oxygen for combustion (by replacement of oxygen with carbon dioxide) of the powder is 13% by volume, what is the minimum flow rate of carbon dioxide which must be added to ensure safe operation?

[Answer: $1.75 \, m^3/s$.]

14.2 A combustible dust has a lower flammability limit in air at 20 °C of 1.2% by volume. A dust extraction system operating at $3 \, m^3/s$ is found to have a dust concentration of 1.5% by volume. What minimum flow rate of additional air must be introduced to ensure safe operation?

[Answer: $0.75 \, m^3/s$.]

14.3 A flammable pharmaceutical powder suspended in air at a concentration within the flammable limits and with an oxygen concentration above the minimum oxygen for combustion flows at $40 \, m/s$ through a tube whose wall temperature is greater than the measured ignition temperature of the dust. Give reasons why ignition does not necessarily occur.

14.4 A fine flammable plastic powder is leaking from a pressurized container at a rate of $0.5 \, litre/min$ into another vessel of volume $2 \, m^3$ and forming a suspension in the air in the vessel. The minimum explosible concentration of the dust in air at room temperature is 1.8% by volume. Stating all assumptions, estimate:

(a) The delay from the start of the leak before explosion occurs if there is no ventilation.

(b) The delay from the start of the leak before explosion occurs if the air ventilation rate in the second vessel is $0.5 \, m^3/h$.

(c) The minimum safe ventilation rate under these circumstances.

[Answer: (a) 1.2 h; (b) 1.43 h; (c) $1.67 \, m^3/h$.]

14.5 Using the information in Table 14.E5.1, calculate the heat released per unit volume for each of the fuels at their lower and upper flammability limits. Comment on the results of your calculations.

Table 14.E5.1 Combustion data for various fuels

Substance	Lower flammability limit (vol% of fuel in the air) (20 °C and 1 bar)	Upper flammability limit (vol% of fuel in air)	Standard enthalpy of reaction (MJ/kmol)
Cyclohexane	1.3	8.4	−3953
Toluene	1.27	7	−3948
Ethane	3	12.4	−1560
Propane	2.2	14	−2220
Butane	1.8	8.4	−2879

14.6 Based on the information for propane in Table 14.E5.1, produce a plot of heat released per m^3 of mixture as a function of fuel concentration for propane–air mixtures at atmospheric pressure and 20 °C. In the light of this, explain why fuels have an upper flammability limit.

EXERCISES – HEALTH EFFECTS OF FINE POWDERS

14.7 Determine, by calculation, the likely fate of a 20 µm particle of density 2000 kg/m³ suspended in the air inhaled by a human at a rate giving rise to the following velocities in parts of the respiratory system:

Part	Number	Diameter (mm)	Length (mm)	Typical air velocity (m/s)	Typical residence time (s)
Mouth	1	20	70	3.2	0.022
Pharynx	1	30	30	1.4	0.021
Trachea	1	18	120	4.4	0.027
Two main bronchi	2	13	37	3.7	0.01

14.8 Given the following information, in which region of the respiratory tract is a 3 µm particle of density 1500 kg/m³ most likely to be deposited and by which mechanism? Support your conclusion by calculation.

Part	Diameter (mm)	Length (mm)	Typical air velocity (m/s)	Typical residence time (s)
Trachea	18	120	4.4	0.027
Bronchioles	2	20	0.6	0.032
Terminal bronchioles	0.7	5	0.2	0.023
Alveolar ducts	0.8	1	0.0023	0.44
Alveoli	0.15	0.15	0.00004	4

14.9 Compare the Stokes numbers for 2, 5, 10 and 40 µm particles of density 1200 kg/m³ in air passing through the nose. What conclusions do you draw regarding the likelihood of deposition of these particles in the nose?

Data:
Characteristic velocity in the nose: 9 m/s
Characteristic diameter of the airway in the nose: 6 mm
Viscosity of air: 1.81×10^{-5} Pa s
Density of air: 1.21 kg/m³

14.10 Carrier particles are used in dry powder inhalers. What is a carrier particle? What is the role of a carrier particle? Why are carrier particles needed in these inhalers?

14.11 With reference to the control of dusts as a health hazard, explain what is meant by the *hierarchy of controls*.

14.12 The required dose of a particulate drug of particle size 3 µm and particle density 1000 kg/m³ is 10 µg. Estimate the number of particles in this dose and the volume occupied by the dose, assuming a void fraction of 0.6.

[Answer: 7×10^5; 0.25 mm³.]

Notation

a	acceleration	m/s^2
a	surface area of particles per unit bed volume	m^2/m^3
A	area	m^2
Ar	Archimedes number $\left[Ar = \dfrac{\rho_f(\rho_p - \rho_f)gx^3}{\mu^2}\right]$	—
b	dimension	m
$B_{nuc}(v)$	rate of granule growth by nucleation	$\text{m}^{-6}\,\text{s}^{-1}$
c	concentration	kg/m^3
c	contact radius	m
c	particle suspension concentration	kg/m^3
C	concentration	$\text{m}^3/\text{m}^3 \text{ or kg/m}^3$
C_D	drag coefficient	—
C_{fL}	lower flammability limit	$\text{m}^3 \text{ fuel/m}^3$
C_{fU}	upper flammability limit	$\text{m}^3 \text{ fuel/m}^3$
C_g	specific heat capacity of gas	J/kg K
c_{JKR}	contact radius according to JKR theory	m
C_p	specific heat capacity	J/kg K
D	diameter	m
D	diffusion coefficient for molecular diffusion	m^2/s
e	coefficient of restitution	—
E	Young's modulus	Pa
Eu	Euler number	—
f	fractional supersaturation	difference in chemical potential
f^*	friction factor for packed bed flow	—
F	force	N
F_D	total drag force	N
ff	hopper flow factor	—
f_g	Fanning friction factor	—
F_{gw}	gas-to-wall friction force per unit volume of pipe	Pa/m
f_p	solids-to-wall friction factor	—
F_{pw}	solids-to-wall friction force per unit volume of pipe	Pa/m
F_s	drag force due to shear	N

Introduction to Particle Technology, Third Edition. Martin Rhodes and Jonathan Seville.
© 2024 John Wiley & Sons Ltd. Published 2024 by John Wiley & Sons Ltd.
Website: www.wiley.com/go/rhodes/particle3e

F_{vw}	van der Waals force	N
g	acceleration due to gravity	m/s^2
G	shear stiffness	Pa
G	solids mass flux = M_p/A	kg/m$^2\cdot$s
$G(x)$	grade efficiency	—
h	boundary layer thickness	m
h	heat transfer coefficient	W/m$^2\cdot$K
H	height	m
h_a	measure of roughness of granule surface	m
h_{gc}	gas convective heat transfer coefficient	W/m$^2\cdot$K
h_{gp}	gas-to-particle heat transfer coefficient	W/m$^2\cdot$K
h_{max}	maximum bed-to-surface heat transfer coefficient	W/m$^2\cdot$K
H_{mf}	height of bed at incipient fluidization	m
h_{pc}	particle convective heat transfer coefficient	W/m$^2\cdot$K
h_r	radiative heat transfer coefficient	W/m$^2\cdot$K
j	reaction order	—
J	nucleation rate	number/s m^3
k	Janssen constant	—
k	reaction rate constant per unit volume of solids	mol/m$^3\cdot$s
K_C	interphase mass transfer coefficient per unit bubble volume	s^{-1}
k_g	gas conductivity	W/m$\cdot$K
$K_{i\infty}^*$	elutriation constant for size range x_i above TDH	kg/m$^2\cdot$s
k_n	normal stiffness, in DEM	N/m
k_t	tangential stiffness, in DEM	N/m
K_{ih}^*	elutriation constant for size range x_i at height h above distributor	kg/m$^2\cdot$s
L_{max}	maximum Feret length	m
L_{min}	minimum Feret length	m
L_{mean}	mean Feret length	m
m	mass	kg
M	mass of solids	kg
m_{B_i}	mass fraction of size range x_i	—
n	exponent in Richardson–Zaki equation	—
N	number	—
N_p	the number of particles in a given system/simulation	—
$N(t)$	total number of granules in the system at time t	—
Nu	Nusselt number ($h_{gp}x/k_g$)	—
$n(v, t)$	number density of granule volume v at time t	m^{-6}
p	pressure	Pa
Pr	Prandtl number ($C_g\mu/k_g$)	—
p_s	pressure difference	Pa
PVC	pigment-volume concentration	—
Q	volume flow rate	m^3/s
r	radial dimension	m
R	particle radius	m
Re^*	Reynolds number for packed bed flow [Equation (6.12)]	—
Re_{mf}	Reynolds number at incipient fluidization ($U_{mf}x_{Sv}\rho_f/\mu$)	—

Re_p	single particle Reynolds number	—
R_G	mass-based growth rate	kg/s
R_i	rate of entrainment of solids in size range x_i	kg/s
r_1, r_2	radii of curvature of two liquid surfaces	m
s	granule saturation	—
S	total surface area of population of particles	m^2
S	sphericity: $S = \dfrac{2\sqrt{\pi A}}{P}$	—
S_R	roundness of a particle: $S_R = \dfrac{4\pi A}{P^2}$	—
S_v	surface area of particles per unit volume of particles	m^2/m^3
Stk	Stokes number	—
t	time	s
T	reaction temperature	K
T_g	gas temperature	K
T_{ig}	ignition temperature	K
U	superficial gas velocity ($= Q_f/A$)	m/s
U_B	mean bubble rise velocity	m/s
U_{ch}	choking velocity (superficial)	m/s
U_{mb}	superficial gas velocity at minimum bubbling	m/s
U_{mf}	superficial gas velocity at minimum fluidization	m/s
U_{ms}	minimum velocity for slugging	m/s
U_p	actual particle or solids velocity	m/s
U_r	radial gas velocity	m/s
U_R	radial gas velocity at cyclone wall	m/s
U_{rel}	relative velocity (= $U_{slip} = U_f = U_P$)	m/s
U_{salt}	saltation velocity (superficial)	m/s
U_{slip}	slip velocity ($U_f - U_p$)	m/s
U_T	single particle terminal velocity	m/s
U_T^*	dimensionless terminal velocity	—
U_{Ti}	single particle terminal velocity for particle size x_i	m/s
U_θ	particle tangential velocity at radius r	m/s
v	velocity	m/s
v_Y	yield velocity	m/s
w	average granule volume defined in Equation (12.19)	m^3
w	liquid level on dry mass basis	g liquid/g dry powder
W	force	N
x	granule or particle diameter	m
x^*	dimensionless particle size	—
x_{crit}	critical particle size for separation [Equation (8.20)]	m
x_s	equivalent surface sphere diameter	m
x_{SV}	surface-volume diameter (diameter of a sphere having the same surface/volume ratio as the particle)	m
x_v	volume diameter (diameter of a sphere having the same volume as the particle)	m
x_{50}	cut size (equiprobable size)	m
Y	yield stress	Pa
Y_d	dynamic yield stress of the granule	Pa

y_i	composition of sample number i	—
β	liquid bridge half-angle in Figure 2.20	°
β	coalescence kernel or rate constant	s^{-1}
β	$(U - U_{mf})/U$	—
β_0	coalescence rate constant [Equation (12.18)]	—
$\beta_1(u, v)$	coalescence rate constant [Equation (12.18)]	—
$\beta(u,v - u,t)$	coalescence rate constant for granules of volume u and $(u - v)$ at time t	s^{-1}
γ	surface energy	J/m^2
δ	effective angle of internal friction	deg
Δc	concentration driving force	kg/kg solution
δ_{ij}	overlap between two interacting particles i and j, in DEM	m
Δp	pressure drop	Pa
ε	strain	—
ε	void fraction	—
ε_a	agglomerate or granule void fraction	—
ε_B	fraction of bed occupied by bubbles	—
ε_B	porosity or void fraction of powder bed	—
ε_{min}	minimum granule porosity or void fraction reached during granulation	—
θ	contact angle	°
θ	dynamic contact angle of the liquid with the powder	°
μ	friction coefficient	—
μ	liquid viscosity	Pa s
μ_r	rolling friction coefficient	—
μ_s	sliding friction coefficient	—
μ_t	torsion friction coefficient	—
μ_W	friction coefficient at the wall	—
ν	Poisson ratio	—
ξ	tangential overlap between two interacting particles, in DEM	m
ρ_B	powder bulk density	kg/m^3
ρ_f	fluid density	kg/m^3
ρ_g	granule density	kg/m^3
ρ_P	particle density	kg/m^3
ρ_s	solid density	kg/m^3
ρ_1	liquid density	kg/m^3
σ	normal stress	Pa
σ_{rr}	radial stress	Pa
σ_v	vertical component of stress	Pa
σ_y	unconfined yield stress	Pa
σ_{zz}	stress acting in the z direction on a plane orthogonal to the z direction	Pa
τ	shear stress	Pa
τ_R	Rayleigh timestep	s
τ_W	wall shear stress	Pa
Φ_w	kinematic angle of wall friction	°
Ψ_a	dimensionless spray flux	—
Ω	cohesive energy density, in DEM	J/m^3

References

Abrahamsen, A.R. and Geldart, D. (1980). Behaviour of gas fluidized beds of fine powder. Part 1. Homogenous expansion. *Powder Technol.* **26**: 35.

Adams, M.J., Briscoe, B.J., Gorman, D.M., Hollway, F. and Johnson, S.A. (1999). The boundary lubrication of glass-glass contacts by mixed alkyl alcohol and cationic surfactant systems. *Tribol. Ser.* **36**: 49.

Adetayo, A.A. and Ennis, B.J. (1997). Unifying approach to modelling granule coalescence mechanisms. *AIChE J.* **43** (4): 927–934.

Allen, T. (1990). *Particle Size Measurement*. Springer Dordrecht [Republished as an eBook 2012].

Amblard, B., Bertholin, S., Bobin, C., and Gauthier, T. (2015). Development of an attrition evaluation method using a jet cup rig. *Powder Technol.* **274**: 455–465.

Andrews, M.J. and O'Rourke, P.J. (1996). The multiphase particle-in-cell (MP-PIC) method for dense particulate flows. *Int. J. Multiphase Flow* **22** (2): 379–402.

Baeyens, J. and Geldart, D. (1974). An investigation into slugging fluidized beds. *Chem. Eng. Sci.* **29**: 255.

Barnes, H.A., Hutton, J.F., and Walters, K. (1989). *An Introduction to Rheology*. Amsterdam: Elsevier.

Baumann, H. and Tillman, A.-M. (2015). *The Hitch Hiker's Guide to LCA*. Lund, Sweden: Studentlitteratur.

Bemrose, C.R. and Bridgwater, J.A. (1987). Review of attrition and attrition test methods. *Powder Technol.* **49**: 97–127.

Beverloo, W.A., Leniger, H.A., and Van de Velde, J. (1961). The flow of granular solids through orifices. *Chem. Eng. Sci.* **15** (3-4) September 1961: 260–269.

Bi, H.T. and Grace, J.R. (1995). Flow regime diagrams for gas-solid fluidization and upward transport. *Int. J. Multiphase Flow* **21** (6): 1229–1236.

Bi, H.T. and Grace, J.R. (1999). Flow patterns in high-velocity fluidized beds and pneumatic conveying. *Can. J. Chem. Eng.* **77**: 223–230.

Boerefijn, R., Gudde, M.J., and Ghadiri, M. (2000). A review of attrition of fluid cracking catalyst particles. *Adv. Powder Technol.* **11** (2): 145–174.

Botterill, J.S.M. (1975). *Fluid Bed Heat Transfer*. London: Academic Press.

Botterill, J.S.M. (1986). Fluid bed heat transfer. In: *Gas Fluidization Technology* (ed. D. Geldart). Chichester: Wiley Chapter 9.

Bröckel, U., Meier, W., and Wagner, G. (ed.) (2007a). *Product Design and Engineering, Vol. 1 Basics and Technologies*. Weinheim, Germany: Wiley-VCH.

Bröckel, U., Meier, W., and Wagner, G. (ed.) (2007b). *Product Design and Engineering, Vol. 2 Raw Materials, Additives and Application*. Weinheim, Germany: Wiley-VCH.

Introduction to Particle Technology, Third Edition. Martin Rhodes and Jonathan Seville.
© 2024 John Wiley & Sons Ltd. Published 2024 by John Wiley & Sons Ltd.
Website: www.wiley.com/go/rhodes/particle3e

Brone, D., Alexander, A., and Muzzio, F.J. (1998). Quantitative characterization of mixing of dry powders in V-blenders. *AIChE* **44** (2): 271–278.

Butcher, J.C. (2016). *Numerical Methods for Ordinary Differential Equations*. Chichester, UK: Wiley.

Cahyadi, A.H. Neumayer, C.M. Hrenya, R.A. Cocco, J.W. Chew (2012). Comparative study of transport disengaging height (TDH) correlations in gas–solid fluidization. *Powder Technol.* **275**: 220–238.

Campbell, C.S. (2006). Granular material flows–an overview. *Powder Technol.* **162** (3): 208–229.

Chambers, A.J. and Marcus, R.D. (1986). Pneumatic conveying calculations. In: *Proceedings of the Second International Conference on Bulk Materials Storage and Transportation Wollongong, Australia*, 49–52.

Chambers, A.J., Keys, S., and Pan, R. (1998). The influence of material properties on conveying characteristics. In: *Proceedings of the 6th International Conference on Bulk Materials Storage, Handling and Transportation*, 309–319.

Che, H., M. Al-Shemmeri, P. J. Fryer, E. Lopez-Quiroga, T. K. Wheldon and K. Windows-Yule (2023). PEPT validated CFD-DEM model of aspherical particle motion in a spouted bed. *Chem. Eng. J.* **453**: 139689.

Chen, H., Xiao, Y., Liu, Y., and Shi, Y. (2017). Effect of Young's modulus on DEM results regarding transverse mixing of particles within a rotating drum. *Powder Technol.* **318**: 507–517.

Clift, R. (1986). *Hydrodynamics of bubbling fluidized beds. In: Gas Fluidization Technology* (ed. D. Geldart). Chichester: Wiley Chapter 4.

Clift, R., Grace, J.R., and Weber, M.E. (1978). *Bubbles, Drops and Particles*. London: Academic Press (Republished: Dover Publications, December 2013).

Clift, R., Ghadiri, M., and Thambimuthu, K.V. (1981). *Progress in Filtration and Separation*, **Vol. 2** (ed. R.J. Wakeman). Amsterdam: Elsevier.

Coulson, J.M. and Richardson, J.F. (2019). *Chemical Engineering, Volume 2A: Particle Technology and Separation Processes*, 6the (ed. R. Chhabra and M.G. Basavara). Butterworth-Heinemann.

Dalby, R.N., Tiano, S.L., and Hickey, A.J. (1996). Medical devices for the delivery of therapeutic aerosol to the lungs. In: *Inhalation Aerosols: Physical and Biological Basis for Therapy* (ed. A.J. Hickey), 441–473. New York: Marcel Dekker.

Danckwerts, P.V. (1952). Definition and Measurement of some characteristics of Mixtures. *Appl. Sci. Res. A* **3**: 279–296.

Darcy, H.P.G. (1856). Les fontaines publiques de la ville de Dijon. In: *Exposition et applications à suivre et des formules à employer dans les questions de distribution d'eau*. Paris, France: Victor Dalamont.

Darton, R.C., La Nauze, R.D., Davidson, J.F., and Harrison, D. (1977). Bubble growth due to coalescence in fluidised beds. *Trans. Inst. Chem. Eng.* **55**: 274.

Davey, R. and Garside, J. (2000). *From Molecules to Crystallizers, Oxford Chemistry Primers*. Oxford Science Publications.

Davidson, J.F. and Harrison, D. (1963). *Fluidised Particles*. Cambridge University Press.

Di Renzo, A., Napolitano, E.S., and Di Maio, F.P. (2021). Coarse-grain DEM modelling in fluidized bed simulation: a review. *Processes* **9** (2): 279.

Dixon, G. (1979). The impact of powder properties on dense phase flow. In: *Proceedings of International Conference on Pneumatic Transport, London*.

Edwards, M.F. and Instone, T. (2001). Particulate products - their manufacture and use. *Powder Technol.* **119**: 9–13.

Ennis, B.J. and Litster, J.D. (1997). *Particle Size Enlargement in Perry's Chemical Engineers' Handbook* (ed. D.W. Green and M.Z. Southard). New York: McGraw-Hill.

Ennis, B.J. and Litster, J.D. (2018). *Particle Size Enlargement in Perry's Chemical Engineers' Handbook*, 9the (ed. D.W. Green and M.Z. Southard), 21–89. McGraw-Hill.

Ennis, B.J., Tardos, G., and Pfeffer (1991). A microlevel-based characterization of granulation phenomena. *Powder Technol.* **65**: 257–272.

Ergun, S. (1952). Fluid flow through packed columns. *Chem. Eng. Prog.* **48**: 89–94.

Evans, I., Pomeroy, C.D., and Berenbaum, R. (1961). The compressive strength of coal. *Colliery Eng.* **75–81** (123–127): 173–178.

Fell, J.T. and Newton, J.M. (1970). Determination of tablet strength by the diametral-compression test. *J. Pharm. Sci.* **59**: 688–691.

Filtvedt, W.O., Javidi, M., Holt, A., Melaaen, M.C., Marstein, E., Tathgar, H. and Ramachandran, P.A. (2010). Development of fluidized bed reactors for silicon production. *Sol. Energy Mater. Sol. Cells* **94**: 1980–1995.

Francis, A.W. (1933). Wall effects in falling ball method for viscosity. *Physics* **4**: 403.

Franks, G.V. and Lange, F.F. (1996). Plastic-to-brittle transition of saturated alumina powder compacts. *J. Am. Ceram. Soc.* **79** (12): 3161–3168.

Franks, G. V., Zhou, Z., Duin, N. J. and Boger, D. V. (2000), Effect of interparticle forces on shear thickening of oxide suspensions. *J. Rheol.*, **44** (4), 759–779.

Geldart, D. (1973). Types of gas fluidization. *Powder Technol.* **7**: 285–292.

Geldart, D. (ed.) (1986). *Gas Fluidization Technology*. Chichester: Wiley.

Geldart, D. (1990). Estimation of basic particle properties for use in fluid–particle process calculations. *Powder Technol.* **60**: 1.

Geldart, D. (1992). *Gas Fluidization Short Course*. Bradford: University of Bradford.

Geldart, D. and Jones, P. (1991). Behaviour of L-valves with granular powders. *Powder Technol.* **67**: 163–174.

Geldart, D., Cullinan, J., Gilvray, D., Georghiades, S. and Pope, D. J. (1979). The effects of fines on entrainment from gas fluidized beds. *Trans. Inst. Chem. Eng.* **57**: 269.

Gilvary, J.J. (1961). Fracture of brittle solids. I. Distribution function for fragment size in single fracture. *J. Appl. Phys.* **32**: 391–399.

Goldsmith, W. (1960). *Impact*. London: Arnold.

Govender, N., Wilke, D.N., and Kok, S. (2015). Collision detection of convex polyhedra on the NVIDIA GPU architecture for the discrete element method. *Appl. Math Comput.* **267**: 810–829.

Grace, J.R. (1986). Contacting modes and behaviour classification of gas solid and other two-phase suspensions. *Can. J. Chem. Eng.* **64**: 353–363.

Grace, J.R., Avidan, A.A., and Knowlton, T.M. (ed.) (1997). *Circulating Fluidized Beds*. London: Blackie Academic and Professional.

Grace, J.R., Bi, X.-T., and Ellis, N. (ed.) (2020). *Essentials of Fluidization Technology*. Weinheim: Wiley-VCH.

Green, D. and Southard, M.Z. (ed.) (2018). *Perry's Chemical Engineering Handbook*, 9th Edition. New York: McGraw-Hill.

Guigon, P., Simon, O., Saleh, K., Bindhumadhavan, G., Adams, M.J., Seville, J.P.K., Salman, A.D., Ghadiri, M. and M.J. Hounslow (2007). Roll pressing. In: *Granulation* (ed. A.D. Salman, M.J. Hounslow, and J.P.K. Seville), 255–288. Amsterdam: Elsevier.

Guo, Y. and Curtis, J.S. (2015). Discrete element method simulations for complex granular flows. *Annu. Rev. Fluid Mech.* **47**: 21–46.

Haider, A. and Levenspiel, O. (1989). Drag coefficient and terminal velocity of spherical and non-spherical particles. *Powder Technol.* **58**: 63–70.

Hapgood, K.P., Litster, J.D., and Smith, R. (2003). Nucleation regime map for liquid bound granules. *AIChE J.* **49** (2): 350–361.

Hapgood, K., Litster, J.D., White, E.T., Mort, P.R. and Jones, D.G. (2004). Dimensionless spray flux in wet granulation: Monte-Carlo simulations and experimental validation. *Powder Technol.* **141**: 20–30.

Hapgood, K.P., Iveson, S.M., Litster, J.D., and Liu, L. (2007). Granulation rate processes. In: *Granulation*, (ed. A.D. Salman, M.J. Hounslow, and J.P.K. Seville), 897–977. Amsterdam: Elsevier.

Hardgrove, R. M. (1932) 'Grindability of Coal', *Trans. ASME, Fuels and Steam Power*, 54 FSP-54-5, 37–46.

Harnby, N., Edwards, M.F., and Nienow, A.W. (1992). *Mixing in the Process Industries*, 2nd Edition, London: Butterworth-Heinemann.

Herald, M.T., Sykes, J.A., Werner, D., Seville, J.P.K., and C. Windows-Yule (2022). DEM2GATE: Combining discrete element method simulation with virtual positron emission particle tracking experiments. *Powder Technology* **401**: 117302.

Hiemenz, P.C. and Rajagopolan, R. (1997). *Principles of Colloid and Surface Chemistry*, 3rd Edition, New York: Marcel Dekker.

Hinkle, B.L. (1953) *PhD Thesis*, Georgia Institute of Technology.

Hirama, T., Takeuchi, T., and Chiba, T. (1992). Regime classification of macroscopic gas-solid flow in a circulating fluidized-bed riser. *Powder Technol.* **70**: 215–222.

Horio, M., Taki, A., Hsieh, Y.S., and Muchi, I. (1980). Elutriation and particle transport through the freeboard of a gas–solid fluidized bed. In: *Fluidization* (ed. J.R. Grace and J.M. Matsen), 509. New York: Engineering Foundation.

Hukki, R.T. (1961). Proposal for Solomonic settlement between the theories of von Rittinger, Kick and Bond. *Trans. AIME* **220**: 403–408.

Hunter, R.J. (2001). *Foundations of Colloid Science*, 2nd Edition. Oxford: Oxford University Press.

Inglis, C.E. (1913). Stress in a plate due to the presence of cracks and sharp corners. *Trans. Inst. Naval Arch.* **55**: 219–230.

Israelachvili, J.N. (2011). *Intermolecular and Surface Forces*, 3rde. London: Academic Press Imprint.

Iveson, S.M., Litster, J.D., Hapgood, K.P., and Ennis, B.J. (2001). Nucleation, growth and breakage phenomena in agitated wet granulation processes: a review. *Powder Technol.* **117** (1–2): 3–39.

Janssen, H.A. (1895). Versuche über Getriededruck in Silozellen. *Z. Ver. Dtsch. Ing.* **39** (35): 1045–1049.

Jenike, A.W. (1964). Storage and flow of solids. *Bull. Utah Eng. Exp. Station* **53** (123): 26.

Johnson, K.L. (1985). *Contact Mechanics*. Cambridge: Cambridge University Press.

Johnson, S.B., Franks, G.V., Scales, P.J., and Healy, T.W. (1999). The binding of monovalent electrolyte ions on alpha alumina. II. The shear yield stress of concentrated suspensions. *Langmuir* **15**: 2844–2853.

Johnson, S.B., Franks, G.V., Scales, P.J., Boger, D.V. and T.W. Healy. (2000). Surface chemistry – rheology relationships in concentrated mineral suspensions. *Int. J. Mineral Process* **58**: 267–304.

Jones, M.G. (1988). *The Influence of Bulk Particulate Properties on Pneumatic Conveying Performance*, Doctoral dissertation. UK: Thames Polytechnic.

Jones, M.G. and Williams, K.C. (2008). Predicting the mode of flow in pneumatic conveying systems—a review. *Particuology* **6**: 289–300.

Jones, D.A.R., Leary, B., and Boger, D.V. (1991). The rheology of a concentrated colloidal suspension of hard spheres. *J. Colloid Interface Sci.* **147**: 479–495.

Kay, J.M. and Nedderman, R.M. (1985). *Fluid Mechanics and Transfer Processes*. Cambridge: Cambridge University Press.

Keleb, E.I., Vermeire, A., Vervaet, C., and Remon, J.P. (2004). Twin screw granulation as a simple and efficient tool for continuous wet granulation. *Int. J. Pharm.* **273**: 183–194.

Kendall, K. (1978). The impossibility of comminuting small particles by compression. *Nature* **272**: 710.

Kendall, K. (2001). *Molecular Adhesion and its Applications*. New York: Kluwer Academic/Plenum Publishers.

Khan, A.R. and Richardson, J.F. (1989). Fluid–particle interactions and flow characteristics of fluidized beds and settling suspensions of spherical particles. *Chem. Eng. Commun.* **78**: 111.

Khan, A.R., Richardson, J.F., and Shakiri, K.J. (1978). Heat transfer between a fluidized bed and a small immersed surface. In: *Fluidization, Proceedings of the Second Engineering Foundation Conference* (ed. J.F. Davidson and D.L. Keairns), 375. Cambridge University Press.

Knight, P.C. (2001). Structuring agglomerated products for improved performance. *Powder Technol.* **119**: 14–25.

Knowlton, T.M. (1986). Solids transfer in fluidized systems. In: *Gas Fluidization Technology* (ed. D. Geldart). Chichester: Wiley Chapter 12.

Knowlton, T.M. (1997). Standpipes and non-mechanical valves', Notes for the Continuing Education Course. In: *Gas Fluidized Beds: Design and Operation*. Department of Chemical Engineering, Monash University.

Kodam, M., Bharadwaj, R., Curtis, J., B.C. Hancock and C. Wassgren (2009). Force model considerations for glued-sphere discrete element method simulations. *Chem. Eng. Sci.* **64** (15): 3466–3475.

Konno, H. and Saito, S.J. (1969). Pneumatic conveying of solids through straight pipes. *Chem. Eng. Jpn.* **2**: 211–217.

Konrad, K. (1986). Dense phase conveying: a review. *Powder Technol.* **49**: 1–35.

Kunii, D. and Levenspiel, O. (1991). *Fluidization Engineering*, 2nd Edition. Chichester: Wiley.

Kuo, H.P., Knight, P.C., Parker, D.J., Tsuji, Y., Adams, M.J. and Seville, J.P.K. (2002). The influence of DEM simulation parameters on the particle behaviour in a V-mixer. *Chem. Eng. Sci.* **57**: 3621–3638.

Lacey, P.M.C. (1954). Developments in the theory of particulate mixing. *J. Appl. Chem.* **4**: 257.

Lambourne, R. and Strivens, T.A. (1999). *Paint and Surface Coatings*, 2nde. Cambridge, UK: Woodhead Publishing Limited.

Leung, L.S. (1980). Vertical pneumatic conveying: a flow regime diagram and a review of choking versus non-choking systems. *Powder Technol.* **25**: 185–190.

Litster, J.D. (2016). *Design and Processing of Particulate Products*. Cambridge, UK: Cambridge University Press.

Litster, J.D., Hapgood, K.P., Michaels, J.N, Sims, A., Roberts, M., Kameneni, S.K. and T. Hsu. (2001). Liquid distribution in wet granulation: dimensionless spray flux. *Powder Technol.* **114**: 32–39.

Luding, S. (2004). Molecular dynamics simulations of granular materials. In: *The Physics of Granular Media* (ed. H. Hinrichsen and D.E. Wolf), 297–324. Weinheim: Wiley-VCH.

Lumay, G., Boschini, F., Traina, K., Bontempi, S., Remy, J-C., Cloots, R. and Vandewalle, N. (2012). Measuring the flowing properties of powders and grains. *Powder Technol.* **224**: 19–27.

Lunn, G. (1992). *Guide to Dust Explosion, Prevention and Protection, Part I*. Rugby: Venting, Institution of Chemical Engineers.

Mainwaring, N.J. and Reed, A.R. (1987). Permeability and air retention characteristics of bulk solid materials in relation to modes of dense phase pneumatic conveying. *Bulk Solids Handling* **7** (3): 415–425.

Martin, T.W., Seville, J.P.K., and Parker, D.J. (2007). A general method for quantifying dispersion in multiscale systems using trajectory analysis. *Chem. Eng. Sci.* **62** (13): 3419–3428.

Meyer, H., Rosdahl, G., Saarnak, A., Säberg, O. (1997) *Notes for the Bachelor Course on Färger og Lacker, Nordiska Ingenjörsbyrån för Färg AB*.

Mills, D. (1990). *Pneumatic Transport Design Guide*. London: Butterworth.

Molerus, O. (1982). Interpretation of Geldart's type A, B, C and D powders taking into account interparticle cohesion force. *Powder Technol.* **33**: 81–87.

Molerus, O. (2000). Fluid mechanics and heat transfer in fluidized beds. *Kona* **18**: 121–130.

Naveen Tripathi, N., Sharma, A., Mallick, S.S., and Wypych, P.W. (2015). Energy loss at bends in the pneumatic conveying of fly ash. *Particuology* **21**: 65–73.

Nedderman, R.M. (1992). *Statics and Kinematics of Granular Materials*. Cambridge University Press.

Neveu, A., Francqui, F., and Lumay, G. (2022). Measuring powder flow properties in a rotating drum. *Measurement* **200**: 111548.

Norton, J.E., Fryer, P.J., and Norton, I.T. (ed.) (2013). *Formulation Engineering of Foods*. Oxford, UK: Wiley-Blackwell.

Ogarko, V. and Luding, S. (2012). A fast multilevel algorithm for contact detection of arbitrarily polydisperse objects. *Comput. Phys. Commun.* **183** (4): 931–936.

Oko, C.O.C., Diemuodeke, E.O., and Akilande, I.S. (2010). Design of hoppers using spreadsheet. *J. Agric. Eng. Res.* **56** (2): 53–58.

Orcutt, J.C., Davidson, J.F., and Pigford, R.L. (1962). Reaction time distributions in fluidized catalytic reactors. *Chem. Eng. Prog. Symp. Ser.* **58** (38): 1.

Palmer, K.N. (1990). Explosion and fire hazards of powders. In: *Principles of Powder Technology* (ed. M.J. Rhodes), 299–334. Chichester: Wiley.

Pan, R. (1992). *Improving Scale-Up Procedures for the Design of Pneumatic Conveying Systems, Doctoral dissertation,*. Australia: University of Wollongong.

Pan, R. (1999). Material properties and flow modes in pneumatic conveying. *Powder Technol.* **104**: 157–163.

Pan, R. and Wypych, P.W. (1998). Dilute and dense phase pneumatic conveying of fly ash. In: *Proceedings of the sixth international conference on bulk materials storage and transportation, Wollongong, NSW, Australia*, 183–189.

Podlozhnyuk, A., Pirker, S., and Kloss, C. (2017). Efficient implementation of superquadric particles in discrete element method within an open-source framework. *Comput. Part. Mech.* **4**: 101–118.

Poole, K.R., Taylor, R.F., and Wall, G.P. (1964). Mixing powders to fine-scale homogeneity: studies of batch mixing. *Trans. Inst. Chem. Eng.* **42**: T305.

Punwani, D.V., Modi, M.V. and Tarman, P.B. (1976) Generalized correlation for estimating choking velocity in vertical solids transport in *Proc. Int. Powder and Bulk Solids Handling and Processing Conference*, Powder Advisory Center, Chicago, IL. USA.

Randolph, A.D. and Larson, M.A. (1971). *Theory of Particulate Processes*. London: Academic Press.

Reh, L. (1971). Fluidized bed processing. *Chem. Eng. Prog.* **67** (2): 58–63.

Rhodes, M.J. (1989). The upward flow of gas/solid suspensions, part 2: a practical quantitative regime diagram for the upward flow of gas-solid suspensions. *Chem. Eng. Res. Des.* **67**: 30–37.

Rhodes, M.J., Sollaart, M., and Wang, X.S. (1998). Flow structure in a fast fluid bed. *Powder Technol.* **99**: 194–200.

Richardson, J.F. and Zaki, W.N. (1954). Sedimentation and fluidization. *Trans. Inst. Chem. Eng.* **32**: 35.

Rizk, F. (1973). *Dr-Ing*. Dissertation. Karlsruhe, Germany: Technische Hochschule.

Rosato, A.D. and Windows-Yule, C. (2020). *Segregation in Vibrated Granular Systems*. London: Academic Press.

Salman, A.D., Hounslow, M.J., and Seville, J.P.K. (2007a). *Granulation, Handbook of Powder Technology*, **Vol. 11**. Amsterdam: Elsevier.

Salman, A.D., Ghadiri, M., and Hounslow, M.J. (2007b). *Particle Breakage, Handbook of Powder Technology*, **Vol. 12**. Amsterdam: Elsevier.

Sanchez, L., Vasquez, N., Klinzing, G.E., and Dhodapkar, S. (2003). Characterization of bulk solids to assess dense phase pneumatic conveying. *Powder Technol.* **138**: 93–117.

Sastry, K.V.S. (1975). Similarity of size distribution of agglomerates during their growth by coalescence in granulation or green pelletization. *Int. J. Miner. Process.* **2**: 187.

Sastry, K.V.S. and Fuerstenau, D.W. (1970). Size distribution of agglomerates in coalescing disperse systems. *Ind. Eng. Chem. Fundam.* **9** (1): 145.

Schofield, C. (1985) *Guide to Dust Explosion, Prevention and Protection, Part I*, Venting, I. Chem. E., Rugby.

Schofield, C. and Abbott, J. A. (1988) *Guide to Dust Explosion, Prevention and Protection, Part II, Ignition Prevention, Containment, Suppression and Isolation*, I. Chem. E., Rugby.

Seville, J.P.K. (ed.) (1997). *Gas Cleaning in Demanding Applications*. London: Blackie/Chapman & Hall.

Seville, J.P.K. and Wu, C.-Y. (2016). *Particle Technology and Engineering*. Oxford: Butterworth-Heinemann.

Seville, J.P.K., Tüzün, U., and Clift, R. (1997). *Processing of Particulate Solids*. Glasgow: Blackie A&P/Chapman & Hall.

Seville, J.P.K., Fryer, P.J., and Norton, I.T. (2007). Formulation of structured chemical products. In: *Product Design and Engineering, Vol 2: Raw Materials, Additives & Applications* (ed. U. Bröckel, W. Meier, and G. Wagner). Weinheim, Germany: Wiley-VCH.

Shaw, D.J. (1992). *Introduction to Colloid and Surface Chemistry*, 4the. Oxford: Butterworth/Heinemann.

Simons, S.J.R. (2007). Liquid bridges in granules. In: Granulation, (ed. A. D. Salman, M. J. Hounslow and J.P.K. Seville), 1257–1316. Amsterdam: Elsevier.

Smith, S.J. and Bernstein, J.A. (1996). Therapeutic use of lung aerosols. In: *Inhalation Aerosols: Physical and Biologic basis for Therapy* (ed. A.J. Hickey), 233–269. New York: Marcel Dekker.

Squires, A.M., Kwauk, M., and Avidan, A.A. (1985). Fluid beds: at last, challenging two entrenched practices. *Science* **230**: 1329–1337.

Svarovsky, L. (1981). *Solid–Gas Separation*. Amsterdam: Elsevier.

Svarovsky, L. (1986). Solid–gas separation. In: *Gas Fluidization Technology* (ed. D. Geldart), 197–217. Chichester: Wiley.

Svarovsky, L. (1990). Solid–gas separation. In: *Principles of Powder Technology* (ed. M.J. Rhodes), 171–192. Chichester: Wiley.

Tardos, G.I., Khan, M.I., and Mort, P.R. (1997). Critical parameters and limiting conditions in binder granulation of fine powders. *Powder Technol.* **94**: 245–258.

Taylor, K.M.G. and Aulton, M.E. (ed.) (2022). *Aulton's Pharmaceutics*, 6the. Oxford, U.K: Elsevier.

Thornton, C., Cummins, S.J., and Cleary, P.W. (2013). An investigation of the comparative behaviour of alternative contact force models during inelastic collisions. *Powder Technol.* **233**: 30–46.

Toomey, R.D. and Johnstone, H.F. (1952). Gas fluidization of solid particles. *Chem. Eng. Prog.* **48**: 220–226.

Tu, W.-D., Ingram, A., and Seville, J.P.K. (2009). Regime map development for continuous twin screw granulation. *Chem. Eng. J.* **145**: 505–513.

Turton, R. and Levenspiel, O. (1986). A short note on the drag correlation for spheres. *Powder Technol.* **47**: 83.

Walton, O.R. and Braun, R.L. (1986). Viscosity, granular-temperature, and stress calculations for shearing assemblies of inelastic, frictional disks. *J. Rheol.* **30** (5): 949–980.

Ward, I.M. and Sweeney, J. (2004). *An Introduction to the Mechanical Properties of Solid Polymers*, 2nde. Wiley-Interscience.

Weibel, E.R. (1963). *Morphometry of the Human Lung*. Berlin: Springer-Verlag.

Wen, C.Y. and Yu, Y.H. (1966). A generalised method for predicting minimum fluidization velocity. *AIChE J.* **12**: 610.

Wensrich, C. and Katterfeld, A. (2012). Rolling friction as a technique for modelling particle shape in DEM. *Powder Technol.* **217**: 409–417.

Wesselingh, J.A., Kiil, S., and Vigild, M.E. (2007). *Design and Development of Biological, Chemical, Food and Pharmaceutical Products*. Chichester, UK: Wiley.

Willett, C.D., Johnson, S.A., Adams, M.J., and Seville, J.P.K. (2007). Pendular capillary bridges. In: *Granulation* (ed. A.D. Salman, M.J. Hounslow, and J.P.K. Seville). Amsterdam: Elsevier.

Williams, J.C. (1990). Mixing and Segregation in Powders. In: *Principles of Powder Technology* (ed. M. J. Rhodes). Chichester: Wiley Chapter 4.

Wills, B.A. and Finch, J. (2015). *Wills' Mineral Processing Technology - An Introduction to the Practical Aspects of Ore Treatment and Mineral Recovery*, 8th Edition. Amsterdam: Elsevier/Butterworth-Heinemann.

Windows-Yule, C. and Neveu, A. (2022). Calibration of DEM simulations for dynamic particulate systems. *Pap. Phys.* **14**: 140010–140010.

Windows-Yule, C., Seville, J., Ingram, A., and Parker, D. (2020). Positron emission particle tracking of granular flows. *Annu. Rev. Chem. Biomol. Eng.* **11**.

Wolf, B., Scirocco, R., Frith, W.J., and Norton, I.T. (2000). Shear-induced anisotropic microstructure in phase-separated biopolymer mixtures. *Food Hydrocoll.* **14** (3): 217–225.

Wong, Y.S. (2004). *Experimental and numerical investigations of fluidization behaviour with & without the presence of immersed tubes*, PhD Thesis. University of Birmingham.

Yagi, S. and Muchi, I. (1952). *Chem. Eng., (Jpn)* **16**: 307.

Yerushalmi, J., Turner, D.H., and Squires, A.M. (1976). The fast fluidized bed. *Ind. Eng. Chem. Process. Des. Dev.* **15**: 47–51.

Zabrodsky, S.S. (1966). *Hydrodynamics and Heat Transfer in Fluidized Beds*. Cambridge, MA: MIT Press.

Zeng, X.M., Martin, G., and Marriott, C. (2001). *Particulate Interactions in Dry Powder Formulations for Inhalation*. London: Taylor & Francis.

Zenz, F.A. (1949). Two-phase fluid-solid flow. *Ind. Eng. Chem.* **41**: 2801–2806.

Zenz, F.A. (1964). Conveyability of materials of mixed particle size. *Ind. Eng. Fund.* **3** (1): 65–75.

Zenz, F.A. (1983). Particulate solids—the third fluid phase in chemical engineering. *Chem. Eng.* **90**: 61–67.

Zenz, F.A. and Weil, N.A. (1958). A theoretical–empirical approach to the mechanism of particle entrainment from fluidized beds. *AIChE J.* **4**: 472.

Zhou, Z., Scales, P.J., and Franks, G.V. (2001). Chemical and physical control of the rheology of concentrated metal oxide suspensions. *Chem. Eng. Sci.* **56**: 2901–2920.

Zhou, Z., Gan, Y., Wanless, E.J., Jameson, G. J. and G.V. Franks (2008). Interaction forces between silica surfaces in aqueous solutions of cationic polymeric flocculants: effect of polymer charge. *Langmuir* **24** (19): 10920–10928.

Index

Note: Page numbers followed by "n" denote footnotes.

Introduction to Particle Technology, Third Edition. Martin Rhodes and Jonathan Seville.
© 2024 John Wiley & Sons Ltd. Published 2024 by John Wiley & Sons Ltd.
Website: www.wiley.com/go/rhodes/particle3e